STEAM POWER ENGINEERING
Thermal and hydraulic design principles

Many important advances have occurred in the use of steam power in the last fifteen years. Large power plants as well as small-capacity boilers have become more efficient and reliable, and their design and operation have been enhanced by the application of modern computational techniques.

This book provides a timely review of the latest developments in steam power engineering. It discusses thermohydraulic principles and processes, with an emphasis on practical problems of steam power plant design and operation. Among topics covered are historical analysis, cycle design, thermal and hydraulic design of heating surfaces, pollutant control, and flow instability, from which are deduced strategies for further development. Each chapter reflects the author's considerable experience in teaching, research, and consulting.

Intended for both students and engineers, the book treats realistic problems and offers a wealth of valuable insight into the design of large and small steam power plants.

Seikan Ishigai is Professor Emeritus at Osaka University, Japan.

T0192483

STEAM POWER ENGINEERING

Thermal and hydraulic design principles

Edited by

SEIKAN ISHIGAI

x

CAMBRIDGE
UNIVERSITY PRESS

CAMBRIDGE UNIVERSITY PRESS
Cambridge, New York, Melbourne, Madrid, Cape Town, Singapore,
São Paulo, Delhi, Dubai, Tokyo

Cambridge University Press
The Edinburgh Building, Cambridge CB2 8RU, UK

Published in the United States of America by Cambridge University Press, New York

www.cambridge.org
Information on this title: www.cambridge.org/9780521135184

First published 1999
This digitally printed version 2010

A catalogue record for this publication is available from the British Library

Library of Congress Cataloguing in Publication data

Steam power engineering: thermal and hydraulic design principles/
edited by Seikan Ishigai.
      p.   cm.
   Includes bibliographical references.
   ISBN 0-521-62635-8
   1. Steam power plants.   I. Ishigai, Seikan.
TJ400.S73   1999
621.31′2132 – dc21   98-15350
                          CIP

ISBN 978-0-521-62635-4 Hardback
ISBN 978-0-521-13518-4 Paperback

Contents

Contributors

Koji Akagawa
Professor Emeritus
Kobe University
Nada, Kobe 657-8501
Japan

Terushige Fujii
Department of Mechanical
Engineering
Kobe University
Nada, Kobe 657-8501
Japan

Seikan Ishigai
Professor Emeritus
Osaka University
Suita, Osaka 565-0871
Japan

Shigeyasu Nakanishi
Department of Mechanical and
System Engineering
Ryukoku University, Seta
Otsu 520-2194
Japan

Eiichi Nishikawa
Engineering Systems Department
Kobe University of Mercantile
Marine
Higashinada, Kobe 658-0022
Japan

Mamoru Ozawa
Department of Mechanical
Engineering
Kansai University
Suita, Osaka 564-8680
Japan

Preface

Steam power has been at the frontier of engineering since the beginning of the 18th century due to its importance in both power and control engineering. Progress in steam power has been notable in the last 15 years, both in fossil-fuel-fired plants and in nuclear plants. During this period, nuclear plants of sound design have constantly been improving their safety. Fossil-fuel-fired units have successfully increased their efficiency, operability, and reliability by bringing into operation sliding-pressure supercritical units for daily start and stop operation with a wide load range down to 10%. A drastic reduction in the weight and size of gas-fired boilers to nearly one-half is within reach and some small-capacity boilers are now proving the reliability of the advanced design concept in actual operation.

Computerization of the process of design and operation of steam power plants has certainly helped such advancement. However, computerization has also brought the essence of computation into a black box. Understanding the fundamental principles of processes to be computed is indispensable for design engineers and operation engineers, as well as for research staff who supply them with basic data and computer programs.

The thermo-hydraulic process is a fundamental process in steam power and permeates every part of the plant irrespective of its fuel, type, and scale. The basic principles are the same in both nuclear plants and fossil-fuel plants of different scales, and must be applied to each particular plant with due consideration to its particular nature.

This book is intended to help practicing engineers and graduate students build a basic frame of ideas with which to treat practical problems of steam power plants. Basic principles are extended by practical applications so that the reader can absorb the experiences of successful engineers and scientists.

Each contributing author of this book has been engaged in research and development in some field of steam power engineering and also in consulting services for design of steam power plants. The authors held meetings in the course of writing to form a common general view on the frontier of steam power engineering and also to make clear presentations of fundamental theories in this rapidly changing field. It would give us great pleasure if the essence of our experience of research and development is shared by careful readers.

The authors are grateful to Dr. George Hayward for his help in producing this English edition, and to Dr. Ian N. Guy for editing the manuscript. Without their help, this book could not have survived the Kobe earthquake of 1995.

1

Historical development of strategy for steam power

SEIKAN ISHIGAI

Professor Emeritus, Osaka University

1.1 The field of steam power

Four main types of heat engine are now in use: the internal combustion reciprocating engine, the gas turbine, steam power, and the rocket. The internal combustion reciprocating engine is predominant on land and sea in vehicles such as the automobile, bulldozer, combine, and the motor ship. The gas turbine, in the form of the jet engine, is predominant among aircraft propulsion engines, and the rocket is the sole motor for space vehicles. Steam power is predominant in electricity-generating thermal power stations of both the fossil-fuel-firing type and the nuclear type. Each of the four types of heat engine has thus its own proper field of application, and the boundaries between these fields are stable within a short span of several years but are flexible over a long period of history because of interactions between the four types and competition with other power sources such as the water turbine and the fuel cell.

The ever-growing demand for electricity in both developed and developing nations raises various energy problems. Because of the position of steam power in electricity generation, the choice of strategy for steam power development is an important issue. In this chapter a historical method is proposed for generating or amending the general strategy for energy technology and for steam power in particular.

1.2 Fundamental characteristics from a technological viewpoint

From the viewpoint of natural science, the fundamental characteristics of steam power are simple. The steam power unit is (1) a type of heat engine, (2) having water as the working fluid, (3) which is heated and cooled externally through surfaces that divide the working fluid from the environment.

These characteristics are certainly fundamental but do not directly explain the history of steam power, since this is a branch of production technology that is

1

driven to development not by nature but by society. The power demand comes from society, which at the same time supplies both the social and the natural resources necessary for the development. Particular laws of development are thus observed in the history of power technology as well as of total technology.

This relation is similar to the emergence of physiological laws among living creatures that obey the laws of physics, and the emergence of ecological laws among animals and plants that obey the laws of physiology. Similarly, laws of development of production technology emerge therein, while every component of production technology obeys the laws of nature and society. Three notable facts are presented in the following as evidence of the existence of laws of development of technology beside the general laws of nature and society.

First, thermodynamics and other branches of thermal science were established many decades after the extensive use of steam power in industry. The first practical steam engine was built by Thomas Newcomen (1663–1729) and went into successful operation in 1712. About 100 units of the same type of engine were reported to be in operation in northern England in 1769 (Dickinson 1963), whereas the establishment of the first and second laws of thermodynamics by Clausius, Lord Kelvin, and others did not occur until around 1855.

This does not mean that thermal science is useless. Trial and error is an excellent method of technical development, but it is certainly exhausting. Formulated laws of nature help to develop technology by giving an insight into natural processes that take place in the branch of technology under development. However, it should be understood that there are many examples of successful development of technology without knowledge of natural laws, or even with belief in a mistaken theory of nature such as alchemy, which was once prevalent.

Second, no technology can start to develop unless it is economically feasible and safe. Persistent efforts had been devoted by many inventors before Newcomen. However, all of them failed until the maximum unit output, or the maximum output of one unit of the then available power sources, including human bodies, horses, windmills, and waterwheels, grew to several kilowatts at the beginning of the 18th century. Then came an extensive search worldwide into the hitherto unknown field of getting power out of heat. This is an example of the importance of unit output as the most fundamental parameter of power source technology.

Third, every technology has not only a shining side but also a dark side of faults and harmful effects, which take time to overcome. In the case of steam power, all inventors before Newcomen used boilers at higher pressures than atmospheric, and their boilers were sure to leak or explode sooner or later. This defect was overcome by Newcomen, who used boilers at atmospheric pressure and completely avoided the risk of boiler explosion.

These three points indicate that the development of steam power is neither directly nor solely controlled by laws of nature. It is jointly controlled by natural and social laws so that characteristics particular to steam power are generated, and such characteristics can lead to the formulation of laws of development of steam power and finally to technology in general. Three of these characteristics are briefly presented below.

The most fundamental characteristic of steam power is its aptitude for large-output units. Steam power has helped society to expand production, which in turn has expanded not only its total power demand but also its unit power output. Although it is now generally forgotten, the earliest automotive road vehicles, invented by N.J. Cugnot (1725–1804) and R. Trevithick (1771–1833), were steam-driven. In the course of development, steam power handed over such smaller units as automotive engines and small-factory engines to the internal combustion engine and the electric motor, and has always been a prime mover toward development of higher unit output and better efficiency and automation. Steam power is now specializing in central electricity-generating stations and nuclear naval vessels. Thus the most important characteristic of steam power is its large-output units.

Common sense might lead one to think that the weight and volume of the boiler and the condenser penalize the steam engine in comparison with the diesel engine, which does not need them. The situation is not that simple, since, at around the borderline unit output, steam is lighter and less voluminous than diesel. This merit in machinery weight and volume did balance to some extent the inferiority in fuel consumption, and hence the above-mentioned borderline was formed.

Common sense might also lead one to believe that the steam engine can burn coal or nuclear fuel and has an inherent domain in land use in comparison with the internal combustion engine, which needs liquid or gaseous fuels. By and large this is true if due consideration is paid to the unit capacity. There are exceptional cases, even in marine power plants such as for natural-gas carriers, in which liquid- or gaseous-fuel-burning steam power is preferred to the internal combustion engine or gas turbine if the unit capacity is large enough.

Thus, the ability for large unit capacity has been the most fundamental characteristic of steam power up to the present time, although even the steam engine should have an upper limit to its unit output, since no type can fulfil all purposes and each has its proper domain of application.

The second important characteristic of steam power is its use of low-temperature heat sources. The thermodynamic properties of water are such that the theoretical cycle of the low-pressure steam engine is very close to the Carnot cycle, the theoretical efficiency of which amounts to 18% for very moderate temperatures, for example, 100°C for steam and 30°C for the condenser. This is probably the most important thermodynamic reason why the steam engine attained practical success as early as 1712 with the first Newcomen engine, why Watt's engines were welcomed by industries in the Industrial Revolution, and also why steam turbines are now increasingly accepted in the bottoming cycle that receives exhaust gas from diesel engines, gas turbines, and other processes.

Steam power's third important characteristic is its ability to combine with industrial and air-conditioning uses of steam. In the ideal case, in which all extracted or exhausted steam from a steam power plant is consumed in the combined industrial or air-conditioning plants, the theoretical thermal efficiency of the steam power plant is equal to the boiler efficiency and can be 90% or over. This old method of combining power with heat is, even now, still the most efficient method of power generation.

1.3 A brief history of steam power development

All technologies that have a future begin in fields to which they are not ideally suited, simply because their main spheres of activity are not yet ready at the time of their infancy. Steam power is no exception; it started as pumping engines made by Newcomen.

Steam power met the then badly needed social demand for power sources that could develop enough horsepower to pump out water from mines. Demands for such unit-output power sources were already there on land and sea, and were being supplied by means of windmills and waterwheels on land and sails on sea. Steam helped the industry of the time to build up its productive ability, which in turn created demands for more power in various fields. Quantitative growth and qualitative diversification thus went hand in hand.

When the patent by James Watt expired in 1800, steam engines were greatly helping to promote the Industrial Revolution but were at a trial stage for ships and carriages. However, demands for mobile engines were already in existence since sailing ships were cruising on oceans, horse carriages were traveling on roads – some even running on wooden rails – and horse ploughs were working on farms.

Steam met the demands for mobile power sources as well as stationary power sources. Marine steam engines and steam locomotives for vehicles, steam tractors for farming, and of course stationary steam engines for factories greatly helped society to expand production and traffic on land and sea and to prepare the future market for electric motors and internal combustion engines.

In 1882, Pearl Street Station, constructed by Thomas A. Edison as the first central electricity-generating station, opened the era of electricity with its three steam-driven 150-kW generators. In 1883 Gottlieb Daimler invented the gasoline engine and in 1885 made his first engines of about 0.5 kW output mounted on a motorcycle and an automobile, opening the era of automobiles. In 1893, Rudolf Diesel obtained patents for his engine, and in 1912 two diesel engines, each 920 kW, were mounted on the ship *Selandia*.

Electric motors and internal combustion engines grew very rapidly in use by opening up new horizons and also by taking the place of steam power for small-output uses. Steam lost its place in automobiles in very early days, in industrial vehicles such as tractors in early days, in factories gradually, on railways more recently, and on ships still more recently. Note that steam was replaced in the earliest days in the smallest unit output field, that is, the automobile, and more recently as the necessary unit output grew larger, irrespective of its use. This is evidence of the special importance, mentioned above, of the unit output among various parameters of power sources in the history of development of power technology.

When this replacement was going on, some people suspected that steam had lost its future, but it did not do so. Steam had actually found its right place among prime movers such as in electricity-generating stations, in power houses for blast furnaces, and in ships propelled by nuclear energy.

Electric lamps and electric motors created a rapidly increasing demand for electricity, which always needed more powerful generating stations than those already in existence, since, first, the growth of the total electricity demand was very rapid (doubling in ten to twenty years), and, second, larger-capacity stations meant lower cost per kilowatt-hour sold. Every year, at around the turn of the 20th century, the world record for the largest-output steam engine or turbine was broken by a new installation.

In 1929 the famous State Line Station with its 208-MW turbine was commissioned. The unit output of the steam turbine was nearly 1400 times as large as that of the Pearl Street steam engine, and the growth rate during the 47 years since the Pearl Street Station was equal to doubling every 6.59 years.

The world record for the largest steam turbine output in 1996 is 1300 MW for a nuclear power plant, and even today steam power has merit in scale: The larger the machine's unit capacity, the better is the efficiency and the less is the cost per kilowatt-hour.

Note in this short description of the history of steam power that a clear presentation of the history is possible by using only the unit output without recourse to other factors such as pressure, temperature, fuel types, material characteristics, and positioning of services. This is evidence that unit output forms the basis of steam power. Definition and further analysis of the basic factor in technology development is found in the literature (Ishigai 1989).

In this short description, factors other than unit output are not ignored but are represented implicitly. They are backstage but are ready to appear on the main stage if a more detailed presentation is to be made. If all factors are treated equally, the result is simply confusion. Distinction between the foundation and other conditions is necessary in understanding history (Ishigai 1989).

At the turn of the 21st century, not only steam power but also the whole of production technology seems to be facing crises, the only exception being information-processing technology represented by the computer. Environmental problems caused by waste products from power generation such as CO_2, NO_x, and radioactive ash are believed to be endangering the earth, while expansion of power production is still necessary to sustain the world economy. It was once claimed that the introduction of nuclear energy would ultimately solve the problem of energy resources, but now there are doubts about its future. A drastic change in the strategy of energy development is obviously necessary. In the following is proposed a method of forming such a strategy.

1.4 Strategy at transition to the new technological era

It is generally believed that the world is stepping into a new era of civilization now that three centuries have passed since the Industrial Revolution of the 18th century, and that the transition to the new era is being promoted by emerging new technologies. A safe and happy passage over the transition is in no way predetermined. Since there have already been many shipwrecks due to mis-steering over the tideway, and also since energy technology that includes nuclear power is among the most controversial currents of this tideway, strategy for a safe passage should be carefully studied.

It might well be argued that correct foresight into the future of such a drastic transition is impossible since the new era is totally unexperienced. Every transition to a new era in history is certainly unique and happens just once. However, it is also true that all such era transitions have some aspects and factors in common, which are interesting to study and understanding of which is proven to be useful in actual management of business and living, with inevitable limitations due to the circumstances and capabilities of the persons involved.

The proposed set of laws of development of technology (Ishigai 1989) has sometimes been misunderstood as a type of determinism, but it is not. An example of the distance between finding governing laws and finding their particular solutions for practical problems is the case of nonlinear differential equations, which can yield many solutions with jumps between them depending on initial and boundary conditions, and in which seemingly slight differences can yield tremendous differences in the solutions. Social events have one more factor that contributes to indeterminism, that is, the free will of humans. Another example is the distance between planning on maps and making actual explorations. Still, maps and laws of development of technology are certainly helpful.

The following sections of this chapter are devoted to introducing a historical method of strategy construction for tiding over this period of transition to a technological era, not by oracle or fortune telling but by reasoning and reconnoitering.

1.5 Three tools of strategy making for technology

1.5.1 Intrinsic laws of technology development

The intrinsic laws of technology development (Ishigai 1989) are very briefly introduced below.

Advanced apes such as chimpanzees and gorillas can make and use primitive tools. However, even the most primitive tools made by the earliest human, known as *Homo habilis*, are qualitatively different from those made by apes in the following two respects:

1. Apes handle the raw material for the tool directly with their forelegs and teeth to make it into tools, whereas the most primitive humans handled the raw material indirectly through tools in their hands when making paleolithic tools. In other words, the earliest humans created tool-made tools.
2. Even the most primitive paleolithic tools are not results of random trials but products of intentional labor. In other words, the product belongs to a particular type since it is a materialization of a more or less clear image in mind.

The simple use of means of labor by the earliest humans, between their hands and the objects of their labor, is the origin of the huge system of technology as it is today, and contains all moments of technology development through its history. Each paleolithic tool, as well as the tool system for making it, consisted of a simple one-piece solid body, but the use of means of labor gave both the maker and the user the freedom to select each particular form, material, and size from a wide variety of possible forms, materials, and sizes at his will in accordance with his purpose. In other words, the humans obtained a type of means of labor with one degree of freedom of selection.

Note here that the logic in the idea of the tool-made tool is circular. It is logically impossible to make the first one since existence of itself is the prerequisite for its production, and hence apes cannot make it. When the contradiction was solved by the trial-and-error method and became fixed by biological evolution, which created the yet unknown human ancestors by forging their brain through natural selection, the unknown ancestors of humans started to depart from the ape stage and became human.

Note also that the physical means of labor is so closely connected with the mental system of knowledge in making and using it that the knowledge can even be estimated by analyzing the physical body, as is usually done in archeology. On the basis of recognition of this close relation, technology is here defined as the system of means of labor.

The history of technology as defined above can be divided into eras by taking the type of direct means of labor, represented by its degree of freedom (df), as the sole criterion, as follows:

0. Prehistory (df = 0). The age of natural tools used by apes.
1. Era of the one-piece tool (df = 1). The freedom of selecting a one-piece solid body to be inserted between the worker and his objects of labor is the first degree of freedom obtained.
2. Era of the compound tool (df = 2). The additional new degree of freedom is relative motion in the direct means of labor as seen in the bow and arrow, wheel and shaft, mill, and many others.
3. Era of the machine (df = 3). The additional new degree of freedrom is the mechanism.
4. Era of the computer (df = 4). The additional new degree of freedom is information processing.

The history of power technology is similarly divided into eras, also consistently by a sole criterion, if the type of the maximum-unit-output power source is taken as the criterion, as follows:

1. Era of manpower (in which the human body is the power source that delivers maximum unit output among all available power sources)
2. Era of cattle power
3. Era of waterwheels and windmills
4. Era of heat engines (inclusive of the present day)

Note that such names as the era of electricity, the coal age, or the nuclear era are automatically discarded by this criterion. These names are expected to appear in some subdivisions. This is an important point of this era division and is not a fault, as is made clear later.

1.5.2 Transition period theory

There are many books on particular cases of era transition such as the rise and fall of a civilization, a country, or a hero. Some general properties common in these era transitions are noted below.

1. A transition period as a whole is a crisis to the changing body. A smooth passage over the transition is not assured beforehand. However, the same period is sure to be recalled as a chance occurrence by those who successfully navigate the transition. It is at the same time a chance and a crisis because it is changeable.
2. In an era transition, the dominant idea or system changes from the preceding era to the succeeding one. The time of transition is far shorter than the span of the preceding or succeeding era. Even so, several stages are also noticed in the transition period

such as the preparatory stage, the beginning stage, the steady progress stage, and the completion stage.

3. The era transition proceeds by solving hitherto unexperienced problems insoluble by traditional and prevalent methods. The trial-and-error method is therefore inevitable and the result is initial diversification in the type of emerging new quality.

4. An example of initial diversification is the rush of new types of steam turbine at the turn of the 20th century; that is, the De Laval single-stage impulse turbine in 1883, the Parsons reaction turbine in 1885, the Curtis velocity-compounded turbine in 1891, the Rateau pressure-compounded turbine in 1898, the Zoelly version of the pressure-compunded turbine in 1903. The diesel engine has a similar initial diversification, and the boiler abounds in many examples of initial diversification that, in the long run, gradually undergo selection by society to a final convergence.

5. Hence not all variations in the initial diversification have a true future, and some of them survive long but finally perish due to flaws that are not evident in their early days. An example is Edison's direct-current electricity distribution system. Screening at the earliest possible time can be carried out by applying the laws of development of technology. A simple knowledge of dependence of the suitable type on size may be helpful. An example is the DC distribution system in automobiles.

6. Prevalent ideas and systems before the transition are seen as assuredly mistaken, curious, or unsuitable from the post-transition viewpoint. Two systems fighting against each other in the pre-transition period might expect that only one will survive the transition, but both may perish if they miss the new key quality that becomes dominant after the transition. An example is the struggle between the open-country policy by the feudal Tokugawa regime and the closed-country policy by the new pro-emperor group in Japan in the middle of the 19th century. What controlled the post-transition era in Japan known as the Meiji Age was an open-country policy by the imperial regime.

7. New technologies that have a future are nourished and trained by special applications that are inherently not the most suitable major applications. An example is electric power supply technology, which started as electric light supply technology.

8. A new technology that has a future is healthy only when it remedies in practice its initially hidden faults and dangers that appear in the course of scaling up to its maturity.

1.5.3 Recognition of the present state of technology

Correct foresight into the future depends on correct recognition of the present state. The basis of such recognition is to find the right place of the present stage in the historical era sequence. In this study the historical sequences given in Section 1.5.1 are presupposed.

The assumption of other era sequences will certainly lead to different recognitions, and the acceptability of the recognition derived thus is the test of correctness or credibility of the presupposed era sequence. Those given in Section 1.5.1 have successfully passed several such tests up to now.

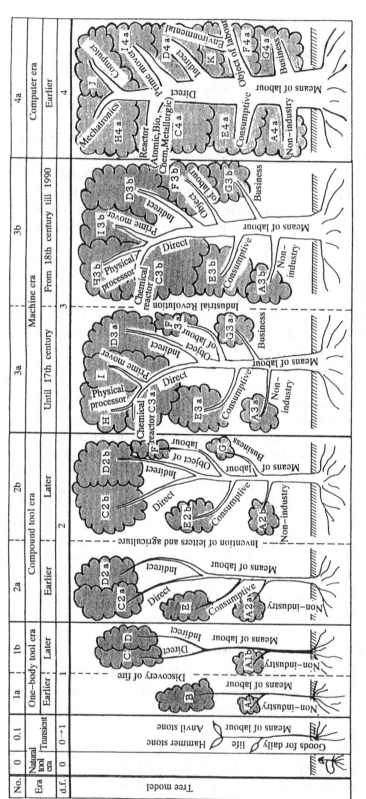

Figure 1.1 Historical eras of total technology and a tree-model presentation of progress in its differentiation. Time flows to the right. In each subdivision of the technical era, total technology is represented as a tree having a trunk identified as a means of labor after the definition of total technology given in the text. Vertical thick lines are borders between two successive eras with one exception, between the natural-tool era and the one-body-tool era, which is represented by an area as a reminder of the particular importance of this transient period that created the origin of total technology. (d.f.) degrees of freedom; (A) nonindustrial technology; (B) implements for tool production; (C) direct means of labor; (D) indirect means of labor; (E) consumptive technology; (F) technology embodied in objects of labor such as cattle, mines, and nautical routes; (G) business technology; (H) technology for processing, such as machine tools; (I) prime movers.

Another test of credibility or correctness is the acceptability of results obtained by going further into details. One result is presented in Figure 1.1 as a tree model, but without further explanation for the sake of brevity. A model or analogy is not a proof but is a good help for grasping the total image. Details shown in the figure are believed to be acceptable, and the era sequences presented above are presupposed in the following sections.

1.6 Position of present technology in historical eras

1.6.1 Division of total technology into eras

A division of the history of technology as a whole – total technology – has already been presented and is reproduced here for easy reference:

- prehistory
- the one-piece tool era
- the compound tool era
- the machine era
- the computer era

The single and consistent criterion for the above era sequence is, as also explained before, the type of direct means of labor represented by the number of degrees of freedom.

In Figure 1.1, time flows from left to right. The thick straight lines are borders between historical eras of technology, with one exception: The transition from prehistory or the natural tool era to the one-piece tool era is given not as a line but as an area, so as to express the particular importance of this transition, which created the earliest human by overcoming the logical contradiction of creating the tool-made tool.

The present time is postulated as being directly after the border from 3b (the latter machine era) to 4a (the earlier computer era) but still in the transient period from 3b to 4a.

The new degree of freedom in the computer age is information processing. This has always been exercised by the human brain through history, including today, and such instruments as the speed governor of Watt's engine and the mechanical calculator in the preceding machine era were prototypes of its materialization. The electronic computer, invented in the preceding era, has grown up very rapidly and has now taken over the position of the most dominant type of direct means of production, which is the common criterion of era division.

The electronic computer is dominant today for the following reasons:

1. The computer industry is very powerful, as indicated by the quantity of production and the number of employees. The computer is shown as a new but thick branch J in the model tree in Figure 1.1.

2. Just as important, the computer is the most important component of practically all devices of both labor and daily life, such as machine tools, washing machines, and even game machines.

1.6.2 Division of power technology into eras

The division of the history of power technology into eras is also reproduced here for easy reference:

- the manpower era
- the cattle power era
- the waterwheel and windmill era
- the heat engine era (inclusive of the present day)

Power technology is a part of technology as a whole. As such, it is under control of the basic factor of total technology; hence, the manpower era corresponds roughly to prehistory and the one-piece tool era, the cattle power era to the compound tool era, and the waterwheel and windmill era plus the heat engine era to the machine era.

1.6.3 Present power technology in historical eras

Power technology is a subsystem with its own basic factors that are dependent on the basic factors of total technology. Hence its era sequence is not in rigorous but in rough correspondence to that of total technology, as shown in Figure 1.2. Thus the compound tool era started while manpower was in use; this gradually brought in the cattle power era and also the early waterwheel and windmill era, which fostered the machine era to the stage that necessitated the invention of the heat engine, which in turn promoted total technology to the higher era of today, that is, the computer era.

Thus, the new technical era of the computer has already started by making use of the power technology now available. The evidence of the excellence of the new era of total technology is the very rapid increase in world population.

As mentioned before, history has proven that the tremendous progress made by entering a revolutionary new era of total technology requires a corresponding new type of power technology. In other words, there is a mismatch between total technology that is already in the computer era and power technology that is still in the traditional heat engine era, as indicated in Figure 1.2. This mismatch is evidently the true but hidden cause of the global environmental troubles that can eventually transform the earth into an uninhabitable planet.

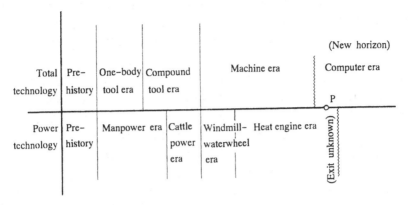

Figure 1.2 Comparison of era divisions of total technology with those of power technology. Time flows to the right. The present time is positioned at the point P at the beginning of the computer era and at the same time at the end of the heat engine era. The dominant type of power source in the future is not yet known. It may be a direct electricity generator, a hydrogen generator, or something else.

1.7 Proposed strategy for power technology development

The above analysis indicates that the emerging computer era of total technology is currently in a mismatch with the old heat engine era of power technology. This postulate is also supported by the contrast between the bright prospect of information-related industries and the cloudy prospect of leading power industries such as the electricity generation industry, which is suffering from difficulties in finding sites for power plants to meet its responsibility of supplying electricity to satisfy the ever-increasing demand.

Since initial diversification followed by selection to final convergence is generally seen in every new technology that has a future, as pointed out in Section 1.5.2, the most fundamental strategy for power technology development at the present time should be multidirectional.

At present it is very difficult to form a clear image of the post-heat engine era. It may be the electricity era, in which natural energy is directly converted to electricity by means of the fuel cell, the solar cell, or some others. There might be a transitional period in which the heat engine is still the most important power source, especially for vehicles that are supplied with hydrogen, methanol, hydrogen-enriched natural gas, or some other substance produced by nuclear, chemical, or solar reactors. When the transitional era is over, the selection process is nearly over and the result of the final convergence to one or two dominant types of means of power generation will be clearly recognized by everybody. However, the present time seems to be the beginning of a transitional period in which reconnaissance is important. This is the reason for the multidirectional strategy mentioned above.

Nuclear power is certainly a great invention but cannot be an exception. In its very early days of development up to around 1950, initial diversification was certainly there. However, the effect of the Cold War was so strong as to make the dual-purpose system most attractive, producing both plutonium for weapons and electricity for industry and life. The thorium–uranium fuel cycle that does not produce plutonium was hastily discarded and the light-water reactor line based on the uranium–plutonium fuel cycle was selected.

It is worth recalling that David E. Lilienthal, the first chairman of the U.S. Atomic Energy Commission, wrote of the danger of too-early convergence to the light-water reactor, urging that initial diversification as seen in the car industry is a necessary process (Lilienthal 1980). He recalls that the automobile industry started with the steam carriage and that the number of gasoline-driven cars produced was for some years behind that of steam cars, which continued to be produced until about 1926.

The light-water reactor is very suitable for the dual purposes of power generation for peaceful uses and fissile material production for nuclear weapons. However, the end of the Cold War in 1989 and later developments have changed the value of weapons-grade plutonium. It is now estimated that "50 or more metric tons of plutonium on each side are expected to become surplus to military needs, along with hundreds of tons of highly enriched uranium" (National Academy of Sciences 1994). The time has come to place nuclear power at its right place in the world economy.

All nuclear reactors now in use for electricity generation operate on the uranium–plutonium fuel cycle and generate by their operation more or less plutonium, which is a very important material for nuclear weapons and itself can be used as an explosive. Disposal of fissile materials from excess nuclear weapons by the thorium–uranium fuel cycle is attractive, since the cycle generates less radioactive nuclear waste, allows quicker plutonium disposal, and even allows the breeding of ^{233}U (Furukawa 1994) if necessary. This is not to say that further improvements in light-water reactors are unnecessary, since they are now extensively used for electricity generation and their replacement by a more suitable system in the post–Cold War period needs some lead time.

The thermal efficiency of light-water nuclear power plants improves if the turbine inlet steam pressure is raised. Several different approaches are now in progress (Japan Society of Mechanical Engineers 1996) and are welcome from the viewpoint of multidirectional strategy. The success of coal-fired supercritical-pressure power plants is encouraging since it proves the existence of practical solutions to problems associated with flow and heat transfer of supercritical water as the working fluid of steam power plants for electricity generation.

Another approach is to utilize nuclear heat for generating hydrogen or for converting fossil fuels to clean gaseous fuels. In this case, steam power is used either as the component that works on the heat from the gas-turbine exhaust gas, as in the integrated coal gasification combined cycle (IGCC) (Japan Society of Mechanical Engineers 1996; Kaneko 1996), or as the topping component that has an exhaust gas turbine. Still another approach is to use the molten-salt homogeneous reactor that sends high-temperature molten salt to a supercritical steam power plant (Furukawa 1964).

There should be a clear understanding as to the meaning of initial divergence and final convergence. It must be admitted that there is currently competition among several possible processes.

1.8 Fossil-fuel-fired steam power technology development

Among fossil energy resources, natural gas is the most promising because:

1. it is the "youngest" among fossil energy resources and is expanding its share while increasing its expected life, that is, the known amount of the resource divided by the annual consumption;
2. it is the cleanest among the fossil energy resources;
3. it paves the way to direct electricity generation by means of the fuel cell; or, in other words, its pipeline systems help to expand application of the fuel cell;
4. natural-gas-fired steam power plants cost less to construct and are more efficient than steam power plants fired by any other fuel, owing to the excellent properties of natural gas;
5. natural-gas-fired combined-cycle power plants have a better efficiency than pure steam plants;
6. natural-gas-fired co-generation plants in industrial or urban districts combine electricity generation with heat supply and bring about a high efficiency that is not attainable from pure electricity generation (the electricity-generating part can be fuel cells as well as heat engines);
7. other innovative ways of power production from natural gas may still be expected and should be explored.

The eastern coast of the Asian continent is the most rapidly developing part of the world and is already suffering from environmental problems caused by primitive coal-burning equipment. Coal supplies 60% of primary energy in this area, excluding Japan, whereas the world average share of coal is 25%. The exceptionally high dependence on coal there is a problem not only for this area but also for the earth's environment and should be remedied quickly. A master plan for a system supplying natural gas to this area is urgently needed.

In selecting the strategy for steam power technology development, its three fundamental characteristics mentioned in Section 1.2, that is, its aptitude for (1) large-output units, (2) low-temperature heat sources, and (3) combining power generation with process heat supply should be taken into account. It is impossible to go into details in this book, and the following paragraphs simply give a few hints.

Simultaneous development toward both larger and smaller unit output has always been apparent in the history of power technology. While a variety of miniature power sources is currently being developed, the largest unit output power source throughout the 20th century has been the steam turbine. One reason for this dominance is its adaptability to any kind of fuel, including low-grade coal and nuclear energy. Steam is even now responsible for creating advanced means of large-scale coal-burning at better efficiency and with less environmental contamination. Pressurized combustion and pressurized coal gasification are inevitable targets. Efforts toward higher steam pressure and temperature should be accelerated (Valenti 1996).

Progress in waste heat recovery and in extracting electricity from low-pressure steam has reached a point that pulp mills, steel mills, and many other industrial fuel consumers as well as incinerators can generate more electricity than they consume by carefully recovering their waste energy. Integration of such industrial byproduct electricity into electric utility grids is a good method of improving overall social energy efficiency.

Since the heating load and other fuel-burning loads are fairly independent of electric load, energy storage equipment is an important element in integrating fuel consumers' byproduct electricity into electric utility grids and is worthy of extensive R&D effort.

The transient period is at the same time an opportunity and a crisis. Bold and careful reconnoitering under well-thought-out strategies is necessary for successfully tiding over the period.

1.9 Emergence of intrinsic laws of boiler development through the interplay of natural and social laws of production

1.9.1 Introduction

In the preceding sections, the intrinsic laws of total technology development and power technology development were presented without going into the mechanism of their emergence, for the sake of brevity. This section demonstrates the mechanism of emergence of intrinsic laws of boiler development through the interplay of natural and social laws in the field of production.

Technology has always been developing. Hence it always contains young as well as old elements. The problem is that not all young elements have a future, nor do all old elements lack a future. An important key for correct judgment on the future state of total technology, as well as of any component of it, is the set of intrinsic laws of its development (Ishigai 1989).

Era divisions of both the history of total technology and the history of its important subsystem, power technology, were presented in Section 1.5.1, taking as the sole criterion, respectively, the type of direct means of labor and the type of maximum-unit-output power source. It was demonstrated that the era divisions embody intrinsic laws of their development.

Similarly, the era division of the history of the boiler presented in the following takes as its sole criterion the type of water circulation (in a broad sense), which is defined later in the last part of Section 1.9.2. The classification has three proper types and four intermediate types, as indicated in Figure 1.3, and goes as follows:

Transition from prehistory to the submergence principle, or the era of the primitive boiler

1. Era of the submergence principle class, or the **cylindrical boiler**

    (1–2) Transition from submergence to circulation principle, or the era of the submerged water-tube boiler

2. Era of the circulation principle class, or the **water-tube boiler**

    (2–3) Transition from circulation to through-flow principle, or the era of the forced-circulation boiler

3. Era of the through-flow principle class, or the **once-through boiler**

    (3–4) Transition from through-flow to single-phase flow principle, or the era of the nonboiling heater.

The era division embodies intrinsic laws of boiler development, since:

1. the sole criterion is water circulation in a broad sense, which is the most basic factor of the boiler. This does not mean that other factors such as fuel, combustion, and heat transfer are unimportant. They are certainly important, but they are conditional factors and not the basis of the boiler.

2. The era sequence of the history of the boiler is determined by qualitative evolution of the water circulation in a broad sense, as proven in this section.

It should be noted here that the postulation of the existence of intrinsic laws of development is not deterministic, since dependence of the process of development on its history is admitted. The actual mechanism of the emergence of intrinsic laws of boiler development is presented in the following sections.

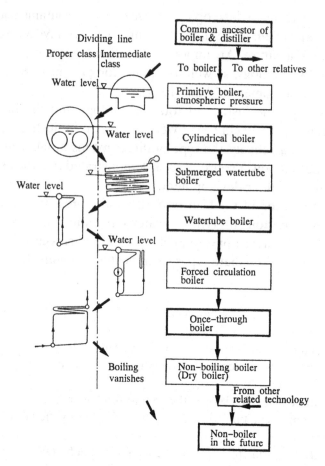

Figure 1.3 Technological era sequences of dominant classes of the boiler (Ishigai 1983). Note that the boiler emerged from nonboilers and that the boiler will someday fuse into some other nonboiler. Each class has its domain of excellence, depending on the scale of the boiler. Every proper class turns into the next proper class via an intermediate class.

1.9.2 From the ancestors to emergence of the primitive boiler

The histories of cooking and fabric processing, such as dyeing and washing, are very old. Hence the history of utilization of hot water and steam is older by far than the history of steam power. The first industrially successful steam engine by Newcomen in 1712 was helped by long-accumulated experience of using steam and hot water in such processes. Since 1712, the most advanced boiler with the largest steam generation in ton/hour (t/h), hereafter referred to as the unit capacity of the boiler, has always been used for power generation. Thus the prehistory of the boiler ended in 1712 when the boiler for the first Newcomen engine began the proper history of the boiler.

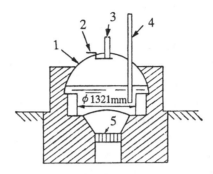

Figure 1.4 The primitive boiler for an atmospheric engine by Newcomen, 1715. (1) Boiler shell; (2) steam valve; (3) steam pipe; (4) pipe for inserting a float, 100 mm diam.; (5) firing grate (Dickinson 1963).

All steam engine inventors before Newcomen tried to use steam at above-atmospheric pressure. Because of the low quality of materials and fabrication technology at that time, their boilers were sure to leak or even explode sooner or later. The essential point of Newcomen's invention was to use steam only for vacuum generation, and hence his boiler worked at atmospheric pressure to totally avoid the possibility of boiler explosion. His boiler for his first engine is well described and illustrated by Dickinson (1963) and is diagrammatically sketched in Figure 1.4. The shape of the copper shell is strange from the modern standpoint, but it must have been familiar to engineers of his day. The shell contains water and is heated from outside. Since the shell is closed, the inner steam pressure is not naturally atmospheric but is dependent on the energy balance of the boiler, although it was easy to keep this at atmospheric pressure by controlling the firing rate because of the large amount of water in the shell in comparison with its unit capacity. Such a large water content in the boiler is a shortcoming by modern standards, but it made Newcomen's engine easy to control.

Since the shell was closed, a steam valve and a feedwater valve were indispensable. Some device was necessary to indicate the invisible water level in the shell in order to control it at a predetermined level by manipulating the amount of feed water supply, so that all the heating surface was submerged in water to prevent overheating. As steam generation progressed, solute in the feed water was mostly left in the boiler water, increasing its concentration, so that a water blow-off valve was necessary to control the concentration. A manhole was also necessary to get inside the shell to clean the contaminated surface and to take out sediment and scale.

Thus the boiler already had the minimum necessary devices for a boiler but the steam pressure was controlled at atmospheric level. Hence this boiler type is referred to as a primitive boiler and is postulated as the transition from its non-boiler ancestors to modern proper boilers that generate steam or water at any desired pressure and temperature.

Note that steam generated at the heating surface floats up naturally as bubbles by its own buoyancy to the water surface, transferring to the steam space without any help from outside as long as the water level is kept high enough to have the heating surface submerged in water. The function of removing steam from, and at the same time supplying water to, the points of steam generation distributed on the heating surface is referred to as the water circulation (in a broad sense),* the type of which is the most basic criterion of boiler classification that embodies its intrinsic laws of development. The primitive boiler is thus postulated to work on the "submergence principle," which the boiler has in common with its ancestors the kettle and the distiller.

1.9.3 Scale-up of the primitive boiler

The Newcomen engine met the long-awaited demand for a means of pumping water out of mines; its number of installations as well as its unit output rapidly increased. The first Newcomen engine is estimated, from the amount and head of the pumped water, to have delivered 4.1 kW. The largest engine, built in 1769, is similarly estimated to have delivered 76.5 kW, which is 19 times the output of the first engine. The steam consumption of the engine may be estimated at about 10 times that of the first Newcomen engine, since Smeaton is reported to have improved the Newcomen engine by cutting its steam consumption rate by about one-half.

James Watt is reported to have built roughly 500 engines before expiration of his patent in 1800, and the average unit output of his engine was about 11 kW, with a few 60 kW engines (Dickinson 1963), which is smaller than the largest Newcomen engine of 1769. The rate of growth of unit output from 1712 till 1800 is roughly 1.5 times in 10 years or 50 times in 100 years.

The social demand for larger unit output was usually met first by the engine or turbine and then propagated to the boiler. The earliest Newcomen engine had one boiler for its engine. The largest engine in 1769 had four boilers, of which one was a spare (Dickinson 1963). Early in the 20th century, the largest turbines each received steam from more than 20 boilers. The boiler caught up gradually and one-to-one coupling became a common practice after the Second World War.

---

* The phrase "in a broad sense" is added here since both "submersion" and "once-through flow" are not specifically included in "circulation." In some cases, "water circulation" is used in a narrow sense to indicate the particular case of the natural-circulation water-tube boiler, in which the boiler water actually circulates through a loop consisting of the upflow tubes, the drum for steam–water separation, the downflow tubes, and the lower drum for connecting the downflow tubes to the upflow tubes. The definition in the text is given here following tradition in boiler engineering.

In a series of geometrically similar boilers, the heating surface area is proportional to the square of the linear size and the water content is proportional to its cube. The equivalent depth of water, which is the water content divided by the heating surface area and is the virtual depth of water if the boiler consists of a vertical cylinder heated only from its bottom (as is the kettle for home use), increases linearly with size. Hence excessive scale-up makes any type of boiler thick-walled, and too long a time is necessary, for example, to bring it to boiling after igniting the boiler from cold.

An excellent method of reducing the equivalent water depth is to install heating surfaces as flues or smoke tubes submerged in the water. Smeaton is reported to have built some such boilers.

If the largest Watt engine in 1800 is estimated to have delivered 60 kW, this is 15 times that of the first Newcomen engine but is still smaller than the largest Newcomen engine of 1769. If the steam consumption per kilowatt output of Watt's engine is estimated at one-third that of Newcomen's first engine, which is roughly correct, then the necessary unit output of the boiler for the largest Watt engine is five times that of the first Newcomen boiler. If two boilers were used for one Watt engine, then the necessary boiler unit–output ratio reduces to 2.5. This small ratio tells us that Watt's policy of using the primitive boiler was sound.

1.9.4 Emergence of the cylindrical boiler as the first proper class of boiler

Although Watt opened the way to raising the boiler pressure to above atmospheric by separating the condenser from the power cylinder, he was reluctant to raise the pressure, for good reasons. There were, however, some inventors such as Richard Trevithick in England and Oliver Evans in the United States who were waiting for Watt's patent to expire and then started to build engines that worked at a steam pressure of several atmospheres. They made the boiler shell cylindrical to withstand the internal pressure and installed a flue for each boiler in its water space, observing the submergence principle. An example by Trevithick is reproduced as Figure 1.5. Thus a new boiler type, referred to as the cylindrical boiler, was born that works on the traditional submergence principle, generating steam from submerged heating surfaces in the water space.

The Industrial Revolution in the 18th century created demands for powerful prime movers that generate rotary motion. Spinning and weaving mills driven by water wheels were increasing in mountainous districts at the end of the 18th century, and steam helped such textile industries to locate their sites in plains. Sailing ships navigating the oceans were replaced by steamships. In the 19th century, railroads using steam locomotives expanded very rapidly. Blast furnaces

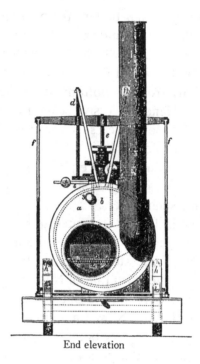

End elevation

Figure 1.5 Trevithick's Camborne road carriage, 1802 (Dickinson and Titley 1934).

were candidates for large land engines. The number and unit output of steam engines increased continuously.

In the meantime, Cornish engines were built according to the traditions set by Watt in which boiler pressure was raised, but very cautiously, so as to enhance fuel economy. Engine manufacturers started by supporting the flat parts of traditional primitive boilers with various stays, but they gradually switched to the cylindrical boiler mentioned above.

Numerous types of boilers were developed that worked on the submergence principle but with very different configurations. Initial diversification took place to meet expanding demand from land and sea. Saving in the weight and size of the cylindrical boiler is largest when the coal-firing grate is installed in a flue set in the unheated shell to eliminate heavy and bulky brickwork, as in the locomotive boiler and in the marine Scotch boiler. These two boilers were operating successfully as early as 1860 and prospered until about 1950, and the locomotive boiler survives even today on local railways, helped by progress in supporting technologies such as boiler water treatment, proving the toughness of proper types of the cylindrical boiler as against the weakness of intermediate types of the primitive boiler.

There are drawbacks to the expanded use of the cylindrical boiler, that is, the possibility of boiler explosion due to the raised pressure and the difficulty of cleaning the water side of the heating surface due to its reduced accessibility in comparison with the plain cylindrical shell. As countermeasures to boiler explosion, Trevithick mounted two safety valves on the shell and one fusible plug on the flue; the latter was a lead rivet that stays there while the flue is sufficiently cooled by the boiler water, but melts down to save the flue if the flue gets overheated. These methods became standard practice in boiler manufacture.

As usual in the process of evolution, there were several blind alleys in boiler development. An example is the elephant boiler, which did not have any flue or smoke tube in the shell but consisted of multiple horizontal cylindrical shells joined together to increase its unit capacity. Such a configuration retained an important merit of the primitive boiler, that is, easy access to all heating surfaces from inside, but lacked the merit of the cylindrical boiler, that is, reduction of weight and size. Since it was not cost-effective, it disappeared.

The final convergence in the cylindrical boiler is still in progress, along with improvements in performance such as efficiency, weight and size, NO_x emission, and automation. It is to be noted that even now the cylindrical boiler is the most dominant in the small unit-output range from 2 to 20 t/h.

The case of the cylindrical boiler mentioned above proves the following propositions of the intrinsic laws of boiler development:

1. There is no all-conquering type in anything; every type has its limited domain of excellence.
2. Each proper type in the history of technology has its own domain of excellence mainly determined by its scale, which in the case of the boiler is the unit output.
3. The first proper class of the boiler is the cylindrical boiler, which works on the submergence principle. It could not emerge directly from its ancestor, but emerged via the intermediate type, that is, the primitive boiler.
4. Almost every intermediate type perishes after performing its historical task of preparing the soil for the incoming proper type. The primitive boiler was no exception.

1.9.5 Transition to the natural-circulation water-tube boiler via an intermediate type

Ever since the 18th century, the social demand for more powerful prime movers than the one available then pushed up the unit output of the steam engine and turbine, as presented in Figure 1.6, which, in turn, pulled up the boiler unit capacity.

The thermodynamic properties of steam ensure that the most efficient steam pressure for a given steam power plant rises with the engine/turbine unit capacity, and the pressure in turn sets a practical limit to each boiler type. The rise in

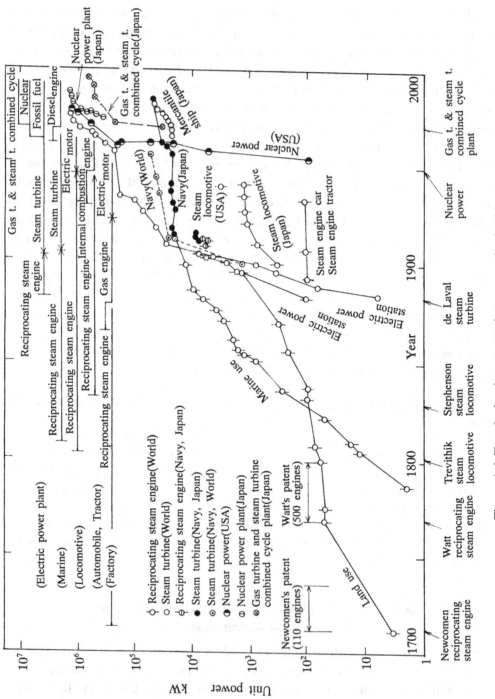

Figure 1.6 Trend of unit output of steam engine/turbine.

steam pressure from atmospheric to about 0.5 MPa was the direct cause of the transition to the cylindrical boiler, which has its upper limit at about 2 MPa. This type was in turn taken over by the water-tube boiler, which has its suitable pressure range from about 0.5 to 15 MPa. And the water-tube boiler was taken over by the once-through boiler, which can go up to above the critical pressure (22.120 MPa) but whose practical lower limit is about 8 MPa.

The cylindrical boiler met the demand for larger unit capacity at higher pressure than the primitive boiler by installing flues and/or smoke tubes in the shell as well as by enlarging the shell that contains them. Since the necessary wall thickness of the shell is proportional to the product of the pressure and the diameter, there is a practical upper limit to the shell size in spite of the improvement in materials from copper to cast iron, wrought iron to steel. There is thus a limit to the contained heating surface area and consequently to the unit capacity of the cylindrical boiler.

An excellent method of evading the limitation on wall thickness is to use for the heating surface only small-diameter water tubes heated from outside, whereas the flues and smoke tubes are heated from inside. It should be remembered here that, in the early days of the steam engine in 1712, steel was known but was very expensive. The Bessemer converter and Siemens–Martin open hearth, which revolutionized steel production, were invented only around 1860, and Mannnesmann's method of seamless steel-pipe production was patented in 1885. Numerous trials of very diversified designs of water-tube boiler extended over more than a hundred years until about 1875 (Dickinson 1963), when the initial diversification common to every important technology ended by social selection and the final convergence to two types started.

One type referred to in this book is the submerged water-tube boiler, and the other is the natural-circulation water-tube boiler, which, however, is generally referred to simply as the water-tube boiler. The former has its water level controlled at some height in the water tubes, as indicated in Figure 1.7, and works on the submergence principle, while the latter has its water level controlled in the drum, as indicated in Figure 1.8, and works on the natural circulation principle, which was an important innovation in the mechanism of supplying water to the boiling points on the heating surface.

To maintain the water level in the submerged water-tube boiler, as shown in Figure 1.7, the water in the tubes must remain stationary like the water in the cylindrical boiler shell, while the generated steam must pass through the supposed water surface in the tube. The steam velocity through the water surface must be slow enough not to violate the in-tube surface too much, so that the indicated level in the connected level indicator is stable and correct. This new limiting condition for the submerged water-tube boiler is not present in the natural-circulation water-tube boiler, since steam–water separation is not a function of the water

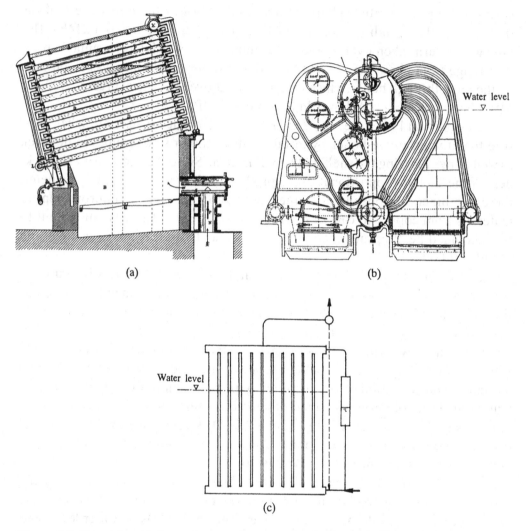

(a) (b)

(c)

Figure 1.7 Examples of the submerged water-tube boiler. (a) Howard's boiler, 1871 (Dickinson 1963); (b) Thorneycroft boiler for *Daring*, a British destroyer, 1893 (Smith 1937); (c) modern Japanese design, 1995.

tubes as in the submerged water-tube boiler, but a function of the downstream drum, which is similar to the shell of the cylindrical boiler.

Both types of water-tube boiler could meet the need for the increasing pressure and unit output of the time. The earlier winner of the tough competition was the submerged water-tube boiler. An example at its peak of prosperity is the French-origin Belleville boiler. This was built by Vickers in the United Kingdom in 1902 for the Japanese battleship *Mikasa*, which had 25 Belleville boilers to supply her 11 MW engines with 2.1 MPa of steam.

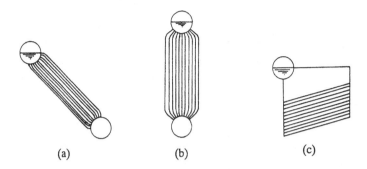

Figure 1.8 Examples of the natural circulation water-tube boiler. (a) Inclined tube type; (b) vertical tube type; (c) sloped tube type.

In the meantime, practical experience gradually revealed the favorable effects of water circulation on boiler tube reliability by reducing corrosion and scale accumulation that eventually caused tube explosion. If the boiler unit capacity is small and the water tubes are short, the driving force of natural circulation is small, and the difference in safety and reliability between the submersion-type water tube and the circulation-type water tube is insignificant. However, progress in steam engines reached the stage in which larger unit capacity than the submerged water-tube boiler could provide became a necessity. The need was evident from the large number of 25 boilers coupled with the two propulsion engines of the *Mikasa*. The earlier winner of the competition perished all of a sudden at the transition to the 20th century and was overtaken by the natural-circulation water-tube boiler, which reached 700 t/h, 200 MW, and 18 MPa in the 1950s.

A new problem for the natural-circulation water-tube boiler was to generate a reliable flow of the steam–water mixture through the water tubes into the drum, since local flow stagnation brings about a local accumulation of steam, which is a poor heat conductor and, if accumulated in stagnation, causes overheating, corrosion, and eventually tube explosion. The solution of this new problem was difficult and took time, since, first, measurement of velocity and quality (i.e., the mass ratio of steam in a steam–water mixture) of the two-phase flow in tubes under high pressure is difficult even today, and, second, the actual flow of circulating water through many parallel tubes with unequal steam generation is quite complex. A simplified loop having unheated downcomer tubes, as indicated in Figure 1.9, is today a standard practice for high-pressure boilers. However, this configuration was impossible in the days of its initial diversification. The final version was realized shortly before the Second World War.

The reason is that the latent heat of water diminishes with the pressure and finally vanishes at its critical pressure, 22.120 MPa. In its early days when the pres-

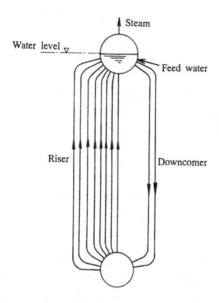

Figure 1.9 Principle of the natural-circulation water wall.

sure was low, the latent heat made up nearly all of the heat absorbed by the water. As the pressure rises, heat absorption by the superheater, reheater, and econo-mizer other than the evaporator increases, and finally only a part of the radiant heating surface in the furnace suffices to supply the latent heat. Thus the simple loop design, as in Figure 1.9, was the result.

In Figure 1.9, steam generated in heated water tubes mixes with the flowing water in the tube. The steam–water mixture is lighter in specific weight than the water in the unheated downcomer, and the difference is the driving force for the water circulation through the loop. Note that no mechanical help from outside is necessary as long as in the drum the water surface is at the designed position. Feedwater that compensates for the mass of the outgoing steam is supplied into the drum to control the water level by regulating its flow rate.

An important parameter of the flow of boiling water through tubes is its mass velocity, which is equivalent to the speed of water at the tube inlet. It is very close to zero in the submergent water-tube boiler, since liquid water is supposedly sta-tionary and the steam is supposed to float up as bubbles through the stationary water with a small amount of entrained water.

The mass velocity is higher in the natural-circulation water-tube boiler because the steam generated in the tube is designed to entrain several times more mass of water or still more through the circulation loop. The measured inlet speed is around 0.7 m/s or more in well-designed high-pressure natural-circulation water walls, that is, furnace walls made up with water tubes. It is interesting to note that

0.7 m/s is a very rough but good measure of sufficient inlet water velocity for all three boiler types, the natural-circulation, forced-circulation, and once-through boiler.

Progress of the natural-circulation water-tube boiler since 1900 brought about a "previously undreamt-of rise in pressure and [unit] output of boilers of the orthodox design" (Dickinson 1963, p. 238). In 1900, 2 MPa was regarded as high pressure, and 8 MPa was a reality in 1920 at Edgar Station (Orrok 1930). The success of natural circulation itself indicated that further strengthening of water circulation by use of a pump would have a future, and this proved true. The natural-circulation boiler ceded its dominance around 1938 to the forced-circulation boiler, which was the intermediate type and was in turn replaced by the once-through boiler very soon after the Second World War.

It should be added here that, even in the postwar period, the natural-circulation boiler was competitive enough to reach its peak pressure of 17.4 MPa in about 1970. It is still the most prosperous in the unit-output range from 10 to 200 t/h, proving that any proper type has a lasting strength in its proper domain, as the one-piece tool like the knife or hammer has a history just as long as that of humankind.

The case of the water-tube boiler treated in this section is again a proof of the intrinsic laws of boiler development as stated at the end of Section 1.9.4.

1.9.6 Initial divergences for meeting the new era of the once-through boiler

The third class of proper boilers listed in Section 1.9.1 is the once-through boiler, relying on the once-through principle. The logical sequence in the development of boiler technology after the era of the natural-circulation water-tube boiler is, from the viewpoint of the intrinsic laws of boiler development as shown in Figure 1.3, a transition to the once-through boiler via an intermediate type. History has developed exactly in this way. It is interesting to look into some details of the process of emergence of the intrinsic laws through the interplay of physical and social laws of production.

As usual in every emerging new technology that later becomes dominant, the once-through boiler started as an initial divergence in the 1920s. At that time, it was generally admitted that the difficulty of natural circulation had to be surmounted, because high pressure means less volume of steam and consequently less driving force of natural circulation.

The actual situation is, however, not so simple, since less steam volume means less velocity of the steam–water mixture in the evaporating tubes, which in turn means less flow resistance that helps natural circulation. Thus there were also good reasons for trials to improve natural circulation. All the following possibilities were actually tried:

(a) improved natural circulation;
(b) forced circulation;
(c) once-through flow;
(d) other methods (indirect steam generation, revolving heating surface, use of fluids other than water, etc.).

Category (a) was very successful, as proven by its later development toward its final version, in which evaporation takes place only in the water wall and consequently the differences shown in Figure 1.9 disappear. The natural-circulation water-tube boiler was dominant until about 1943, when it was overtaken by the forced-circulation boiler (b).

Category (b) involves the addition of a circulation pump in the downcomer from the drum, as shown schematically in Figure 1.3. Since both (b) and (c) depend on the pump for the water flow through the steam-generating tubes, they were usually both categorized at that time as the forced-flow boiler, which certainly identifies an important aspect. However, from the viewpoint of the intrinsic laws of boiler development, such classification masks the truth and does not explain important differences in their development such as the very short dominance of (b) in comparison with the much more lasting dominance of (c).

They in fact demonstrate different stages of development, since the one works on the same circulation principle as the natural-circulation water-tube boiler, with a drum for steam–water separation, whereas the other works on the different once-through principle without a drum for steam–water separation. Evidently (b) is the transition from the proper type (a) to the next proper type (c), and behaved exactly as such. The crucial point is not the drum itself but its reason for being there. Classification is thus an important key to finding intrinsic laws of development.

Category (d) includes many trials, including the Loeffler boiler. No fewer than 23 Loeffler boilers were actually constructed at 12 MPa pressure for electricity generation by the end of the Second World War. They are certainly examples of initial divergences seen in every technological era transition.

1.9.7 Challenges by forced-flow groups

1.9.7.1 Number of parallel tubes and other characteristic numbers

A bird's-eye view of the relative position of each domain of the three competing types, that is, the natural-circulation boiler, the forced-circulation boiler, and the once-through boiler, is given by combining the mass velocity (0.7 m/s in Section 1.9.5) with other characteristic numbers presented herewith.

One is the tube diameter for a water flow of 1 t/h at 0.7 m/s. A simple calculation yields 22.48 mm as the diameter, which is very close to one inch. Among the others is the circulation ratio, which is the mass ratio of flowing water to the steam flow generated in the tube.

The actual water circulation in the usual types of low- and medium-pressure natural-circulation boilers sketched in Figure 1.8 is very complex. Because these boilers have so many water tubes in parallel, most of them, other than water-wall tubes, are located in passages of relatively low-temperature gas to cool the gas down to the stack gas temperature, or to the gas temperature at the inlet to the air preheater or the economizer, if any. In such evaporating tubes, the speed and direction of the boiling water flow are not stable but are incessantly fluctuating.

For example, if the water flow is slow, the generated steam gradually accumulates in the tube to form large cylindrical bubbles that suddenly leave the tube and pass into the drum when the bubble head reaches the tube outlet. Then water flows into the tube from both ends, fills up the tube, remains stationary for a while, and the process repeats. Parallel tubes make the situation more complex. Such flow instability has been proven safe if the pressure is not very high and the circulation loop is designed with due care. Most intensely heated are the water-wall tubes that have unheated downcomers, as in Figure 1.9, where the inlet velocity is around 0.7 m/s.

An important type of forced-circulation boiler is the La Mont boiler, which was patented in the United States in 1925 but was developed in Germany about 1928 (Dickinson 1963). The inlet water velocity of the evaporating tubes, which represents the mass velocity, ranges from 0.3 to 1.2 m/s, averaging around 0.7 m/s, similar to well-designed water walls of natural-circulation boilers. The circulation ratio is selected at about 4 to 12 depending on the situation, and this again coincides with well-designed natural-circulation high-pressure water walls.

The circulation ratio indicates that 4–12 parallel tubes of 25 mm diameter are enough to generate each ton per hour of steam. To construct water walls that cover all furnace walls with such a small number of tubes, the tubes must go meandering along the furnace wall or in the flue. If 50 mm is selected as the tube diameter, the number decreases to 1–3.

The number of parallel tubes is even fewer in once-through boilers since the circulation ratio is unity and all water inflow is sent out as steam. The earliest once-through boiler, by Sulzer, was actually made up of one continuous tube. This is the origin of its trade name "Monotube Boiler," although several or more parallel tubes are necessary in boilers of unit capacity larger than about 10 t/h.

Another interesting characteristic number is the equivalent water depth of water tubes, defined as the volume of liquid water in the tube divided by the heating surface area of the tube (see Section 1.9.3). If a water tube is filled with water, the equivalent water depth is equal to one-fourth of its diameter, for example, 19 mm for 75 mm diameter and only 6 mm for 25 mm diameter.

Since the large volume of generated steam displaces water in the evaporating tubes, the equivalent depth of natural-circulation water tubes is about one-half to one-fourth of the 19 or 6 mm mentioned above, that of forced-circulation water tubes is about one-tenth of the natural-circulation water tube, and in once-through water tubes it is reduced by a further factor of five. This number is a measure of sensitivity. Again, regarding sensitivity to disturbances, the forced-circulation water tube falls between the natural-circulation and the once-through water tube.

It is well known that the high-pressure steam power plant has always been a pioneering user of instrumentation and automatic control. In its early days, the once-through boiler was feasible only for base-load stations even when the best automatic control system then available was used, while the development of advanced control systems was supported by the wide market for controllers for traditional boilers. Today, such systems not only allow the most advanced supercritical once-through boiler to undergo a daily start-and-stop, but also provide automatic frequency control of a group of stations that have quite a number of supercritical boilers.

1.9.7.2 Flow distribution and flow instability

Other important factors in designing forced-flow evaporating tubes are flow distribution and flow instability. An effective countermeasure to both is the installation of resistance at the entrance in single-phase liquid flow. Two-phase downward flow should be avoided as much as possible and prohibited if the volume velocity is so low as to cause phase separation. Two-phase flow in parallel channels has various modes of flow oscillation, which can be avoided, however, by careful design of the channel.

Since the equivalent water depth in tubes, and consequently the liquid content, relative to the natural-circulation boiler, is less in the forced-circulation boiler and least in the once-through boiler, the danger of overheating increases in the same order. At the same time, however, damage accompanied by tube rupture decreases in the same order due to the differences in the equivalent water depth and the diameter of the tube.

1.9.7.3 Challenge by the forced-circulation group

Initial diversification of the water-tube construction, including water walls, was naturally there in the early history of development toward high pressure. A common problem for all forced-flow groups was the meandering coil configuration, due to the smaller number of tubes than in the natural-circulation boiler. It was an important departure from the natural-circulation boiler practice in which nearly straight tubes were preferred because it was easy to clean the inner surface. However, chemical protection from corrosion and fouling was, in any case, necessary in all boiler types and was eagerly pursued. The resultant progress

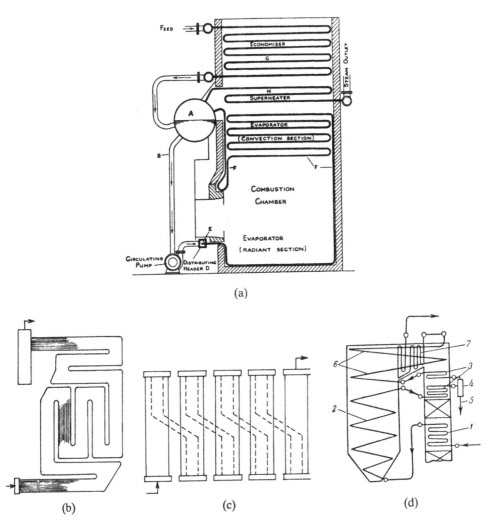

Figure 1.10 Scheme of the La Mont boiler and the three major once-through boilers: (a) La Mont; (b) Monotube; (c) Benson; (d) Ramzin.

of water chemistry improved the reliability of every type of boiler at any steam pressure.

The La Mont boiler (Fig. 1.10a) was a typical forced-circulation boiler that was actually an intermediate type between the natural-circulation boiler and the once-through boiler. As such, it could compete with the natural-circulation boiler in all pressure ranges and in all fields of application. Even vertical up-and-down meandering was applied in practice to boilers having a quickly swinging load since the water circulation was established irrespective of the boiler load. This merit was attractive in some waste-heat recovery units, although it did not mean much for ordinary applications of boilers since there are many other factors that restrain the effects of very quick load swing.

1.9.7.4 Challenge by the once-through group

Three major types that were successfully constructed in the 1930s after passing the experimental stage as large-capacity high-pressure boilers for electricity generation were the Monotube boiler, Benson boiler, and Ramzin boiler. They worked on the same once-through principle but each had its own particular sub-principle for water-wall construction, as shown schematically in Figure 1.10b–d.

The Monotube boiler had each economizer tube connected to each evaporation tube continuously without insertion of a header. By this configuration, the pressure drop in the economizer was utilized as the inlet resistance for flow stabilization. Meandering of water-wall tubes in the furnace was horizontal immediately after their entrance into the furnace, then vertical, both up and down, after the velocity of the steam–water mixture became high enough to carry the buoyant steam downward along with the water, and finally horizontal. The selected water-wall tube diameter was the largest among the three types of once-through boiler, probably for the purpose of reducing the number of parallel tubes that had individual controllers for flow adjustment.

Each tube had a flow control valve at the inlet and a temperature sensor at the outlet. The total feedwater flow to all evaporation tubes was controlled by the saturated steam flow at the outlet from the separator drum that joined the evaporator tubes. The flow to each tube was adjusted by its outlet temperature to keep the quality of steam at the outlet at about 97%. Fouling of the tube inner surface was most liable to happen at the part where the final 3% of wetness was evaporated. Hence the evaporator tube ended its evaporation at about 97% quality and discharged into the separator drum to blow off the remaining liquid water.

Until about 1945, the Benson boiler adopted vertical meandering for its water wall. The vertical upflow leg was constructed with water-wall tubes and the unheated downflow leg was placed outside the furnace to send the steam–water mixture to the neighboring uprisers, as shown in Figure 1.10. Thus the most intensely heated water walls consisted only of vertical uprisers. The instability problem of the downward two-phase flow shifted thus to unheated downcomers and was solved by higher flow velocity in the downcomers.

The steam–water mixture from the final water-wall section was then brought into the inlet header of the final evaporator located in the flue gas passage from the superheater. There the mixture was heated moderately to a slightly super-heated state and any impurities were left there safely as deposits, which were occasionally cleaned out. Unlike the Monotube boiler, the drain-separating drum was seldom installed.

The water-wall tube size of the Benson boiler was smaller than that in the Monotube boiler, and consequently a greater number of parallel water-wall tubes were used in the Benson boiler. Numerous headers and downcomers in the evaporating section served as mixers that absorbed nonuniformity among parallel tubes in each water-wall section.

The water-wall configuration of the Ramzin boiler was totally different from that of the Monotube boiler and the Benson boiler. Parallel water-wall tubes wound around the furnace wall helically at a small angle, starting from the bottom. Downward flow of the steam–water mixture was thus completely avoided, at the expense of some difficulty in field construction of the water wall.

The Ramzin boiler was also constructed for lower pressures than the Benson or the Monotube boiler, apparently for the purpose of evading the costly drum necessary for natural-circulation boilers. In such medium-pressure once-through Ramzin boilers, horizontal evaporating tubes arranged in a configuration similar to the economizer were necessary since the latent heat is larger than that of high-pressure water.

The differences in these three types were thus very significant in the 1930s, when 10 MPa was regarded as very high.

1.9.7.5 The situation at around 1943

In 1943, 12 MPa was a reality. Peculiar experiments such as revolving boilers and two-stage evaporation boilers had already been discarded. However, the Loeffler boiler, the Mercury boiler, and some others were still being constructed and were in the course of further development. The natural-circulation boiler was predominant, but the forced-circulation boiler and the once-through boiler were increasingly accepted by the market, proving their ability to be developed as more advanced boilers of 300 t/h or more unit capacity at 12 MPa and 510°C or higher steam conditions. Block coupling of one boiler with one turbine was gaining acceptance.

1.9.8 Transition to the once-through boiler via an intermediate type

Water at a constant pressure above the critical gradually changes its state by heat absorption without boiling, changing from liquid to vapor as the temperature passes through a narrow zone beyond its critical temperature, 374.15°C. This means that the natural-circulation principle cannot work and also that steam–water separation in the drum is impossible. Hence neither the natural- nor the forced-circulation principle can work. The once-through principle also suffers from a sudden degradation of heat transfer when passing through the near-critical zone, if the mass velocity is insufficient.

Intensive fundamental research and accumulation of practical experience were necessary to endow the supercritical boiler with the necessary reliability and operability, as was clear from the construction of the two pioneering units in 1956: the Huels unit at 88 MW and 30 MPa, and the Philo unit at 125 MW and 31 MPa.

The transitional process leading to the dominance of the once-through principle via the intermediate forced-circulation boiler is presented below in Section 4.7, and only comments on the following three points are added here.

An interesting point, from the viewpoint of the intrinsic laws of boiler development, is that the three types that were originally so different from each other have now practically fused together to a final convergent type, especially since stepping into the supercritical region. The sliding-pressure type, which can be operated safely at a very low load of 15% by sliding the steam pressure down to 7 MPa while keeping the superheated steam temperature constant, is an important achievement that has expanded its application to daily start-and-stop uses under control of the automatic grid-frequency control system. This is also interesting as another proof of dependence of the optimum boiler pressure on the unit output.

The strong position held at present by the natural-circulation boiler in the domain from 5 to 300 MW is also interesting. An example is the case of the main boilers for LNG carriers equipped with 5–30 MW steam turbines for propulsion (Marine Engineering Society of Japan 1996). One important reason for its dominance is that the supercritical steam condition is seldom adopted in this domain and the natural-circulation boiler consumes less auxiliary power because the circulation pump is unnecessary in comparison with the forced-circulation boiler, or because its boiler feed pump consumes less power than the once-through boiler. There are many other physical and social reasons, and the interesting point is that the intrinsic laws of development emerge as the result of the interplay of natural and social laws in the field of production.

The fate of the forced-circulation boiler is typical of the intermediate type. It climbed to its most dominant position when it served as the bridge from the era of the water-tube boiler to that of the once-through boiler. In such a large-unit-output and high-pressure domain, the forced-circulation boiler is no longer used since, on the one hand, the circulation principle cannot work in the supercritical domain, and, on the other hand, the water circulation pump is the weak point in comparison with the natural-circulation boiler in the subcritical domain. It now has an established field of application only in some particular waste heat recovery units. Its fate is similar to that of the lung fish, which was once very prosperous when it bridged the gap between the marine animals and the land animals, but now survives only in very limited areas since it is defeated by purely marine animals in water and by purely air-breathing animals on land.

However, the enterprise that gave rise to the La Mont boiler in 1928 did not perish with the fall of the forced-circulation boiler. It created a new type of natural-circulation water-tube boiler known as the Cornertube boiler, which is based on experience in developing the La Mont boiler. A similar situation is also seen in the United States, which once specialized in the natural-circulation water-tube boiler but has successfully switched to the once-through boiler. These are examples of the categorical difference between business and technology.

1.9.9 Concluding remarks

The concluding remarks given in the last part of Section 1.9.4 hold as they stand there. The following remarks are given as additions to the foregoing remarks.

5. The second and third proper classes of the boiler are the natural-circulation water-tube boiler, which works on the circulation principle, and the once-through boiler, which works on the through-flow principle.
6. Each proper class has its own domain of excellence dependent mainly on its scale, which is the unit output in the case of the boiler. (Point (2) is repeated here for the sake of consistency.)
7. The criterion of the classification of the boiler mentioned above is water circulation in a broad sense as defined in Section 1.9.2. This classification embodies the intrinsic laws of boiler development as the due result of the property of this criterion, which is the most basic factor of the boiler, not simply one of the important factors of a boiler such as the fuel, material, or use.
8. It is to be noted here that the driving force that washes away the generated steam from, and supplies water to, its point of generation is strong enough in the submergence principle to make pool-boiling an excellent heat transmitter. It is stronger in the circulation principle, and strongest in the through-flow principle.
9. It is therefore postulated as one of the intrinsic laws of boiler development that the historical era sequence shown in Figure 1.3 was driven by ever-increasing social demand for steam power. The demand was first met by the primitive boiler, which was the intermediate class to the first proper class, or the submergence-principle class. As the social demand for steam power grew further, beyond the suitable domain of the then-dominant working principle, the technological era sequence shown in Figure 1.3 took place to bring in a class of boiler more fitted to the larger unit output.
10. There are many fundamental laws that jointly control the boiler, such as natural laws that include thermodynamics, hydraulics, and strength of materials, and social laws that include economics, ethics, and juristic laws. Interplay of these laws in production and operation of boilers gives rise to the intrinsic laws of boiler development.
11. The boiler is an element of power technology, which is in turn a component of total technology, each having its own intrinsic laws of development. However, all such laws can affect the development of the boiler itself through the intrinsic laws of boiler development.
12. The key to discovery of the intrinsic laws of boiler development was the finding of the boiler classification that includes missing classes between existing proper classes.

13. Such intermediate classes are the primitive boiler, the submerged water-tube boiler, and the forced-circulation boiler.
14. Intrinsic laws of total technics development and of power technics development are derived by applying the same method of deriving the intrinsic laws of boiler development.

1.10 Proposed strategy and suggestions for boiler development

The strategy for boiler development is subordinate to the strategy for fossil-fuel-fired steam power technology development presented in Section 1.8. Since the most fundamental strategy for power technology development at the present time is multidirectional, as pointed out in Section 1.7, the strategy for boiler development must also be multidirectional. In other words, the boiler must dispatch a reconnaissance into the future following a substrategy for itself formed by using the three tools for strategy-making presented in Section 1.5, with due consideration for the properties of the boiler. Some suggestions, however, are presented in the following.

One suggestion is the tube-nested combustor for burning natural gas. Installation of a well-designed tube nest in the burning fireball in a natural-gas-firing combustion chamber, as in Figure 1.11, proved to agitate and cool the burning flame to bring about uniform flame temperature distribution and NO_x reduction. Quenched burning gas in the boundary gas film on the tube surface is carried by Karman vortices into the burning gas to be burnt up, since the tube-nest arrangement is so designed as to generate Karman vortices after all tubes in the tube nest. There is no longer a very intense radiant heat flux from the large fireball that fills the combustion chamber, since the fireball is penetrated by the tube nest (Ishigai 1992).

If the tube-nested combustor is used as the gas turbine combustor for the combined cycle, as in Figure 1.12, excess air supply to the combustor for the purpose of tempering the gas temperature at its exit is unnecessary; thus the compressor flow of the air is reduced to improve the cycle efficiency, although the efficiency gain is mostly lost by the less efficient Rankine cycle that accepts steam from the tube nest in the combustor (Ishigai 1993). It is also possible to cool the tube nest by tempering air, as in Figures 1.13 or 1.14 (Ishigai 1995), or by using steam from the exhaust-gas boiler. At any rate, the overall effect of the tube-nested combustor should be carefully studied, since it certainly provides a new degree of freedom for improving the quality of combustion.

Some words should be added here on other directions of boiler furnace development. There are three hierarchic classes in boiler furnaces that burn fuels and/or wastes. The most basic class is the traditional atmospheric-pressure furnace, which is being developed by improving the firing equipment and by

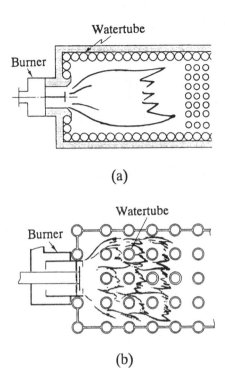

(a)

(b)

Figure 1.11 Natural-gas-firing combustion chamber without (a) and with (b) tube nest in the burning fireball.

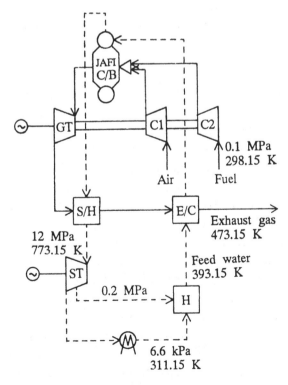

Figure 1.12 Application of the JAFI combustor to a combined cycle.

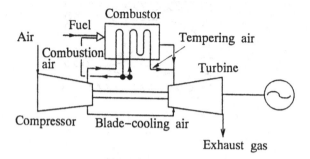

Figure 1.13 Application of air-cooled JAFI combustor to gas turbine.

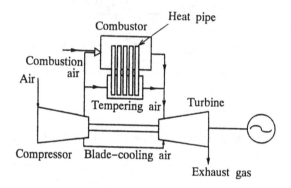

Figure 1.14 Application of heat-pipe-cooled JAFI combustor to gas turbine.

adding gas-cleaning equipment and ash-treatment equipment. Fluidized-bed coal/waste combustion is an example of successful development in this class.

The second class is the pressurized boiler furnace, or the supercharged boiler. It has an exhaust-gas turbine to supercharge the air to the furnace, as in the Velox boiler built since the 1930s, but the Velox boiler gained only limited success. Another variation of the pressurized furnace, for oil-firing marine boilers, was developed after the Second World War, aiming primarily to reduce weight and volume, but it met with little success in the market.

A new development referred to as the pressurized fluidized-bed combined cycle (PFBC) is now ongoing, backed by recent progress of gas turbine and fluidized-bed coal firing. The difference from the earlier supercharged boiler is that the gas turbine generates more power than is necessary for air compression and the excess power is sent out as electricity. It is more efficient than the traditional steam power plant of similar unit-output class, and is environmentally safer.

The third class is characterized by the addition of one more new quality other than pressurization. It might be postulated that coal gasification is a promising new quality, and the IGCC is a typical example. More study is needed on the intrinsic laws of boiler-furnace development.

References

Dickinson, H.W. 1963. *A short history of the steam engine*, 2nd ed. London: Frank Cass & Co.

Dickinson, H.W., and A. Titley. 1934. *Richard Trevithick*. London: Cambridge University Press.

Furukawa, K. 1994. Opening of rational safe thorium utilization way THORIMUS-NES by plutonium-burning and ^{233}U-production. *Ukrainian Chemistry Journal* 60(7): 456–72.

Ishigai, S. 1983. Boiler. In *Encyclopaedia of history of science and technology*, ed. S. Ito, K. Sakamoto, et al. Tokyo: Kobundo, pp. 973–5.

Ishigai, S. 1989. Intrinsic laws of development of technics. Parts 1 & 2. *Technovation* 9: 388–429.

Ishigai, S., H. Kobayashi, Y. Ueda, et al. 1992. Jaggy fireball in tube-nested combustor, an advanced concept for gas-firing and its application to boilers, American Society of Mechanical Engineers, HTD-199, 189–95.

Ishigai, S., H. Kobayashi, Y. Ueda, et al. 1993. Application of jaggy fireball concept to combustor of gas turbine for combined cycle. 20th CIMAC, London, paper no. G04.

Ishigai, S., M. Ozawa, Y. Ueda, et al. 1995. Application of 2nd-generation JAFI concept with heat-pipe tubenest to gas-turbine combustors for gas-firing. 21st CIMAC, Interlaken, paper no. G10.

Japan Society of Mechanical Engineers (JSME). 1996. Session on Future Generation Reactors. 5th National Symposium on Power and Energy Systems, no. 96-3, Kawasaki.

Kaneko, S. 1996. IGCC high efficiency coal firing power plant in 21st century. *Journal of the JSME* 99: 375–8.

Lilienthal, D.E. 1980. Atomic energy, a new start. In *International creative management*. Harper and Row. Japanese ed. (1981). Translated by K. Furukawa. Tokyo: Japan Productivity Center.

Marine Engineering Society of Japan. 1996. Marine engineering progress in 1995. *Bulletin of the Marine Engineering Society of Japan* 24: 76–111.

National Academy of Sciences. 1994. *Management and disposition of excess weapons plutonium*. Washington, D.C.: National Academy Press.

Orrok, G.A. 1930. Central stations. *Mechanical Engineering*, 52: 324–34.

Smith, E.C. 1937. *A short history of naval and marine engineering*. London: Cambridge University Press.

Valenti, M. 1996. Pouring on the steam. *Mechanical Engineering* 118(2): 70–4.

2

Thermodynamic design of the steam power plant cycle

SHIGEYASU NAKANISHI

Ryukoku University

2.1 Thermodynamic fundamentals of assessment of energy conversion

2.1.1 Introduction to exergy

2.1.1.1 What is exergy?

The most general definition of a heat engine, including a steam power plant, is given as a device that extracts thermal energy from some energy source and transforms it to mechanical energy, or work. The quantity of power thus obtained depends not only on the type and performance of the engine but also on the type and properties of the energy source used. The latter are closely connected to the quality of the energy source. From the thermodynamic point of view, the quality of an energy source can be defined as the maximum quantity of mechanical work theoretically obtainable from it. Under the law of energy conservation, that is, under the first law of thermodynamics, this quantity is finite. If an energy source is kinetic, potential, electrical, or magnetic, all the energy can in principle be transformed to mechanical work independently of the surroundings or environment in which the energy conversion proceeds. However, this does not hold for other types of energy, especially thermal energy, as the second law of thermodynamics governs the conversion process.

To begin with, let us examine the simplest and most fundamental case in which thermal energy is supplied by a heat source at constant temperature. A reversible Carnot engine extracts the maximum quantity of mechanical work from the heat source, but requires another heat source because of the second law. It is natural to assume that the second heat source should not have any special value. The environment, such as the atmosphere, river, or sea that surrounds us, is the only candidate for it. The temperature of the heat source under consideration is designated as T and that of the environment as T_0, where the capital letter T implies absolute temperature in degrees kelvin (K). Figure 2.1 shows the energy flow around the Carnot engine where the temperature of the heat source is higher

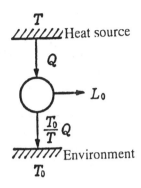

Figure 2.1 Carnot engine.

than that of the environment. Textbooks of thermodynamics teach us that the maximum quantity of mechanical work, L_0, is given by

$$L_0 = \left(1 - T_0/T\right)Q \tag{2.1}$$

where Q is the energy quantity supplied by the heat source, that is, the heat quantity from it. It is necessary to note that the environment participates in the energy conversion and that energy flow exists between it and the engine; the rest of the energy not transformed to mechanical work, $(T_0/T)Q$, is delivered into the environment.

Entropy change occurs at both the heat source and the environment. As their temperatures do not change throughout the conversion, their entropy changes are defined as the heat quantity absorbed divided by the absolute temperature at heat transfer, that is,

entropy decrease of the heat source $= Q/T$

entropy increase of the environment $= \left\{(T_0/T)Q\right\}/T_0 = Q/T$

Therefore, the entropy of the system comprising the heat source, the Carnot engine, and the environment suffers no change. As the system is adiabatic, that is, does not exchange any heat with its surroundings, this is a special case of the "principle of increase of entropy." If Q is considered as an algebraic quantity and its sign is designated as positive for heat flowing out of the heat source, it is easy to confirm that Eq. (2.1) holds also for cases in which the temperature of the heat source is lower than that of the environment $(T < T_0)$; the positive L_0 requires the negative Q. Production of work increases the energy contained in the colder heat source while it decreases that of the hotter environment, which is consistent with the principle of the conservation of energy. This example clearly demonstrates the importance of energy flow to and from the environment when energy conversion involves thermal energy.

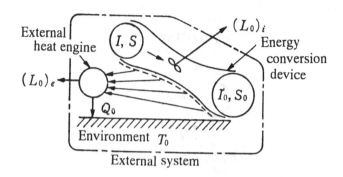

Figure 2.2 Illustration for definition of exergy of substance.

A more general and more complicated situation is the production of work from energy contained in a substance. Ideally the conversion process is to be conducted until the body of the substance completely loses its potential for work production, attaining equilibrium with the environment. Assuming this, we will deduce a formula for the maximum work obtainable in a steady flow system, which classifies most industrial engines, including power plants.

Figure 2.2 represents the thought process for deduction of the formula. A quantity of substance that has enthalpy I and entropy S flows into an energy conversion system to deliver work $(L_0)_i$ and heat at various temperature levels to the external and then flows out with enthalpy I_0 and entropy S_0, which correspond to the state in equilibrium with the environment. The heat delivered to the external is further converted to work $(L_0)_e$ by the aid of a series of external reversible engines that use the environment as another heat source. $(L_0)_i$ is called the inner work and $(L_0)_e$ the external work. The sum of these two is the overall work L_0, which takes its maximum value $(L_0)_R$ when all the processes concerned are reversible. Assuming reversibility, and applying the first law to the extended system comprising the energy conversion system, the external engine, and the environment, we obtain

$$I - I_0 = \left(L_0\right)_R + Q_0 \tag{2.2}$$

where Q_0 is heat quantity absorbed by the environment. As the extended system encircled by a chain line closed curve is reversible–adiabatic, the theorem of constant entropy holds here. As the entropy of the environment increases by Q_0/T_0 while those of the conversion system and the external engines remain unchanged, the resulting equation is

$$S = S_0 + Q_0/T_0 \tag{2.3}$$

Substituting Eq. (2.3) into Eq. (2.2) to eliminate Q_0, we obtain the maximum available work $(L_0)_R$:

$$\left(L_0\right)_{\mathrm{R}} = \left(I - I_0\right) - T_0\left(S - S_0\right) \tag{2.4}$$

The first term of the right-hand side is often misunderstood as the maximum available work if one knows only the first law and, in general, can be smaller or larger than the latter whether the entropy S is larger or smaller than that corresponding to equilibrium with the environment S_0.

From the viewpoint of energy conversion technology, the maximum work given by Eqs. (2.1) and (2.4) is the most important and has to be generalized for all energy sources. So we introduce a new state variable for the energy source defined as *the maximum work obtainable from it under the given environment* and call it "exergy" according to Rant (1956); exergy is also called availability, or available energy, but the latter term is rather out of date now.

2.1.1.2 *Formulae for exergy*

We give here formulae to calculate the exergy value of various energy sources that play an important role in energy conversion processes in steam power plants. In the following we use the symbols E and e for the total quantity of exergy and the specific exergy, respectively.

(a) *Mechanical, electric, and magnetic energies:* As all these energies can be completely converted to mechanical energy in principle, their exergy is identical to the energy quantity.

(b) *Thermal energy* E_q: The exergy of thermal energy is given by

$$E_q = Q\left(1 - T_0/T\right) \tag{2.5}$$

where Q is the heat quantity, T is the absolute temperature of the heat, and T_0 is the environmental temperature.

(c) *General formula of exergy for flowing fluid:* As Eq. (2.4) holds in steady flow systems, the formula for the specific exergy e for a flowing fluid is given by

$$e = i - i_0 - T_0\left(s - s_0\right) \tag{2.6}$$

where i and s are the specific enthalpy and specific entropy of the fluid, respectively, and i_0 and s_0 are the specific enthalpy and entropy, respectively, corresponding to the state under equilibrium with the environment. It should be noted here that the equilibrium state has the following three different definitions:

i. the fluid under consideration is in equilibrium with the environment in temperature only (thermal equilibrium);
ii. the fluid under consideration is in equilibrium with the environment in temperature and pressure (physical equilibrium);
iii. the fluid under consideration is in equilibrium with the environment, including chemical reactions.

The requirement for equilibrium becomes more and more restrictive from (i) to (iii). We have to select the most appropriate one for the system under consideration, taking into account the scheme of the system, the technological level of energy conversion engineering at present, and so on; the equilibrium for a steam power plant should be the thermal one since equilibrium with the environment is established in the condenser.

(d) *Exergy of fuel:* Although it can be defined, in principle, as the quantity calculated from Eq. (2.6) assuming that the equilibrium with the environment is the chemical one, it includes an enormous quantity of work from reversible mixing of the combustion products and the substance comprising the environment, that is, the air, which is impossible to convert to mechanical work by our present technology. We employ the physical equilibrium for the definition of the exergy of fuel, eliminating this component. Moreover, the fuel is usually a complex mixture of chemical compounds with a very complicated structure, apart from gaseous fuels, and so exact estimation of its exergy is practically impossible. However, an accurate value of the exergy of fuel is seldom required in analytical practice. For practical use, there is no problem in using one of the following approximate estimations (Rant 1960):

$$e = H_u + rw, \text{ kJ/kg, for coal}$$
$$e = 0.975H_h, \text{ kJ/kg, for liquid hydrocarbon}$$
$$e = 0.95H_h, \text{ kJ/kg, for gaseous hydrocarbon except methane}$$
$$e = 0.92H_h, \text{ kJ/kg, for methane} \tag{2.7}$$

where H_u is the lower calorific value of fuel, in kJ/kg; H_h is the higher calorific value of fuel, in kJ/kg; r is the latent heat of evaporation of water, in kJ/kg, which can be taken as 254 kJ/kg; and w is the moisture content of coal, in kg/kg.

2.1.2 Principle of assessment of energy conversion process

2.1.2.1 Loss of available work

Flows of energy into and from a heat engine are schematically illustrated in Figure 2.3. The engine is fed the energy flow with exergy E, generates work L, and finally discharges to the environment low potential heat Q_0 and waste containing exergy E_{exh}. The loss of available work is defined as the difference between the exergy feed E and the work output L, and is designated LW. This loss is always positive according to the second law of thermodynamics except for the case where all the processes in the engine and between it and the environment proceed thermodynamically without any loss, or are reversible:

$$LW = E - L \geq 0 \tag{2.8}$$

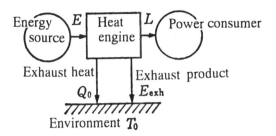

Figure 2.3 Energy flows across a heat engine.

If the same quantity of energy is applied to a reversible heat engine, the work output is equal to the exergy E and the enthalpy of the waste is equal to the enthalpy corresponding to equilibrium with the environment I_0 (see Eq. (2.4)). The equations of energy balance for the reversible and irreversible engines are expressed as

$$E + \left(Q_0\right)_R + I_0 = L + Q_0 + I_{exh}$$

where $\left(Q_0\right)_R$ is the heat discharge of the reversible engine to the environment, and I_{exh} is the enthalpy contained in the exhaust from the irreversible engine. Therefore, we can divide the loss of available work into two components as follows:

$$LW \equiv E - L = \left\{Q_0 - \left(Q_0\right)_R\right\} + \left\{I_{exh} - I_0\right\} \geq 0 \qquad (2.9)$$

The second component corresponds to the exhaust gas loss plus the unburned fuel loss; the former is the thermal loss and the latter is the chemical loss in energy terms. For simplicity, we neglect the chemical loss by waste in this section, and include the thermal loss by waste in the thermal discharge Q_0. Then, Eq. (2.9) is simplified to

$$LW \equiv E - L = Q_0 - \left(Q_0\right)_R \geq 0 \qquad (2.10)$$

This expression means that the loss of available work increases the heat discharge to the environment and, in turn, the entropy of the environment. This entropy increase ΔS^* is due to the irreversibility of the heat engine under consideration. By the definition of entropy,

$$\Delta S^* = \left\{Q_0 - \left(Q_0\right)_R\right\}/T_0 \qquad (2.11)$$

so Eq. (2.10) can be written as

$$LW = T_0 \, \Delta S^* \qquad (2.12)$$

which is a special case of the famous Gouy–Stodola formula.

The exergy efficiency ξ of the heat engine is defined as

$$\xi \equiv L/E = 1 - LW/E = 1 - T_0 \Delta S^*/E \leq 1 \qquad (2.13)$$

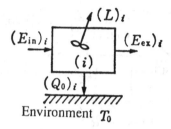

Figure 2.4 Exergy change at system component.

where the exergy of the fuel supplied to the heat engine E is taken as its input and the work delivered to the external L as its output. The thermal efficiency η_{th}, that is, the energy conversion efficiency according to the first law of thermodynamics, is defined as the ratio of the work output L to the energy input U:

$$\eta_{th} \equiv L/U$$

where $U = I - I_0$. The relationship between these two efficiencies is written as follows:

$$\eta_{th} \equiv L/U = (E/U)\xi \qquad (2.14)$$

As $E < U$ for fuels in common use, it is usual that the exergy efficiency is smaller than the thermal efficiency.

To analyze loss generation due to thermodynamic irreversibility in a heat engine in detail, we have to investigate the inflow and outflow of exergy of substance and energy flows to and from each component and to estimate changes of exergy in the engine. Consider the ith component as shown in Figure 2.4, where $(E_{in})_i$ is the sum of exergy flows into the ith component, $(E_{ex})_i$ is the sum of exergy flows from the ith component, $(L)_i$ is the work output of the ith component, and $(Q_0)_i$ is the heat discharge to the environment. As this component can generate the maximum mechanical work equal to $(E_{in})_i - (E_{ex})_i$, the loss of available work attributable to it is equal to

$$(LW)_i = (E_{in})_i - (E_{ex})_i - (L)_i \qquad (2.15)$$

The sum of $(E_{in})_i - (E_{ex})_i$ over all the components gives the exergy quantity E fed to the heat engine and that of $(L)_i$ gives the work output L. Therefore the sum of the local losses of available work $(LW)_i$ gives the loss of available work of the whole engine, LW:

$$LW = \sum_i (LW)_i \qquad (2.16)$$

This formula asserts the summability of the losses of available work, which enables us to identify the intensity of loss generation in each component and to assess its significance in the total system.

2.1.2.2 Exergy analyses of typical processes

(a) *Direct conversion to heat:* When a quantity of energy U is converted to heat at temperature T without work generation, the exergy of the heat obtained is equal to $U(1 - T_0/T)$, according to formula (2.5). This quantity is, of course, less than the exergy corresponding to the energy U, and the difference is the loss of available work LW. As the exergy of electrical energy is identical to its energy quantity, the loss of available work on its direct heat conversion, that is, electric direct or Joule heating, is given as

$$LW = U - U(1 - T_0/T) = U(T_0/T) \qquad (2.17)$$

which teaches us that the higher the temperature, the lower the loss of available work; direct heat conversion is admissible from the energetic point of view only for obtaining a very high temperature.

(b) *Heat transfer with finite temperature difference:* This can be considered as a special case of (a), in which the energy input is also heat. When the temperature of a heat flux changes from T_1 to T_2 through a heat transfer process, its exergy changes according to Eq. (2.5) and, in turn, the loss of available work amounts to

$$LW = Q(1 - T_0/T_1) - (1 - T_0/T_2) = Q(1/T_2 - 1/T_1) \qquad (2.18)$$

where Q is the heat quantity transferred. This is one of the most common and important losses encountered in everyday life and the industrial scene.

(c) *Combustion:* The exergy change on irreversible adiabatic combustion results from entropy change, since enthalpy change does not occur because of the absence of work or heat exchange with the external. The loss of available work is the difference between the exergy of fuel and that of the combustion product, and its quantity per kilogram of fuel, lw, is given by

$$lw = T_0(\Delta s)_c \qquad (2.19)$$

where $(\Delta s)_c$ is the entropy production per kilogram of fuel at combustion. As reversible combustion cannot be realized, even approximately, the loss at combustion always exists and is a relatively large amount. The combustion process in a heat engine using fuel generates most of the loss, which means that exact estimation of the exergy of fuel itself is not required in the exergy analysis of the heat engine.

(d) *Adiabatic expansion:* Adiabatic expansion with thermodynamic loss in a turbine, or more generally in an expander, generates an entropy increase Δs, as shown in Figure. 2.5. The stage efficiency of a turbine stage η_T is defined as follows, not taking into account the kinetic energy of the working fluid:

$$\eta_T = l/\Delta i_a \qquad (2.20)$$

where l = work per kilogram of working fluid on expansion between pressures p_1 and p_2, and Δi_a is the isentropic enthalpy drop. It is clear that the enthalpy of

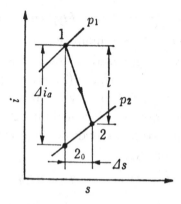

Figure 2.5 Adiabatic expansion process.

the outflow from the stage is higher by $\Delta i_a - l = (1 - \eta_T)\Delta i_a$ than that of an ideal stage where $\Delta s = 0$. Consequently, the exergy of the outflow is larger by $(1 - \eta_T)\Delta i_a - T_0\Delta s$ than that of the ideal stage and therefore the exergy consumption at the stage is reduced by this quantity. Even in this case, the loss of available work is equal to $T_0 \Delta s$, which agrees with the Gouy–Stodola formula, Eq. (2.12).

Next, we will derive the expression for Δs. From the equation of energy conservation,

$$T ds = di - v dp, \quad \text{or} \quad ds = (di - v dp)/T$$

Integrating this from state 2_0 to state 2 along the isobaric line p_2 ($dp = 0$), we obtain

$$\Delta s = \int_{2_0}^{2} \frac{di}{T} = \frac{i_2 - i_{2_0}}{(T_2)_m} = \frac{(1 - \eta_T)\Delta i_a}{(T_2)_m} \tag{2.21}$$

where $(T_2)_m$ is the mean temperature between states 2_0 and 2, and is defined by

$$(T_2)_m = \frac{(i_2 - i_{2_0})}{\displaystyle\int_{2_0}^{2} \frac{di}{T}} \tag{2.22}$$

With this mean temperature, the loss of available work at the turbine stage is expressed as

$$lw = \left\{ T_0/(T_2)_m \right\}(1 - \eta_T)\Delta i_a \tag{2.23}$$

Hence lw depends on the mean temperature at the expansion end as well as the stage efficiency. As it is appropriate to consider the exergy consumption at the stage under examination, that is, $l + lw$, as the input exergy to the turbine stage, the definition of the exergy efficiency of the turbine stage becomes

$$\xi_T \equiv l/(l + T_0 \Delta s) = \eta_T \Delta i_a/(\eta_T \Delta i_a + T_0 \Delta s$$
$$= \eta_T / \{\eta_T + T_0 (\Delta s / \Delta i_a)\} \tag{2.24}$$

Introducing the mean temperature $(T_2)_m$, Eq. (2.22), we obtain the following formula:

$$\xi_T = \eta_T / \{\eta_T + (1 - \eta_T) T_0 / (T_2)_m\}$$
$$= 1 / \{1 + (1/\eta_T - 1) T_0 / (T_2)_m\} \tag{2.25}$$

In the final stage, where the expansion end is connected to the condenser, this mean temperature is identical to the environmental temperature T_0, and, in turn, the exergy efficiency is equal to the stage efficiency:

$$\xi_T = \eta_T$$

In any stage other than the final one, $(T_2)_m > T_0$ holds always, so that

$$\xi_T > \eta_T$$

unless $\eta_T = 1$. If η_T is fixed, an increase of $(T_2)_m$ decreases lw according to Eq. (2.23) and then increases ξ_T, which explains the fact that the Curtis stage in the highest pressure stage does not reduce the overall turbine efficiency in proportion to its low efficiency.

In a condensing multistage turbine without regenerative steam extraction, the formula for the overall loss of available work $(lw)_0$ is obtained by substituting Eq. (2.25) into (2.16) as follows:

$$(lw)_0 = \sum_i \{T_0/(T_m)_i\}\{1 - (\eta_T)_i\}(\Delta i_a)_i \tag{2.26}$$

where $(\eta_T)_i$, $(\Delta i_a)_i$ and $(T_m)_i$ are the stage efficiency, isentropic enthalpy drop, and mean temperature at the expansion end for the ith stage, respectively. As the exergy fed to the turbine is the isentropic enthalpy drop between the turbine inlet and the condenser, $(\Delta i_a)_0$, its overall output per kg of steam, l_0, is given by

$$l_0 = (\Delta i_a)_0 - \sum_i \{T_0/(T_m)_i\}\{1 - (\eta_T)_i\}(\Delta i_a)_i \tag{2.27}$$

The overall exergy efficiency of the turbine is identical to the overall turbine efficiency $(\eta_T)_0$ according to the discussion above and is given by

$$(\eta_T)_0 \equiv l_0/(\Delta i_a)_0 = 1 - \sum_i \{T_0/(T_m)_i\}\{1 - (\eta_T)_i\}(\Delta i_a)_i/(\Delta i_a)_0 \tag{2.28}$$

which gives the relationship of efficiency between the whole unit and individual stages and clearly explains the nature of the reheat factor in existing turbine design theory (Church 1950).

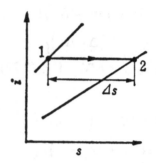

Figure 2.6 Throttling process.

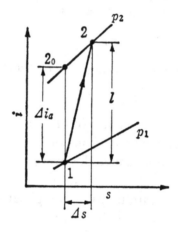

Figure 2.7 Adiabatic compression process.

The adiabatic expansion process without work (Fig. 2.6) is called throttling and has zero exergy efficiency; the loss of available energy lw is equal to $T_0\Delta s$ in this case too.

(e) *Adiabatic compression:* Consider adiabatic compression in a pump or compressor (Fig. 2.7), where the pump efficiency or adiabatic compressor efficiency η_p is defined as

$$\eta_p = \Delta i_a/l \tag{2.29}$$

where Δi_a is the adiabatic enthalpy rise and l is the work input to the pump or compressor. In the compression process, as the enthalpy rise due to its irreversibility is equal to $l - \Delta i_a$ and the entropy rise is Δs, the exergy of the fluid compressed is larger by

$$\Delta i_a / \eta_p - \Delta i_a - T_0 \Delta s = (1/\eta_p - 1)\Delta i_a - T_0 \Delta s$$

compared with that corresponding to isentropic compression. If the purpose of compression is only to pressurize the fluid, the exergy corresponding to the thermal energy, including this, is exhausted to the external and dissipated into the environment as loss of available work unless some device is provided to recover it as a useful energy resource. However, in the case of the heat engine, this thermal energy is recovered in the succeeding components and its exergy is partly utilized. The loss of available energy lw is, therefore, smaller than this excess work consumption $l - \Delta i_a$ and is given by

$$lw = T_0 \Delta s = \left\{ T_0 / (T_2)_m \right\} (1/\eta_p - 1)\Delta i_a \qquad (2.30)$$

where the mean temperature at the end of compression, $(T_2)_m$, is defined by formula (2.22) as in expansion. The exergy efficiency is then defined by

$$\xi_p = 1 - lw/l = 1 - T_0 \eta_p \Delta s / \Delta i_a = 1 - \left\{ T_0 / (T_2)_m \right\} (1 - \eta_p) \qquad (2.31)$$

It should be noted that the premise of this definition is utilization of the thermal exergy of the compressed fluid in the succeeding processes.

(f) *Mixing:* This is a typical irreversible process without work generation and one of the large sources of loss of available work. Adiabatic mixing of fluid flows generates entropy production while maintaining their total enthalpy constant.

(g) *Heat dissipation and exhaust loss:* All the exergy of heat discharged and of material exhausted into the environment becomes loss of available work. However, the temperature level of the discharged heat and exhaust material is usually so low that the ratio of the thermal exergy to the heat quantity is much smaller than unity; the chemical exergy of the exhaust material is very low. When one plans heat and/or exhaust recovery for energy conservation, one should make a deliberate assessment of the exergy.

2.1.2.3 Heat balance analysis and exergy analysis

The exergy analysis of an energy conversion system consists of tracing exergy changes across all the individual components in the system to be analyzed. Calculation of the exergies can be conducted fairly straightforwardly according to the formulae given in the preceding sections. To obtain the quantities required for the calculation, we must first complete the heat balance analysis, which traces enthalpy changes across all the individual components in just the same manner as in the exergy analysis. It should be pointed out here that heat balance analysis demonstrates the energy loss of the system in terms of the energy quantity at

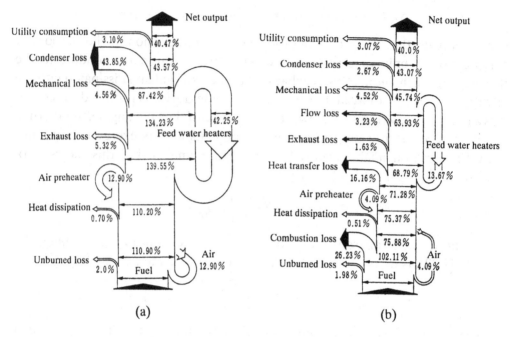

Figure 2.8 Heat (a) and exergy (b) flow diagrams.

its interfaces with the environment, while exergy analysis demonstrates the energy loss in exergy quantity at each component where the loss is generated. The former has no ability to detect degradation of energy or to identify the exact point of loss generation, and gives only the energy outflow from the system without assessment of its energy quality, which often misleads one to consider some components from which degraded energy is exhausted as the real sources of energy loss. Moreover, heat balance analysis overestimates the exhausted heat with a temperature close to that of the environment, especially at the condenser of a steam power plant.

A thermal energy flow diagram (the so-called Sankey diagram) and an exergy flow diagram of a steam power plant (Rant, 1961) are shown in Figure 2.8, which clearly illustrates a large difference in loss distribution. In the thermal energy flow diagram (Fig. 2.8a), the maximum quantity of loss appears at the condenser. On the other hand, in the exergy flow diagram (Fig. 2.8b), most of the loss is attributable to combustion and heat transfer processes in the boiler unit and the loss generation by the condenser itself is relatively small, as is that of the exhaust gas.

Figure 2.9 shows an example of detailed exergy analysis in a 350 MW steam power plant (Kudo et al. 1983).

2.2 Thermodynamic analysis of energy conversion in steam power plants

2.2.1 General discussion of energy conversion in steam power plants

A steam power plant comprises a steam generating unit or boiler, and a turbine unit; the former unit generates steam flow with high specific exergy, consuming the exergy that some energy source provides to it in the form of heat, and the latter unit generates mechanical work using the steam as the working fluid for some steam power cycle. The fundamental arrangement is schematically depicted in Figure 2.10.

The heat flow rate Q and exergy flow rate E of the steam flow generated in the steam generator are, of course, smaller than those provided by the energy source, U_0 and E_0, respectively:

$$U_0 \geq Q, \qquad E_0 \geq E$$

In the case where the energy source is fossil fuel, the energy efficiency of the steam generator, or the boiler efficiency, η_B, is defined as

$$\eta_B = Q/U_0 < 1 \qquad (2.32)$$

Analogously to this, the exergy efficiency of the steam generator, ξ_B, can be defined as

$$\xi_B = E/E_0 < 1 \qquad (2.33)$$

The exergy contained in the steam flow, E, is transformed to mechanical work L by the steam cycle with conversion efficiency ξ_C:

$$L = \xi_C E = \xi_C \xi_B E_0$$

If the overall exergy efficiency of the plant is defined as

$$\xi_0 = L/E_0 \qquad (2.34)$$

then we obtain

$$\xi_0 = \xi_C \xi_B \qquad (2.35)$$

which means that the efficiency of energy conversion by the steam power plant can be divided into two components: ξ_B is that of the heat supply side and ξ_C is that of the steam power cycle. The range of ξ_B is rather wide, while the corresponding energy efficiency η_B hardly changes, taking some value close to unity; the magnitude of the overall efficiency, ξ_0, is governed by ξ_B rather than ξ_C.

Let us examine the flow of exergy at heat transfer in the steam generator (Fig. 2.11). From the heat and exergy balances across the steam generator, we obtain the following relations:

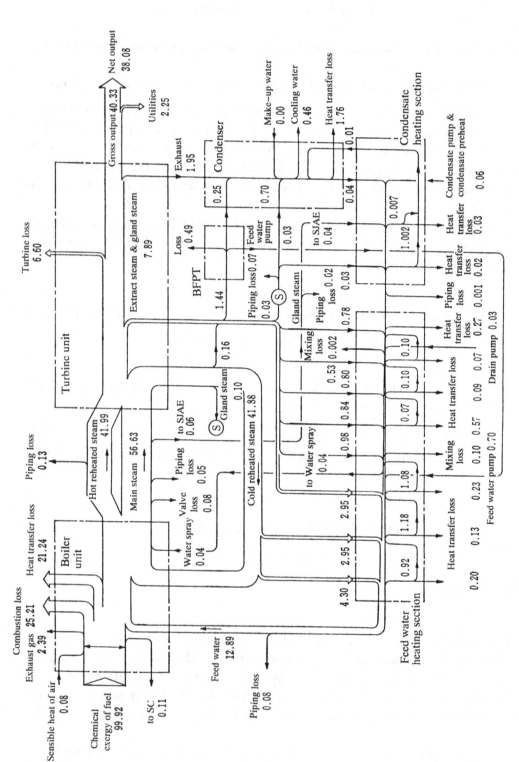

Figure 2.9 Example of exergetic analysis of steam power plant.

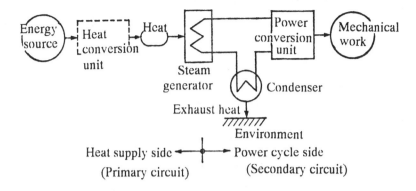

Figure 2.10 Simplified scheme of steam power plant.

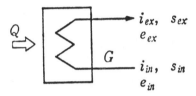

Figure 2.11 Exergy exchange at steam generator.

$$E = G(e_{ex} - e_{in}) \text{ and}$$
$$Q = G(i_{ex} - i_{in})$$

where G is the mass flow rate of steam, e_{ex} and e_{in} are the specific exergies of steam at the exit and inlet of the steam generator, respectively, and i_{ex} and i_{in} are the specific enthalpies at the exit and inlet of the steam generator, respectively. By substituting the definition of exergy, Eq. (2.6), we obtain an expression for E:

$$E = Q(e_{ex} - e_{in})/(i_{ex} - i_{in})$$
$$= Q\{1 - T_0(s_{ex} - s_{in})/(s_{ex} - s_{in})\} \tag{2.36}$$

where s_{ex} and s_{in} are the specific entropies at the exit and inlet of the steam generator, respectively. By introducing the mean temperature of heat transfer, T_m, defined as

$$T_m = (i_{ex} - i_{in})/(s_{ex} - s_{in}) \tag{2.37}$$

we arrive at the exergy formula for transferred heat

$$E = Q(1 - T_0/T_m) \tag{2.38}$$

which is identical to the formula (2.5). The mean temperature T_m can be interpreted as an entropy-averaged temperature because, if the heat transfer is isobaric, the following relation holds:

$$i_{ex} - i_{in} = \int_{s_{in}}^{s_{ex}} T \, ds$$

which implies that

$$T_m = \int_{s_{in}}^{s_{ex}} T \, ds \Big/ \left(s_{ex} - s_{in}\right) \tag{2.39}$$

By substituting Eqs.(2.32) and (2.38) into (2.33), we obtain the final expression for the exergy efficiency of the steam generator:

$$\xi_B = \left(U_0/E_0\right)\eta_B\left(1 - T_0/T_m\right) \tag{2.40}$$

where (U_0/E_0) is the effective ratio of the energy source. As the effective ratio depends only on the type of energy source, ξ_B is determined by η_B and $(1 - T_0/T_m)$ or T_m. η_B is fairly high and its variation is restricted only in a narrow range, as mentioned above, so that the magnitude of ξ_B is largely governed by T_m; the higher the temperature of heat transfer, the higher the exergy efficiency.

The exergy flow transferred to the steam cycle is completely converted to mechanical work only if no thermodynamic irreversibility exists on this side; it should be noted that the exergy efficiency ξ_C is equal to unity while the cycle efficiency η_{th} is smaller than unity. In the state-of-the-art technology, thermodynamic irreversibility in the turbine unit is minimized so that ξ_C is much higher than ξ_B. The cycle thermal efficiency η_{th} can be expressed as follows:

$$\eta_{th} = L/Q = \xi_0 E_0 \Big/ \eta_B U_0 = \xi_C\left(1 - T_0/T_m\right) \tag{2.41}$$

According to the discussion above, it is clear that the governing factor of the cycle efficiency η_{th} is the effective Carnot coefficient $(1 - T_0/T_m)$, which is nearly equal to the exergy efficiency of the steam generator ξ_0; as a rough estimate, the cycle efficiency can be approximated to the Carnot coefficient since ξ_C is close to unity.

2.2.2 Maximization of exergy efficiency of steam generators

As explained in the last section, the overall efficiency of a steam power plant is mainly governed by the exergy efficiency of its steam generating unit. The optimum arrangement of the steam power plant is approximately realized by maximizing the exergy efficiency of the steam generating unit. The strategy for

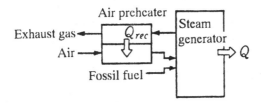

Figure 2.12 Principle of air-preheating at steam generator.

maximization differs according to the type of energy source. We discuss here typical types of energy source.

2.2.2.1 Case where η_B stays constant independently of working fluid temperature

Nuclear fuel belongs to this category. For this type of energy source, the higher the heat transfer temperature T_m, the higher the exergy efficiency. The optimization is attained by making the lowest temperature of heat transmission as high as possible; this temperature corresponds to the state (h_{in}, s_{in}). This limiting temperature is determined by techno-economic factors and, for example, is about 220°C for the light water reactor.

2.2.2.2 Case where η_B changes according to temperature level and arrangement of heating surfaces of the steam generator

(a) *With recirculation of heat at the heat conversion section:* This category includes fossil fuel burnt with preheat of combustion air (Fig. 2.12). In this case the fuel is converted to hot combustion gas, which heats the working fluid. The higher the lowest temperature of heating, the greater the heat loss by the exhaust gas from the system if no special preventive means is devised. This leads to lower η_B and in turn to degraded ξ_B in spite of the higher T_m. This difficulty is avoided if the heat contained in the exhaust gas is recovered by preheating the combustion air and recirculating it to the steam generator, instead of discharging it to the environment. This is the principle of air preheating, which assures low temperature discharge of the exhaust gas from the system and prevents low η_B. The exhaust temperature can in principle be lowered down to the environment temperature, but actually it is limited to about 150°C to inhibit low-temperature corrosion. Although the maximum temperature of combustion increases with the heat quantity of recirculation, Q_{rec}, the upper-limit temperatures of the working fluid flowing in and out are restricted to about 200 and 600°C, respectively, by the limits of material technology.

(b) *Cases where heat recirculation is impossible:* If the energy source is the thermal energy of some hot fluid, we cannot employ the heat recovery system as described above. This category includes steam generators in combined cycle power plants, geothermal plants, Ocean Thermal Energy Conversion (OTEC) plants, heat recovery boilers, and so on. As the maximum temperature is bound to that of the energy source, exergy recovery is optimized by making the temperature of the working fluid passing to the generator as low as possible.

(c) *Radiative heating:* In radiative heat transfer the quantity of heat Q exchanged largely depends on the temperature level of the heating surface and, in turn, on that of the working fluid T_m. An increase of T_m increases the exergy per unit of exchanged heat according to Eq. (2.38) and decreases the heat quantity according to the Boltzmann–Helmholtz law. Clearly an optimum T_m exists, which is determined by the kind of heat source and the arrangement of the heat exchange unit.

2.2.3 Exergy analysis of cogeneration plants

Next, we examine a cogeneration plant that supplies electric power and high temperature heat to consumers outside the plant (Fig. 2.13). The heat balance and exergy balance of this plant are expressed as

$$Q = L + Q_u + Q_0$$

and

$$E = L + Q_u(1 - T_0/T_u) + LW$$

where Q is the energy input to the steam generator, measured as heat quantity; E is the exergy input to the steam generator; L is the power output, mechanical or electric, from the plant; LW is the loss of available work; Q_u and T_u are the heat output from the plant and its temperature, respectively; Q_0 is the heat quantity discharged to the environment; and T_0 is the temperature of the environment. The definition of the total output from the plant is represented as

$$L + Q_u, \text{ based on heat quantity}$$

and

$$L + Q_u(1 - T_0/T_u), \text{ based on exergy quantity}$$

Consequently, we have two efficiencies of the cogeneration plant:

$$\eta_0 = (L + Q_u)/U_0 = \eta_B(L + Q_u)/Q \le \eta_B \tag{2.42}$$

and

$$\xi_0 = \left\{L + Q_u(1 - T_0/T_u)\right\}/E_0 = \xi_B\left\{L + Q_u(1 - T_0/T_u)\right\}/E \le \xi_B \tag{2.43}$$

where η_0 and ξ_0 are the overall thermal efficiency (effectiveness of energy utilization) and overall exergy efficiency, respectively, and U_0 and E_0 are the energy input and exergy input to the plant, respectively.

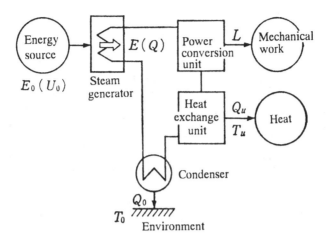

Figure 2.13 Schematic diagram of energy and exergy flow in combined cycle plant.

According to the assessment based on heat quantity, an increase of output heat Q_u produces high η_0 whose limiting value coincides with the boiler efficiency η_B. As η_B is close to unity, η_0 also can approach this high limiting value by making Q_u as large as possible, but this often leads to some misunderstanding or overestimation about the improvement of energy utilization by cogeneration of power and heat. The assessment based on exergy completely removes this possibility; the exergy efficiency ξ_0 never exceeds ξ_B, which cannot differ from that for a steam power plant generating only electric power. The greater Q_u and the lesser Q_0 are, the lower are T_u and the specific exergy $(1 - T_0/T_u)$. The true energy conservation by cogeneration is evaluated by comparing its energy consumption with that of a system comprising a power plant and a heat pump under the same electrical and thermal outputs. The difference between them will not be large. One of the merits of cogeneration is the elimination of loss from the heat pump.

2.2.4 *Exergy analysis of combined cycle plants*

An energy system comprising a number of power-generating subsystems working on different cycles is called a combined cycle power plant. Consider such a plant in which the energy flows in cascade, as shown in Figure 2.14. The work L_i and the loss of available work $(LW)_i$ at the ith subsystem are expressed as

$$L_i = \xi_i \left(E_{i-1} - E_i \right) \tag{2.44}$$

and

$$\left(LW_i \right) = \left(1 - \xi_i \right)\left(E_{i-1} - E_i \right) \tag{2.45}$$

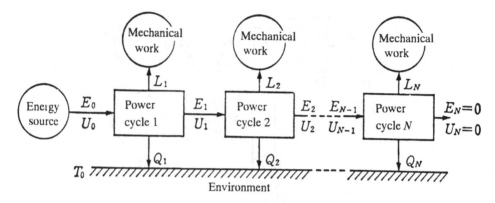

Figure 2.14 Simplified scheme of cogeneration plant.

where E_0 is the exergy input to the whole plant, E_i is the exergy outflow from the *i*th subsystem, and ξ_i is its exergy efficiency. The total output work L and the total loss of available work LW are given as

$$L = \sum_i \xi_i \left(E_{i-1} - E_i \right) = E_0 \sum_i \xi_i \varphi_i \qquad (2.46)$$

and

$$LW = \sum_i \left(1 - \xi_i \right)\left(E_{i-1} - E_i \right) = E_0 \left(1 - \sum_i \xi_i \varphi_i \right) \qquad (2.47)$$

where φ_i is the fraction of exergy consumed at the *i*th subsystem, defined as

$$\varphi_i = \left(E_{i-1} - E_i \right) / E_0$$

The effective utilization of energy sources is limited with any single cycle. For example, a Rankine cycle power plant working with water has a maximum thermal efficiency a little over 40% since the highest temperature of the working fluid does not exceed 600°C. On the other hand, a gas turbine plant and a magnetohydrodynamic (MHD) generator, which are suitable for high-temperature operation, have only low thermal efficiency, as they discharge high-temperature exhaust gas and generate a large exhaust loss. If we combine two cycles that have suitable operating ranges at higher and lower temperatures, respectively, we can expect higher conversion efficiency. A typical example is a gas turbine–steam turbine combined cycle plant, which is considered to attain a conversion efficiency larger than 50% and is now being employed at large-capacity advanced power plants, as will be discussed later.

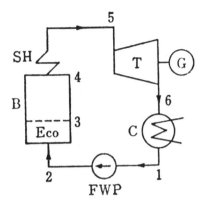

Figure 2.15 Scheme of simple Rankine cycle. (B) Steam generator or boiler; (T) steam turbine; (C) condenser; (FWP) feedwater pump; (G) machine providing power to the external.

2.3 The Rankine cycle

2.3.1 General discussion of the Rankine cycle

In the power generation part of a steam power plant (the cycle side), the working fluid is heated, receiving exergy at the steam generator, and adiabatically expands, generating mechanical work and turning into low pressure vapor in the expander (steam turbine); it is then cooled to be condensed in the condenser, and is compressed in the compressor or pump to be returned to the steam generator. As compression of the working fluid is executed in the liquid phase, the work required for compression is usually insignificant compared with the work output from the expander.

The simplest arrangement of the steam power plant is that without regeneration and reheat, as shown in Figure 2.15. This cycle is named the Rankine cycle after W.J.M. Rankine (1820–1872), who established it as the fundamental cycle for a steam power plant. In the figure, the steam generator B includes an economizer section (Eco) and a superheater section (SH); the former heats up the working fluid to the saturation temperature and the latter superheats its dry saturated vapor. The Rankine cycle comprises an isobaric heating process in B, an adiabatic expansion in T, isobaric cooling in C, and adiabatic compression in FWP; Figure 2.16 shows its cycle diagram in $T–s$ and $i–s$ diagrams that assume the adiabatic processes to be isentropic. From enthalpy or heat balance about unit mass of the working fluid, we express works and heats in terms of specific enthalpies:

$$\left.\begin{array}{ll} q_B = i_5 - i_2, & q_C = i_6 - i_1 \\ l_T = i_5 - i_6, & l_P = i_2 - i_1 \end{array}\right\} \tag{2.48}$$

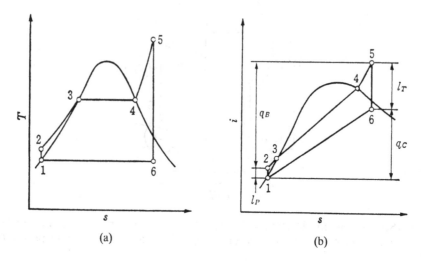

Figure 2.16 Cycle diagrams for simple Rankine cycle: (a) temperature (T)–entropy (s) diagram; (b) enthalpy (i)–entropy (s) diagram.

where q_B and q_C are heat quantities absorbed at B and discharged at C per unit mass of working fluid, respectively; l_T and l_P are works generated in T and consumed in P per unit mass of working fluid; and i_i is the specific enthalpy of working fluid at point i in the figure.

Next, we will conduct an exergy analysis of the cycle. Assume that the working fluid is in equilibrium with the environment at the exit (point no. 1) from the condenser C, and designate the equilibrium temperature and corresponding enthalpy and entropy as T_0, i_0, and S_0, respectively; then

$$i_0 = i_1 \qquad \text{and} \qquad s_0 = s_1$$

Taking into consideration the definition of exergy, Eq. (2.6), we obtain the following expressions of exergy at the node points on the cycle diagram:

$$\left.\begin{aligned} e_2 &= i_2 - i_1 - T_0(s_2 - s_1), & e_5 &= i_5 - i_1 - T_0(s_5 - s_1) \\ e_6 &= i_6 - i_1 - T_0(s_6 - s_1), & \text{and} \quad e_1 &= 0 \end{aligned}\right\} \tag{2.49}$$

where e_i is the specific exergy of the working fluid at point i. Hence the exergy gain per unit mass of the working fluid at the steam generator, Δe_B, is equal to

$$\Delta e_B = e_5 - e_2 = i_5 - i_2 - T_0(s_5 - s_2)$$

The thermal efficiency η_{th} and exergy efficiency ξ_C of the cycle are given by

$$\eta_{th} = \frac{l_T - l_P}{q_B} = \frac{i_5 - i_6 - (i_2 - i_1)}{i_5 - i_2} = 1 - \frac{i_6 - i_1}{i_5 - i_2} \tag{2.50}$$

and

$$\xi_C = \frac{l_T - l_P}{\Delta e_B} = \frac{i_5 - i_6 - \left(i_2 - i_1\right)}{i_5 - i_2 - T_0\left(s_5 - s_2\right)} = 1 - \frac{i_6 - i_1 - T_0\left(s_5 - s_2\right)}{i_5 - i_2 - T_0\left(s_5 - s_2\right)} \qquad (2.51)$$

The above formulae (2.48)–(2.51) are applicable even if thermodynamic irreversibility exists.

The state of the working fluid at the turbine exit is of wet saturated vapor, as explained later, and hence its temperature is equal to the saturation temperature corresponding to the pressure in the condenser, or the environment temperature T_0. As a result, the isobaric line 1–6 in Figure 2.16 becomes the isothermal line corresponding to T_0, that is, the line of the state of environment, on which the exergy is equal to 0. Therefore

$$e_6 = 0$$

and the exergy at the turbine inlet, e_5, is identical with the adiabatic enthalpy drop Δi_a.

For a reversible cycle, the expansion and compression are isentropic:

$$s_5 = s_6 \qquad \text{and} \qquad s_2 = s_1$$

As the state at the turbine exit is wet, the following equation holds:

$$T_0\left(s_5 - s_2\right) = T_0\left(s_6 + s_1\right) = i_6 - i_1 \qquad (2.52)$$

Substituting this into Eq. (2.51), we obtain that the exergy efficiency of this cycle is unity:

$$\xi_C = 1$$

which is no surprise because of the assumption of nonirreversibility.

For cases with irreversibility, Eq. (2.52) is modified to an inequality

$$T_0\left(s_5 - s_2\right) < i_6 - i_i$$

and the exergy efficiency becomes less than unity. More general discussion of cycles with irreversibility is not given in this textbook, but it is simple and easy to evaluate their thermal efficiency η_{th} and exergy efficiency ξ_C by using Eqs. (2.48)–(2.51) with formulae (2.20)–(2.31).

The Rankine cycle has a number of variants apart from the standard cycle given in Figure 2.16. Some of them are as follows (Fig. 2.17):

(a) *Supercritical cycle*: The pressure of the working fluid exceeds the critical pressure at the turbine inlet and no phase change occurs in the steam generator. The difference from the standard cycle is only the absence of a distinct vaporization process and no special problem exists with regard to thermodynamic analysis.

Shigeyasu Nakanishi

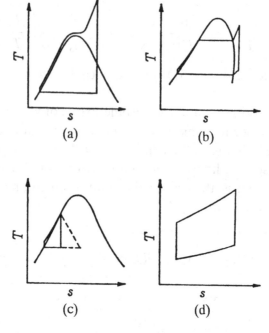

Figure 2.17 Various modified Rankine cycles: (a) supercritical cycle: (b) cycle with super-heated vapor at the turbine exit; (c) hot water cycle; (d) Lorenz cycle.

(b) *Cycle with superheated vapor at the turbine* exit: When some peculiar fluid is employed as the working fluid, the state at the turbine exit can be superheated and the temperature differs from the saturation temperature corresponding to the condenser. This temperature difference generates additional loss of available work due to heat transfer, with a finite temperature difference in the condenser.

(c) *Wet steam turbine* and *hot water expander*: Even if the state of the working fluid is wet or, in a limiting case, a boiling liquid, no special consideration is required in principle. However, there are a number of practical problems given the contemporary level of technology. Erosion of turbine blades by liquid droplets is the most harmful phenomenon in the high-quality region, so the wetness y of the working fluid in the turbine is limited to below 10%. Moreover, wetness lowers the turbine stage efficiency; the following equation estimates this decrease as

$$\eta_{\text{wet}}/\eta_{\text{dry}} = 1 - \alpha_B y \tag{2.53}$$

where η_{wet} and η_{dry} are the stage efficiencies for wet and dry vapor, respectively, α_B is a constant factor called the Bauman coefficient, reported as 0.4 (Kretmeier 1979), and y is the wetness, defined as

$$y = 1 - x_e$$

where x_e is the quality of vapor at the stage exit. The effect of wetness on the stage efficiency is relatively insignificant in the higher-quality region, but can grow enough to deteriorate the thermodynamic performance in the lower-quality region. A limiting case is a hot water expander in which saturated liquid ($x = 0$, i.e., $y = 1$) expands. Its efficiency is as low as 39% and the expansion line largely inclines as the broken line shown in Fig. 2.17c.

(d) *Lorenz cycle*: If the working fluid is a miscible mixture, its saturation temperature for a given pressure changes as vaporization proceeds. In the case where the steam generator is a heat exchanger, that is, the energy source is a high-temperature fluid, the loss of heat transfer can be diminished by employing a suitable mixture as the working fluid and heating it with countercurrent flow heat exchange to minimize the temperature difference between the two fluid flows; this is a variation of the Lorenz cycle. In the theory of the Kalina cycle (Kalina 1983), a distillation process is proposed for the condenser unit to diminish the temperature difference.

2.3.2 Working fluids for the Rankine cycle

As the performance of a thermodynamic cycle largely depends on the properties of the working fluid employed, we will examine here some of those whose effect on performance might be significant. As a performance index, we employ here the thermal efficiency of the cycle η_{th}, corresponding to the plant in which its turbine and pump efficiencies are unity and no heat dissipation loss exists; it is identical to the exergy efficiency of the cycle ξ_0 if $E_0 = U_0$ and $\eta_B = 1$.

(a) *Saturation pressure*: The upper and lower limits of pressure in a cycle are determined by the vapor temperature at the steam generator and the cooling temperature at the condenser, and must be admissible for construction and operation of the plant from the techno-economic viewpoint. Figure 2.18 represents the saturation pressures of substances that are of importance for power generation; circles represent critical points.

(b) *Profile of saturation line*: Figure 2.19 illustrates saturation lines of important substances in the *T–s* diagram. As explained in Section 2.3.1 (cf. Fig. 2.17b), the state of the exhaust from the turbine can be superheated for some peculiar substances. A rough prediction of the appearance of the exhaust superheating is possible through estimation of the gradient $-ds/di$ in the *i–s* diagram. For more precise estimation, the effect of the stage efficiency must be considered and the index should be modified in the following manner.

Assume here an elemental turbine stage with stage efficiency η, as shown in Figure 2.20a. The stage efficiency η is defined as

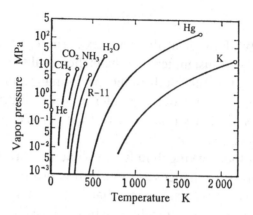

Figure 2.18 Saturation vapor pressure.

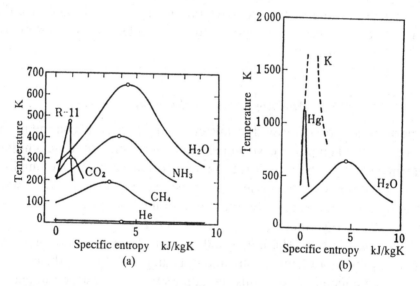

Figure 2.19 Saturation lines: (a) ordinary fluids; (b) liquid metals.

$$\eta = d|i|/di_a$$

where di_a is the adiabatic enthalpy drop for the elemental stage, and $d|i|$ is the enthalpy drop across it, identical with its work generation. Changes of enthalpy and entropy across the stage are given by

$$di = -\eta \, di_a$$

and

$$ds = \left(di_a - d|i|\right)/T = \left(1 - \eta\right)di_a/T$$

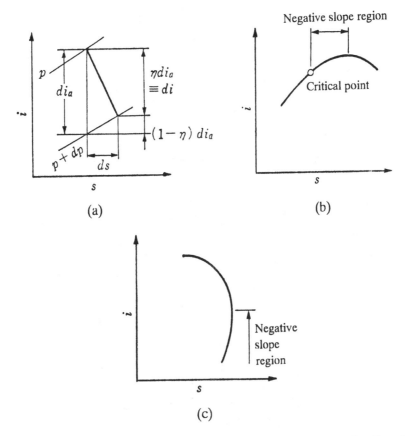

Figure 2.20 Relation between saturated vapor line and turbine expansion line: (a) expansion process in infinitesimal turbine stage; (b, c) two regions of saturation line with negative slope.

where T is the temperature of vapor in the stage. Hence the gradient of the expansion line at this stage is expressed as

$$ds/di = -\left(1/\eta - 1\right)/T$$

Comparison of this with gradient di/ds on the dry saturation vapor line will give information about the possibility of intersection of the expansion line with the saturation line. If the condition

$$\left(Tds/di\right)_{sat} > -\left(1/\eta - 1\right) \tag{2.54}$$

is satisfied for all pressures, the expansion line deviates from the saturation line into the superheated vapor region and never returns to the wet vapor zone as the pressure drops. Hence $(Tds/di)_{sat}$ can be employed as an index for exhaust superheating and its threshold $-(1/\eta - 1)$ is a function only of the stage efficiency η.

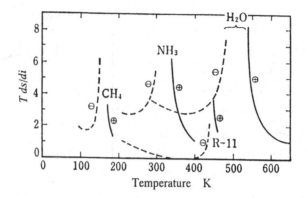

Figure 2.21 Slope of saturated vapor line.

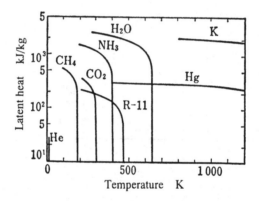

Figure 2.22 Latent heat of evaporation.

Figure 2.21 shows this index $(T\,ds/di)_{sat}$ versus the saturation temperature T for various substances. If the stage efficiency is unity, the threshold is equal to zero. A substance with a positive or small negative value of $(T\,ds/di)_{sat}$ may satisfy condition (2.54). Refrigerant R-11 and methane belong to this class. It should be noted that the positive value region at higher pressure bears no relation to exhaust superheating, which corresponds to what is indicated in Figure 2.20b; that indicated in Figure 2.20c is the relevant one.

(c) *Latent heat of evaporation and specific volume*: Both of these parameters determine mass and volumetric flow rate of the working fluid per output power of turbine and are two of the main factors governing the unit size; Figures 2.22 and 2.23 show the latent heat of evaporation and specific volume, respectively.

(d) *Adiabatic enthalpy drop and Rankine cycle with dry saturation vapor*: The adiabatic enthalpy drop or adiabatic heat drop is identical with the specific exergy of working fluid in a specified state, as explained in Section 2.3.1. Its value for dry

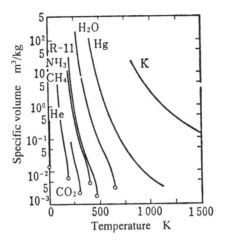

Figure 2.23 Specific volume of dry saturated vapor.

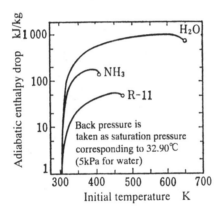

Figure 2.24 Adiabatic enthalpy drop of dry saturation vapor.

saturated steam is shown in Figure 2.24, in which the condenser pressure or back pressure is taken as equal to 5 kPa; the corresponding saturation temperature is 32.9°C. For reference, the thermal efficiency of the ideal Rankine cycle with dry saturated vapor and that of the Carnot cycle are plotted in Figure 2.25. The difference of the working fluid distinctly appears on the thermal efficiency as the critical pressure is approached.

(e) *Rankine cycle with superheated steam*: Water is the most important working fluid and its Rankine cycle for fossil-fuel-fired power plants is the superheated one. Figure 2.26 shows the thermodynamic performance of the ideal Rankine cycle with superheated steam, where no thermodynamic loss exists; thermal efficiency is plotted against pressure at the turbine inlet, with temperature as the

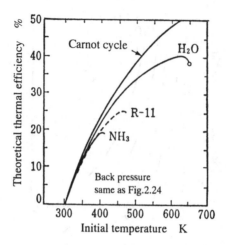

Figure 2.25 Thermal efficiency of ideal Rankine cycle with dry saturated vapor: $\eta_T = \eta_P = 1$.

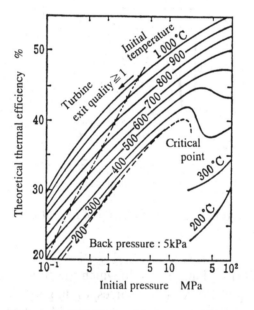

Figure 2.26 Thermal efficiency of ideal Rankine cycle with superheated steam: $\eta_T = \eta_P = 1$.

parameter. The condenser pressure is fixed at 5 kPa. If the state of steam at the turbine inlet enters to the left of the chain line in the figure, the exhaust steam from the turbine is superheated and the cycle generates an intolerable amount of exergy loss as the exhaust temperature is higher than that of the cooling water or the environment. Of course, this situation is never permitted in practice.

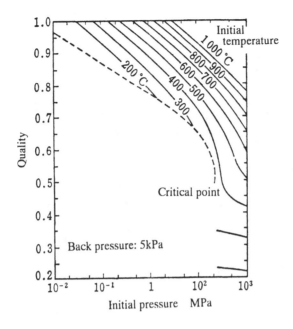

Figure 2.27 Dryness of turbine exhaust with superheated steam: $y_T = 1$.

On the other hand, excess wetness in the turbine exhaust is one of the serious problems of the steam turbine, especially for nuclear power plants, as noted in Section 2.3.1. Nuclear power plants do not or cannot employ the superheated cycle. Figure 2.27 shows the dryness of the turbine exhaust in terms of the condition of the turbine inlet for a condenser pressure of 5 kPa. For simplicity, the cases corresponding to the wet steam turbine are omitted from the figure.

2.3.3 Regenerative and reheat cycles

2.3.3.1 Necessity of introduction of regeneration and reheating

To improve the thermal efficiency of the simple Rankine cycle, one has to increase the area enclosed by its cycle curve in the *T–s* or *i–s* diagram (Fig. 2.16). As the lowest temperature of the cycle is fixed by the condenser temperature or environmental temperature, it is necessary to shift the isobaric heating line 2–3–4 upward and/or to the right. An upward shift corresponds to an increase in the turbine inlet temperature of the working fluid and a rightward shift to that of the turbine inlet pressure. Only from the viewpoint of effective energy conversion is it desirable to use a turbine inlet temperature as high as the energy source can manage. In reality the condition at the turbine inlet is limited, for techno-economic reasons, to about 30 MPa and 600°C even for advanced steam power

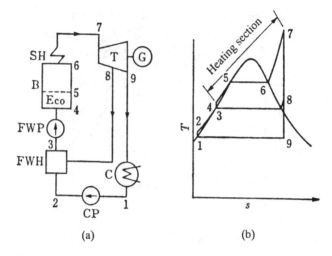

(a)   (b)

Figure 2.28 Principle of regeneration of Rankine cycle.

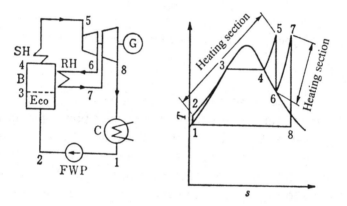

Figure 2.29 Principle of reheating cycle.

plants. In this situation, improvement of the thermal efficiency of the Rankine cycle power plant can be achieved by (i) raising the starting temperature of the heating process in the steam generator, and/or (ii) enlarging the higher temperature sections of the heating process in the steam generator; it should be remembered that the former suffers from the limitation discussed in Section 2.2.2.

The regenerative Rankine cycle achieves (i) in the following way. Some part of the steam flow is extracted during the expansion process in the turbine and heats up the feedwater flow between the condenser and the steam generator, raising its temperature at the inlet of the latter (Fig. 2.28). On the other hand, the reheat Rankine cycle achieves (ii) by introducing a steam flow to the steam generator and reheating it halfway through the expansion process (Fig. 2.29). In addition to improving the thermal efficiency of the ideal cycle, the reheating increases the dryness of the turbine exhaust; the practical merit of this is significant as it largely reduces the problem of erosion of the turbine blades. Regeneration and reheating are employed exclusively on the cycle whose working fluid is water. So, our discussion will be restricted to the Rankine cycle using water steam.

2.3.3.2 Regenerative cycle

Steam for regenerative heating of feed water is extracted at several ports on the turbine. Techno-economic considerations determine the number and positions of the ports for steam extraction; the number of steam extraction points is identical to the number of stages of steam extraction. Heating of feed water by the extracted steam is conducted in feed water heaters that are classified as either the mixing or the surface type (Fig. 2.30). In the former, the feed water is directly heated by injection of extracted steam into it; in the latter, it is indirectly heated via a heating surface by surface condensation. The feed water and the extracted steam and its condensate flow according to one of the following schemes:

i. direct mixing of the two flows by the mixing-type feed water heater, which also works as an apparatus for degassing the feed water and is usually called the deaerator (Fig. 2.30a);
ii. indirect heat transfer by the surface-type feed water heater without mixing (Fig. 2.30b); or
iii. squeezing the extracted steam condensate from the surface-type heater into the feed water flow by an auxiliary pump (Fig. 2.30c).

It should be noted here that schemes (i) and (ii) are equivalent in the energy balance calculation.

The feed water temperature at the exit of the surface-type heater can, in principle, rise to that of the extracted steam, but is usually taken as the saturation temperature corresponding to the extraction pressure. As the pressure in the mixing-type heater cannot exceed that of the extracted steam injected, it is essential to provide a pump after the heater for pressurization of the feed water; the upper limit of the exit temperature is determined from cavitation considerations and is naturally lower than the saturation temperature.

The simplest scheme of the regenerative Rankine cycle is one with only one stage of extraction, and its heater is of the mixing type as shown in Fig. 2.28a. In the regenerative Rankine cycle with multistage extraction, usually one of the feed water heaters is of the mixing type and the rest are of the surface type. Figure 2.31 shows a typical arrangement of a regenerative steam power plant. Given the number of extraction stages, one of the important criteria for selection of the steam extraction points is the maximization of the thermal efficiency, or the ther-

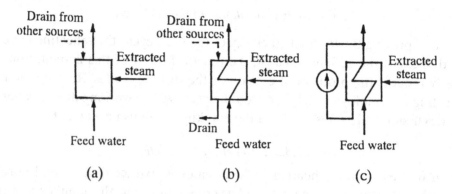

Figure 2.30 Types of feed water heater: (a) mixing type; (b) surface type; (c) mixing of drain flow with main flow by pump.

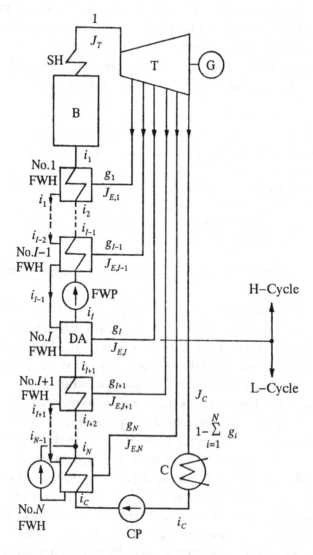

Figure 2.31 Typical scheme of regenerative Rankine cycle. (H–Cycle) Higher pressure cycle; (L–Cycle) lower pressure cycle.

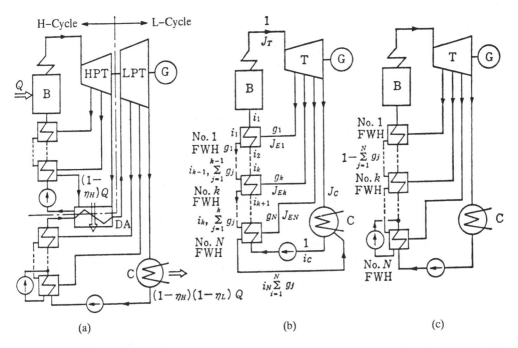

Figure 2.32 Diagrams for analysis of regenerative cycle: (a) division of cycle; (b) H cycle; (c) L cycle.

modynamic optimization of the cycle. Selection methods have long been investigated and proposed by many authors. Among them the best known and the simplest is the one by Laupichler (1926), which is derived for the plant with all mixing-type heaters. It recommends us to determine the positions of steam extraction so as to divide equally the enthalpy drop along the adiabatic expansion line in the turbine, or to divide equally the enthalpy rise of the feed water from the condenser exit to the economizer inlet. This criterion is widely used as a rough estimation, but, strictly speaking, it does not hold because in actual cycles all feed water heaters but one are of the surface type. So, we will analyze the thermodynamic optimization of the regenerative Rankine cycle with a practical feed water heater arrangement in the following (Nakanishi and Kittaka 1984).

The regenerative power plant is modeled as shown in Figure 2.32. It is divided into higher and lower pressure sides at the mixing-type heater, deaerator DA (Fig. 2.32a), and its cycle can be considered as a combination of H cycle (Fig. 2.32b) and L cycle (Fig. 2.32c), joined at the deaerator. It is easily understood that the thermal efficiency of the whole cycle, η_{th}, is determined by those of the component cycles as follows:

$$\eta_{th} = 1 - \left(1 - \eta_H\right)\left(1 - \eta_L\right) \tag{2.55}$$

where η_H and η_L are the thermal efficiencies of the higher and lower pressure cycles, respectively.

To obtain a formula for the thermal efficiency in a closed form, the following assumptions are made, which are commonly used in theoretical analyses of steam power cycles.

i. There are no thermodynamic losses in the steam turbine and the piping. In addition, the work required to operate the pumps is negligible.

ii. The enthalpy difference between any extracted steam and its condensate (boiling water) is assumed to be constant.

iii. The feed water temperature at the exit from the heater is equal to the saturation temperature corresponding to the extraction pressure.

iv. The condensate is discharged in the saturated condition from the surface-type heater.

First, find the formula corresponding to the scheme shown in Figure 2.32b. The starting point is to write heat balance equations for all the heaters. For the no.1 feed water heater (FWH), the equation takes the following form:

$$g_1(J_{E1} - i_1) = i_1 - i_2 \qquad (2.56)$$

where g_k is the flow rate of extracted steam at the kth stage per boiler flow rate, in kg/kg, and J_{Ek} and i_k are the specific enthalpies of extracted steam at the kth stage and its corresponding saturated water, respectively, in kJ/kg. The flow rate g_1 is expressed as

$$g_1 = \frac{(i_1 - i_2)}{(J_{E1} - i_1)} \qquad (2.57)$$

For the FWHs after the second stage, we have to take into consideration the condensate flow from the higher pressure FWHs, and thus obtain a general formula

$$g_k = \frac{i_k - i_{k+1}}{J_{Ek} - i_k} - \frac{i_{k-1} - i_k}{J_{Ek} - i_k} \sum_{j=1}^{k-1} g_j, \quad k = 2, \ldots, N \qquad (2.58)$$

where N is the total number of stages of steam extraction; $k = N+1$ corresponds to the condenser. Thus

$$i_{N+1} = i_C$$

where i_C = specific enthalpy of condensate from the condenser. Introducing a dimensionless parameter

$$a_k = (i_k - i_{k+1})/(J_{Ek} - i_k), \quad k = 1, 2, \ldots, N \qquad (2.59)$$

and using assumption (ii)

$$J_{Ek} - i_k = \text{constant} \qquad (2.60)$$

we transform Eqs. (2.57) and (2.58) into the following system of simultaneous equations:

$$g_1 = a_1 \\ g_k = a_k - a_{k-1}\sum_{j=1}^{k-1}g_j, \quad k = 2, 3, \ldots, N \Bigg\} \tag{2.61}$$

Again, defining a new parameter G_k from g_k,

$$G_1 = 0 \\ G_k = \sum_{j=1}^{k-1}g_j, \quad k = 2, 3, \ldots, N \Bigg\} \tag{2.62}$$

we modify this system to the following simpler one:

$$G_1 = 0 \qquad G_2 = a_1 \\ G_k = a_{k-1} + \left(1 - a_{k-2}\right)G_{k-1}, \quad k = 3, \ldots, N \Bigg\} \tag{2.63}$$

whose solution can be easily obtained by mathematical induction as follows:

$$G_k = a_{k-1} + \sum_{i=1}^{k-2}a_i\prod_{j=1}^{k-2}\left(1 - a_j\right), \quad k = 3, \ldots, N \tag{2.64}$$

The flow rate of extraction steam g_k is found by inversion of Eqs. (2.62):

$$g_1 = a_1 \\ g_2 = a_2 - a_1^2 \\ g_k = a_k - a_{k-1}^2 - a_{k-1}\sum_{i=1}^{k-2}a_i\prod_{j=i}^{k-2}\left(1 - a_j\right), \quad k = 3, \ldots, N \Bigg\} \tag{2.65}$$

This result implies that the arrangement for steam extraction is completely fixed by the set of dimensionless parameters a_k, $k = 1, 2, \ldots, N$, under assumptions (i)–(iv).

The formula for thermal efficiency in terms of g_k, $k = 1, 2, \ldots, N$, is deduced from evaluation of heat quantities absorbed at the boiler and at the condenser:

$$\eta_{\text{th}} = 1 - \frac{\left(J_C - i_C\right)\left(1 - \sum_{j=1}^{N}g_j\right) + \left(i_N - i_C\right)\sum_{j=1}^{N}g_j}{j_T - i_1} \tag{2.66}$$

where J_T, J_C are the specific enthalpies of steam at the inlet and exit of the turbine, respectively. To express this formula in terms of a_k, $k = 1, 2, \ldots, N$, we substitute Eq. (2.65) into the above, and introduce a system parameter A:

$$A = \frac{i_T - i_C}{J_T - J_C} \tag{2.67}$$

where i_T is the specific enthalpy of saturated water at the pressure corresponding to the inlet of the turbine. Finally we obtain the expression

$$\eta_{\text{th}} = 1 - \frac{1 - \sum_{i=1}^{N} a_i \prod_{j=i}^{N} (1 - a_j)}{1 + A - \sum_{i=1}^{N} a_i} \tag{2.68}$$

In transformation, the relationship deduced from Eqs. (2.61) and (2.62) is used:

$$\sum_{j=1}^{k} g_j = a_k + (1 - a_{k-1})G_k \tag{2.69}$$

The other scheme, shown in Figure 2.32c, is different from that of Figure 2.32b only in the connection between the condenser and no. N FWH. Thus the same formula for g_k as for the former holds except for no. N FWH. For the latter, the heat balance around it gives

$$g_N = \frac{a_N}{1 + a_N} - \frac{a_{N-1} + a_N}{1 + a_N} \sum_{j=1}^{N-1} g_j \tag{2.70}$$

So the total flow rate of steam extraction, Σg_j, is easily found through the equation

$$g_j = G_{N-1} + g_N$$

as Eq. (2.64) still holds for $k = N - 1$:

$$\sum_{j=1}^{N} g_j = \frac{a_N + \sum_{i=1}^{N-1} a_i \prod_{j=1}^{N-1} (1 - a_j)}{1 + a_N} \tag{2.71}$$

On the other hand, the general expression for thermal efficiency for the scheme shown in Figure 2.32c is given as

$$\eta_{\text{th}} = 1 - \frac{\left(1 - \sum_{j=1}^{N} g_j\right)(J_C - i_C)}{J_T - i_1}$$

Substituting Eq. (2.71) into the above, we obtain

$$\eta_{\text{th}} = 1 - \frac{1 - \sum_{i=1}^{N-1} a_i \prod_{j=i}^{N-1} (1 - a_j)}{(1 + a_N)\left(1 + A - \sum_{i=1}^{N} a_i\right)} \tag{2.72}$$

Combination of the two efficiency formulae Eqs. (2.68) and (2.72) according to Eq. (2.55) gives the thermal efficiency for the system represented in Figure 2.32a; the cycle is divided at the deaerator DA, or no. I FWH, into the H and L cycles, resulting in division of the parameter A into two parts A_H and A_L:

$$A_H = \frac{i_T - i_I}{J_{EI} - i_I} = A - \sum_{i=I}^{N} a_i \tag{2.73}$$

$$A_L = \frac{i_I - i_C}{J_C - i_C} = \sum_{i=I}^{N} a_i \tag{2.74}$$

The final expressions for η_H and η_L are as follows:

$$\left. \begin{aligned} \eta_H &= 1 - \frac{1 - \sum\limits_{i=1}^{I-1} a_i \prod\limits_{j=i}^{I-1}\left(1 - a_j\right)}{1 + A_H - \sum\limits_{i=1}^{I-1} a_i} \\[2em] \eta_L &= 1 - \frac{1 - \sum\limits_{i=I+1}^{N-1} a_i \prod\limits_{j=i}^{N-1}\left(1 - a_j\right)}{\left(1 + a_n\right)\left(1 + A_L - \sum\limits_{i=I+1}^{N} a_i\right)} \end{aligned} \right\} \tag{2.75}$$

Given the system parameter A, the overall efficiency η_{th} thus obtained is governed by selection of the extraction points and, in turn, is a function of the parameters $a_k, k = 1, 2, \ldots, N$. The thermodynamic optimization of the regenerative cycle is achieved by searching the set of parameters $(a_k; k = 1, 2, \ldots, N)$ that maximizes the thermal efficiency η_{th}. This leads to a system of algebraic equations:

$$\partial \eta_{th} / \partial a_k = 0, \quad k = 1, 2, \ldots, N \tag{2.76}$$

Unfortunately it is impossible to solve this system in a closed form for $N > 2$, and some numerical iteration method is indispensable for its solution. Numerical calculation shows that the difference between the optimal thermal efficiencies obtained by the Laupichler method and the above one is relatively small and is practically identical for stages with number greater than 3, provided values of parameters a_k are not close to each other; the former locates the extraction points at higher pressure than the latter. Generally speaking, the method of equipartition of adiabatic enthalpy drop gives a good approximation for optimization of the ideal regenerative Rankine cycle.

Assumption (ii) does not hold with reasonable accuracy in practical cases, but it is empirically assured that equipartition of enthalpy rise of feed water gives reasonably good results for the extraction pressures if the averaged value of enthalpy difference $(J_E - i)$ is taken.

2.3.3.3 Reheat cycle

Improvement of thermal efficiency is also achieved by reheating steam in the midst of expansion in the turbine. In the scheme shown in Figure 2.29, its thermal efficiency is given by

$$\eta_{th} = 1 - \frac{i_8 - i_1}{i_5 - i_2 + i_7 - i_6}$$

This efficiency is higher than that for the nonreheat cycle:

$$\eta_{th}' = 1 - \frac{i_6' - i_1}{i_5 - i_2}$$

because the following relations hold:

$$i_8 - i_1 = i_8 - i_6' + i_6' - i_1$$

and

$$i_8 - i_6' < i_7 - i_6$$

The improvement by reheating is a natural result from the fact that the reheat cycle is composed of the fundamental nonreheat cycle 1–2–3–4–5–6–6′ and an additional cycle 6–7–8–9–6′; the latter has better efficiency than the fundamental cycle.

In principle, no restriction exists for multistage reheating, but from the practical viewpoint two-stage reheating is employed only in the advanced plant with the largest capacity, and more than three stages is never employed because of difficulty in handling voluminous steam flow at lower pressure. One of the advantages of reheating is reduction of wetness in the lower pressure zone of the turbine. As seen from Figure 2.27, if steam at 600°C expands isentropically from 2.5 MPa to 5 kPa, its final wetness becomes more than 0.1 and the corresponding dryness less than 0.9, which is harmful for the turbine blade because of erosion; in fact, flow friction or irreversibility in the expansion process reduces the wetness a little. In any case, employment of high pressure requires some devices for reducing wetness. Though the wet steam turbine for a nuclear plant employs mechanical separation, the conventional means is reheating.

2.3.3.4 Infinite-stage regenerative and reheat cycles

To ascertain the limitation of improvement by regeneration and reheating, we analyze the infinite-stage regenerative and reheat Rankine cycles (Kittaka and Nakanishi 1985).

First, the subcritical infinite-stage regenerative cycle is considered (Fig. 2.33). The infinite-stage FWH is a series of an infinite number of pairs of infinitesimal mixing-type FWHs and infinitesimal feed water pumps, as shown in Figure 2.34.

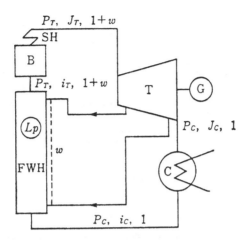

Figure 2.33 Nonreheated cycle with infinite-stage regeneration. (B) Boiler; (C) condenser; (FWH) feed water heater system; (SH) superheater; (T) turbine.

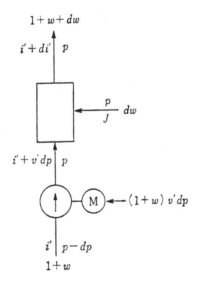

Figure 2.34 Heat balance on infinitesimal feed water heater.

For idealization, the following are assumed:

i. There is no thermodynamic loss in the turbine, pump, piping, and so on.
ii. Mixing of feed water and extracted steam is isobaric and the feed water is heated up to saturation temperature; the state of feed water changes along the saturated liquid line.
iii. The exit pressure of the FWH is equal to the turbine inlet pressure. As seen from the figure, the feed water flow $1 + w$ is compressed from $p - dp$ to p by an infinitesimal

pump and is heated from i' to $i' + di'$ in an infinitesimal heater by injection of the extracted steam dw, whose specific enthalpy is J. The heat balance equation is then expressed as

$$\left(i' + di'\right)\left(1 + w + dw\right) = Jdw + \left(1 + w\right)\left(i' + v'dp\right)$$

where v' is the specific volume of the saturated water. Neglecting infinitesimal terms of more than second order, and applying the thermodynamic first law,

$$T\,ds' = di' - v'\,dp$$

we obtain a differential equation

$$\frac{dw}{1 + w} = \frac{T\,ds'}{J - i'} \tag{2.77}$$

where s' is the specific entropy of the saturated water, w is the total flow rate of extracted steam from the condenser pressure p_C to pressure p

$$w = \int_{p_C}^{p} dw$$

and T is the absolute saturation temperature for p. Consequently

$$1 + w = \exp\left\{\int_{p_C}^{p} \frac{T\,ds'}{J - i'}\right\} \tag{2.78}$$

The total flow rate of extracted steam w_T is given as the integral from the condenser pressure p_C to the turbine pressure p_T:

$$w_T = \exp\left\{\int_{p_C}^{p_T} \frac{T\,ds'}{J - i'}\right\} - 1 \tag{2.79}$$

Finally, we can evaluate the thermal efficiency according to the general formula by substituting w_T obtained above:

$$\eta_{th} = 1 - \frac{J_C - i_C}{\left(1 + w_T\right)\left(J_T - i_T'\right)} \tag{2.80}$$

For supercritical cycles a modification is necessary. Regenerative heating is no longer possible in the usual sense as the phase change does not exist at pressures higher than the critical one. It is therefore assumed that no steam is extracted in this region. Thus the total flow rate of the extracted steam remains at that for the critical pressure p_{cr}:

$$w_T = w_{cr} = \exp\left\{\int_{p_C}^{p_{cr}} \frac{T\,ds'}{J - i'}\right\} - 1 \tag{2.81}$$

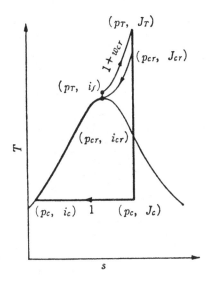

Figure 2.35 Supercritical cycle with infinite-stage regeneration.

Referring to Figure 2.35, we easily obtain the formula for the supercritical cycle:

$$\eta_{th} = 1 - \frac{J_{cr} - i_{cr}}{J_T - i_f} \cdot \frac{J_C - i_C}{(1 + w_{cr})(J_{cr} - i_{cr})} \tag{2.82}$$

Numerical results are shown in Figure 2.36, in which the condenser pressure is taken as 5 kPa. Comparison of this figure with Figure 2.26 will give a clear understanding of the effect of regeneration.

Infinite-stage reheating can also be analyzed in a procedure similar to the above by introducing a series of an infinite number of pairs of infinitesimal turbines and infinitesimal reheaters, but it is somewhat tedious to trace this analysis. Only its result is shown in Figure 2.37, where the thermal efficiency is given for infinite-stage regenerative cycles without reheating, with one-stage reheating, and with infinite-stage reheating.

2.3.3.5 Enthalpy and exergy diagrams of water

The usefulness of thermodynamic diagrams such as the *T*–*s* and *i*–*s* ones for water has been fully demonstrated above, especially in the qualitative discussion of the steam power plant cycle. On the other hand, the significance of the *i*–*e* diagram in cycle analysis has recently become larger. An *i*–*e* diagram of water appears in the Appendix for the convenience of readers.

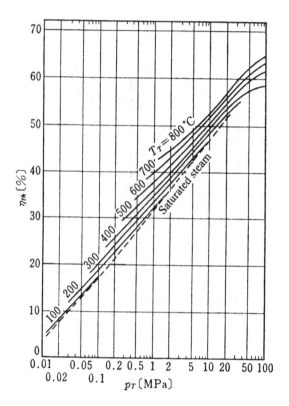

Figure 2.36 Thermal efficiency of nonreheating cycle with infinite-stage regeneration.

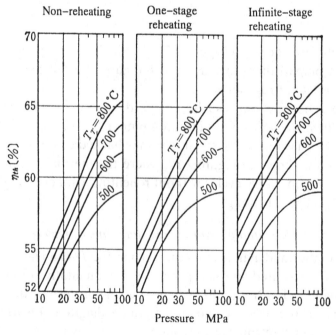

Figure 2.37 Effect of number of stages of reheating on thermal efficiency.

2.4 Steam power plants for practical use

TERUSHIGE FUJII
Kobe University

2.4.1 Classification

Steam power plants have been developed, with the use of fossil fuels as their main heat sources, since Newcomen's steam engine using a saturated steam cycle was first manufactured in the year 1712. However, since about 1900, the superheated steam Rankine cycle has primarily been used when the steam pressure is above 2 MPa, because of the decrease in moisture from the exhaust of the steam turbine and improvements in plant efficiency. However, use of the saturated steam Rankine cycle has increased recently with a variety of fuels and working fluids. There are cases of 1ow-temperature sources using low-boiling-point media in power plants, and those of pressurized water reactors (PWRs) and boiling water reactors (BWRs) using nuclear fuel as their heat source.

These power plants are classified on the basis of differences in the input energy sources, as shown in Table 2.1.

On the other hand, geothermal power plants use geothermal energy resources as input power. These systems are called total flow systems, flash systems, hybrid systems, and so on, according to the characteristics of each system.

In this section, a variety of steam power plants that use the Rankine cycle are classified into saturated steam and superheated steam systems. Next, examples of the combined cycle, the two-fluid cycle, and others are shown.

2.4.2 The saturated steam Rankine cycle

Here, the saturated steam Rankine cycle denotes the Rankine cycle under the turbine inlet condition of saturated steam. In fossil-fuel-burning boilers, the combustion gas temperature is very high and requires superheated steam at high temperature and pressure. However, when using low-temperature heat sources, saturated steam is used. As shown in Table 2.1, a coolant such as Freon, ammonia, propane, methane, or isobutane (see Fig. 2.18), whose saturation pressure is greater than water at the same temperature, is used as the working fluid for heat sources below 200°C. Generally, in such a system, the regeneration process is not used because of the small difference between the maximum and the minimum temperatures.

First, when using water as the working fluid, there is the dry-well-type geothermal power plant. The flowsheet of the 2500 kW output plan of the Oni Kobe Geothermal Power Station is shown in Figure 2.38 (Iida 1975). Steam–water mix-

Table 2.1. *Power generation plants classified by difference
in input energy resources*

Classification	Heat energy resources	Working fluids
Steam power plant	Fossil fuel such as coal, oil, and liquified natural gas	Water
Nuclear power plant	Nuclear fission, nuclear fusion	Water, liquid metal, He, CO_2 gas, etc.
Geothermal power plant	Geothermal water, steam	Water, low-boiling-point media
Liquified natural gas power plant	Liquified natural gas	Water, low-boiling-point media, air
Ocean temperature difference power plant	Ocean	Water, low-boiling-point media
Solar energy power plant	Solar heat	Water
Exhaust heat recovery power plant	Exhaust heat	Water, low-boiling-point media

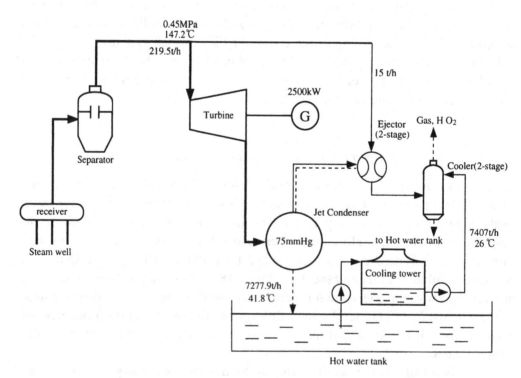

Figure 2.38 Flowsheet of geothermal power generating station (Onikobe, Japan).

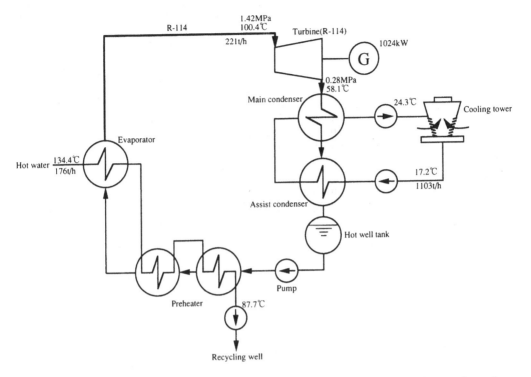

Figure 2.39 Flowsheet of geothermal power generating station (Nigorigawa, Japan).

tures are injected from the steam well. For comparatively dry and good-quality steam, the steam, from which moisture and solids are excluded, enters the turbine as dry saturated steam of quality $\simeq 1$ and is condensed and liquefied in the condenser. Also, uncondensed gases in the steam such as H_2S, CO_2, and H_2O are extracted by the ejector to raise the vacuum of the condenser. This example has been built for a vacuum of 75 mm Hg (that is, 0.1 bar) and a saturation temperature $\simeq 45.8°C$. For such geothermal resources, since fuel cost is not a consideration, the back-pressure-type turbine, which releases the exhaust steam into the atmosphere, is used.

The Rankine cycle is used for applications using low-boiling-point media to effectively utilize geothermal resources (hot water and steam) at 100 to 150°C and also the waste heat from a factory. Low-boiling-point media with a greater saturation pressure than water, such as Freon, isobutane, and ammonia, are evaporated by the indirect contact-type heat exchanger. Figure 2.39 shows the flowsheet of the 1024-kW pilot plant of the geothermal binary-fluid cycle using only hot water, built as a part of the Sunshine Project with the Agency of International Science and Technology, under the Ministry of International Trade and Industry. The working fluid is a low-boiling-point medium, R-114. In general,

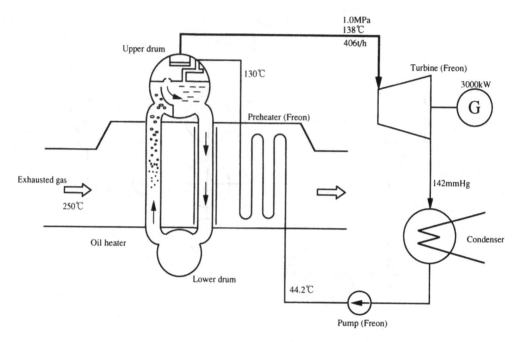

Figure 2.40 Power generation plant operating on low- and middle-temperature heat sources (oil, Freon).

since the specific volume of a low-boiling-point medium is lower than that of steam, the turbine becomes compact. Further, since the heat is relatively great compared with the latent heat, it can take on temperature–heat flux characteristics appropriate to heat recovery, and the efficiency of the heat recovery process increases. Also, both steam and water resources are utilized. For example, the power plant of the saturated steam Rankine cycle (1000 kW) has been tested with isobutane as a working fluid with an upper limit of the explosion ratio of 8.44% and a lower limit of 1.80%. The safety of isobutane was demonstrated by ventilating well and installing a leaked-gas detector (Aikawa and Kawaguchi 1975). Also, an actual experiment on a hot dry rock power plant was carried out by the New Energy Development Organization of Japan, and, for the geothermal resources, the pressurized water was fed into deep ground and warmed to a higher temperature. As shown in Figure 2.40, for low- and middle-temperature resources of 200 to 400°C, from the relationship of the upper-limit temperature of about 160°C for the decay of the working fluids, by mixing the low-cost media of the stable polyolester oil and R-113, both media exchange heat directly in a Freon vapor generator, which is a saturated Rankine cycle.

In the case of power generation by ocean temperature difference, in which the maximum temperature is 30°C at most, the saturated steam Rankine cycle is used

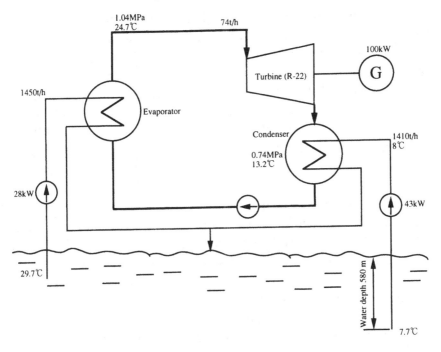

Figure 2.41 Power generation by ocean temperature difference (Naul Republic, plan).

(Uehara 1981). Figure 2.41 shows the flowsheet plan of a demonstration plant with a capacity of 100 kW, which the Tokyo Electric Power Company Ltd. and the Tokyo Electric Designing Company have manufactured in the Naul Republic (Ito et al. 1982). The working fluid is R-22. In 1981 it was demonstrated that the maximum power output was 120 kW and the net output was 31.5 kW.

In general, since solar energy is absorbed by the upper sea layers to a depth of about 10 m and there is a low-temperature difference of about 20°C, the Rankine cycle is used. Because deep sea water is used as the cooling resource, the ratio of in-plant power consumption to total output is above 40%. Efforts are being made to lower this to below 20 to 30% by improving the heat transfer performance of the evaporator and condenser. However, because of its cheap fuel costs and lack of pollution, it is attractive to remote islands.

Also, in relation to the above, there is a method for utilizing a solar pond. If an area of the sea is partitioned off with concrete, the salt concentration in the sea increases as it becomes deeper; for example, the concentration can increase from about 5% in the upper part to about 25% in the lower part. Then, convection is prevented and electricity can be generated by using the heat accumulated in the lower part. In Israel, since 1984, the temperature difference in the water in the Dead Sea, with warm water at 85°C, has been used to generate 500 kW of power.

At present, light nuclear power reactors, that is, PWRs and BWRs, account for more than 80% of the nuclear power plants in the world. Furthermore, the advanced thermal reactor (ATR) in Japan, which is a heavy water reactor, is different only in the form of its fuel relying on nuclear fission. For electric power generation, the saturated steam Rankine cycle is used because of the restriction on the coolant outlet temperature. However, to improve thermal efficiency, the B&W company used the once-through steam generator for the PWR and adopted the superheated steam Rankine cycle (hybrid system; see Section 2.4.6), which generates superheated steam by superheating the saturated steam fossil fuel.

The flowsheets are shown in Figure 2.42a–c. The PWR, which accounts for 70% of the light water reactors in the world, passes heat under a steam condition of about 15 MPa and 300°C in the first circuit to the coolant in the second circuit and feeds it to the turbine as saturated steam. On the other hand, BWRs and ATRs are direct cycle systems in which there is no steam generator, but steam and water are separated, and steam with a moisture fraction of below 0.1%, the moisture being removed by the steam dryer, is fed to the turbine. In particular, the moisture fraction is removed by centrifugal force, utilizing the difference in specific weight between steam and water in expansion. Until the moisture fraction reaches 2%, it is removed by a moisture separator installed in the conduits connecting the high-pressure and low-pressure turbines, or a reheat is done using either the bleeding steam of the high-pressure turbine or the line steam at the exit of the steam generator. A flowsheet of such a bleeding distribution is shown in Figure 2.43. This is an example of a one-stage reheat and six-stage regeneration, 79% being in the low-pressure turbine and 62% in the condenser as regards the evaporation mass flow rate of the steam generator. For the reheat process of moisture removal, 9% of the steam generation flow is used and the drainage is returned back to the no. 6 feed water heater.

2.4.3 The superheated steam Rankine cycle

In modern thermal power plants, marine power plants, liquid metal fast breeder reactors (LMFBR), and so on, the superheated steam Rankine cycle of higher thermal efficiency involving reheat and regeneration is used. Also, since the fast breeder reactors (FBRs) and high temperature gas-cooled reactors (HTGRs) work at a higher temperature than light nuclear reactors, superheat and reheat processes are used.

In general, commercial thermal power plants using fossil fuels are equipped with a supercritical pressure boiler of class 25 MPa and 560°C, foremost among which are the maximum steam conditions of 31.6 MPa and 621/561/538°C in the Eddystone No. 1 power plant. But in recent years, partial load operation of up to

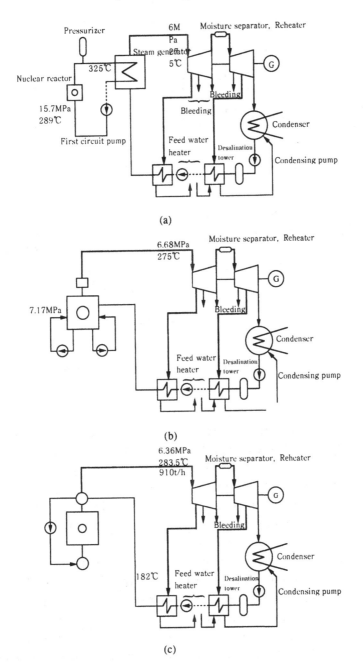

Figure 2.42 Flowsheets of a (a) pressurized water, (b) boiling water, and (c) advanced thermal reactor.

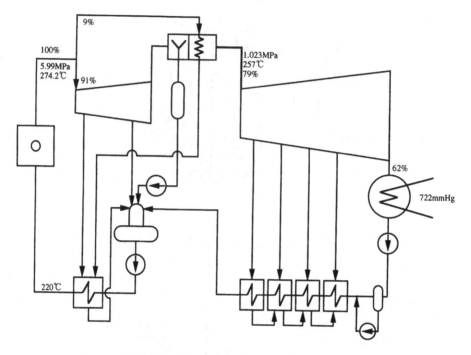

Figure 2.43 Distribution of steam flow rate in PWR.

10%, even in the supercritical pressure boiler, has been required (Nakano 1982). This is due to the spread of the air conditioner, the decline of industries with high electric power consumption like the steel industry, which takes a comparatively flat form of electric power generation, increases in the difference of electric power demand between day and night, and, finally, increases in nuclear power generation. An example of the heat balance diagram involving partial load operation is shown in Figure 2.44. It shows a trial balance sheet of a modern 600-MW power station, where the numerical value in the upper row corresponds to a 100% load and that in the row below corresponds to 15%. However, the recirculation system created by the boiler circulation pump has been refurbished recently for the structure of the boiler (see end of Section 4.6). The plant performance at partial load is improved over that at constant pressure operation because of a decrease in throttle loss provided by the steam control valve, a decrease in the axial power of the feed water pump, and also an increase in the reheated steam temperature due to a rise in the exhaust temperature of the high-pressure turbine, as shown in Figure 2.44b (Iwanaga et al. 1985).

A flowsheet of the No. 2 prototype reactor "Monju" (electric power output 300 MW) of the LMFBR type, which the Power Reactor and Nuclear Fuel Development Corporation is investigating and developing, is shown in Figure

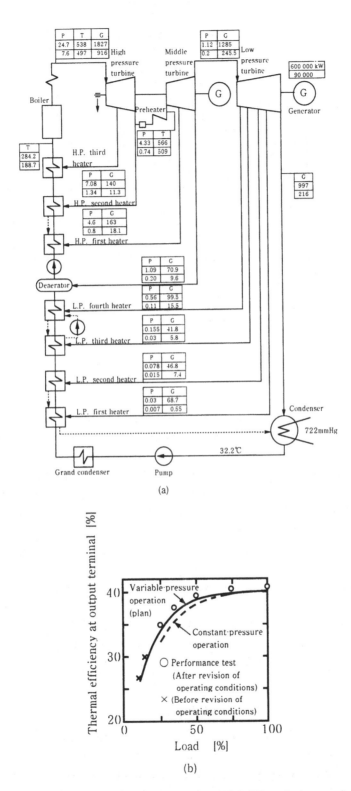

Figure 2.44 Example of a thermal power plant. (a) Heat balance; (b) the thermal efficiency at the generator terminal (overall thermal–electric efficiency).

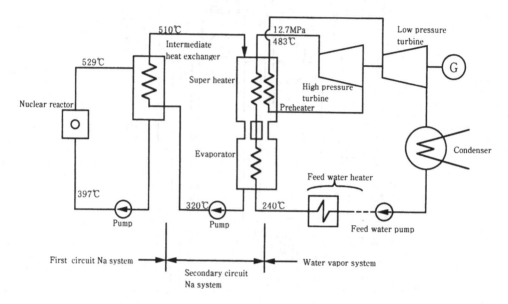

Figure 2.45 Flowsheet of LMFBR "Monju."

2.45. The working fluid of both the first and second cooling circuits is sodium (boiling temperature about 880°C at atmospheric pressure, melting point 98°C, specific weight 0.85). Heat is transferred to water in the steam generator via the middle heat exchanger, and the superheated steam generated at high pressure enters the turbine. A reheat is also carried out. In the HTGR, helium gas at a high temperature is used as the coolant (a chemically stable operation temperature is 750–800°C and over), and electricity is generated.

Furthermore, a combined system is being planned that considers not only electricity generation but also the utilization of the heat processes for manufacturing iron and other purposes, the aim being the development of a multipurpose HTGR.

The gas-cooled fast breeder reactor (GCFR) is expected to be the reactor of the future, producing more fuel than it consumes; but, for the generation of electricity, the superheated steam Rankine cycle is adopted.

2.4.4 The total flow turbine system

This system works simply by flashing highly moist steam and hot water from geothermal heat sources, low- and medium temperature heat sources, and so on. However, the total flow turbine system has received attention, because it saves energy and effectively utilizes unavailable energy since the wet steam and hot water enter the two-phase flow expander as they are and generate electricity. As

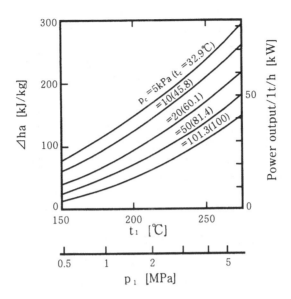

Figure 2.46 Exergy of saturated hot water.

input energy sources such as geothermal become exhausted, other sources such as ocean heat are considered. The available energy, that is, exergy, per 1 kg of saturated hot water is shown in Figure 2.46 for various saturation pressures; in the case of flashing, the steam exergy is lowered to 50% at most. But at present, the two-phase flow expander with higher turbine efficiency has not been developed. This system competes with the saturated steam Rankine cycle using a low-boiling-temperature medium (see Fig. 2.41) and the flash turbine system described below. The relative merits depend strongly on the performance of the expander. When the inlet of the turbine is saturated water, the critical ratio of total-flow turbine efficiency η_T to flash turbine efficiency η_F is as shown in Figure 2.47. It is shown that if the ratio η_T/η_F is greater than the values in the figure, the work of the total-flow turbine is greater and is advantageous in thermal efficiency (Akagawa et al. 1986). At present the impulse type, Heron (reaction) type, helical screw type, and Bunkel type expanders have been investigated. For example, the fineness of liquid droplets can affect performance improvement. The adoption of the steam assist nozzle and the performance improvement due to the insertion of a honeycomb into the nozzle inlet have been considered.

A flowsheet of the hot water power plant that was the first to be manufactured in the world (by Yahata Iron Manufactures of the Shinnihon Iron Manufacturing Company, Japan) is shown in Figure 2.48 (Ikeda and Fukuda 1980). Hot water obtained from the heat exhausted from the blast furnace enters the one-stage impulse turbine. The vapor exhausted from the wet state is separated by a cyclone

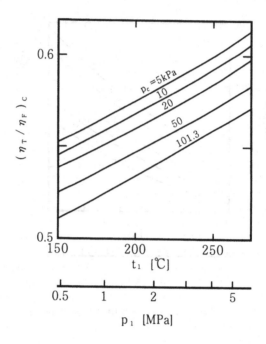

Figure 2.47 Ratio of total-flow turbine efficiency η_T to flash turbine efficiency η_F.

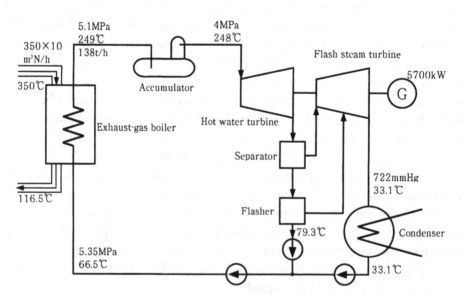

Figure 2.48 Combined use of total-flow turbine and flash turbine systems.

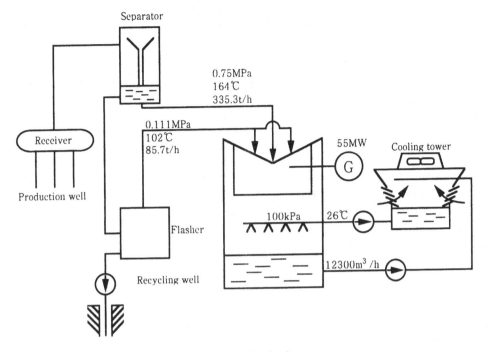

Figure 2.49 Double-flash system.

separator; the steam is fed into the flash turbine; the hot water is flashed in a flasher, and is then fed into the medium stage of the flash turbine. The hot water turbine accounts for 700 kW of the total output of 5700 kW. Also, to absorb the load variation and input variation, the accumulator is increasingly used (Ishigai and Nakanishi 1982).

2.4.5 *The flash turbine system*

In the total-flow turbine system described above, the low dry steam and hot water are depressurized and self-evaporated (flash-evaporation), and then only saturated steam is released into the turbine. This is called a flash turbine system. The separated hot water does not contribute to the power generation and is returned underground. This is possible in both the single-stage and multistage systems, and in some cases a superheater is used externally. A flowsheet of the double-flash system (55 MW) of the Kyushu Electric Company in Haccho-bara is shown in Figure 2.49 (Yoshida and Aikawa 1980). The pressure of the steam–water mixture (ratio approximately 1 to 2) at the turbine inlet is higher, at 0.65 MPa. The steam fed to the power plant from the geothermal well by the gas–water mixture transport method is first separated in the separator. The steam is fed to the first-stage turbine, but the hot water is flashed again in the flasher with the second steam pressure at 0.11 MPa, and is then fed to a mixed-pressure stage on the way to the turbine. The output increases by 15 to 20% compared with that of the first-stage flash system. Furthermore, in the marine field there is an example of a double-flashed power generation system using the exhaust gas of a diesel engine.

2.4.6 The hybrid system

The combination of different heat resources such as fossil fuels, solar heat, and geothermal heat is called the hybrid system, shown in Figure 2.50a and b. Figure 2.50a shows the preheating of the feedwater by geothermal heat in the super-heated steam Rankine cycle, and Figure 2.50b shows the superheater using fossil fuel in the case of geothermal power plants.

2.4.7 The two-fluid cycle

The saturation vapor line of metal vapor is greater in slope; that is, the specific heat is lower than that of the water. As shown in the saturation vapor pressure line of Figure 2.18, the saturation pressure at the same temperature is low and the critical temperature is high. Accordingly, the isothermal absorption process is realized by using metal vapor to increase the upper temperature of the cycle above the critical temperature of water. A combination of two Rankine cycles can be adopted, using metal vapor as the working fluid of the topping cycle and steam as that of the bottoming cycle. A few practical uses of this combination were made in the United States between 1928 and 1950. The inlet temperature of the mercury turbine was 470 to 515°C, and the unit output was 7,500–20,000 kW (Oberly 1950). The flowsheet and T–S diagram are shown in Figure 2.51a and b. The plant efficiency was 34 to 37%, which at the time was high, but at present this cycle is disregarded because the Rankine cycle has increased in efficiency because of improvements in steam conditions from regeneration and reheat, and further because of the corrosion, toxicity, and high cost of steel. But because of requests for improvements in thermal efficiency under conditions of higher tem-perature and higher pressure, the two-fluid cycle of both alkali metal (Na, K, Rb, Cs, etc.) and steam is being reconsidered. For example, an estimation of the two-fluid cycle efficiency is shown in Figure 2.51c (Taniguchi 1979), assuming the boiler efficiency of metal vapor to be 90%, the efficiency of both turbines to be 85%, the temperature difference of heat exchange between the two fluids to be $\Delta T = 10$°C, and the inlet steam temperature of the metal to be 900°C. The thermal efficiency exceeds 50% with the inlet temperature of the steam turbine at 600°C, and increases with higher pressure and temperature, but the effect of the working

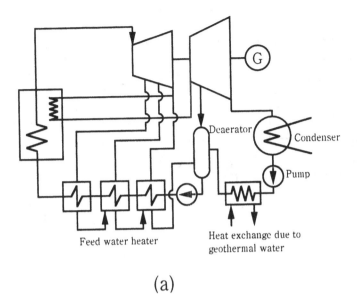

Feed water heater

Heat exchange due to
geothermal water

(a)

Heat exchanger   Fossil fuel burning superheater

Separator

Condenser

Pump

(b)

Figure 2.50 Hybrid system.

fluids on thermal efficiency is low. On the other hand, the mass flow rate ratio of the fluids is Hg $\cong$ 12, Cs $\cong$ 7–8, and Na and K $\cong$ 1–2 for a steam flow rate of 1. But because the specific volume of Na is greater than that of the others, the volumetric flow rate of Na becomes greater inversely and the size of the turbine becomes greater. Recently, K of lower soluble oxygen has received attention because its corroding effect on steel is less than that caused by steam, and it is also comparatively cheap (Mitachi and Saito 1983). However, because the alkali

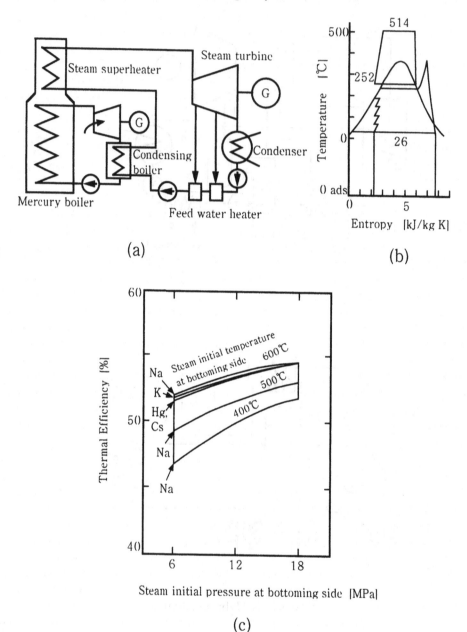

(a)

(b)

(c)

Figure 2.51 Thermal efficiency of the two-fluid cycle. (a) Flowsheet; (b) *T–S* diagram; (c) thermal efficiency.

metal reacts strongly with water and air, it has become very important to establish an appropriate handling technique. Because of the large size of the turbine low-pressure stage due to the increase in unit output, a two-fluid cycle is considered in which steam is used for the topping cycle and a low-boiling temperature medium such as NH_3, SO_2, CO_2, or Freon is used as the working fluid of the bottoming cycle.

2.4.8 The combined cycle

In general, a combination using the Brayton cycle for the topping cycle and the Rankine cycle for the bottoming cycle is called the combined cycle. Between 1930 and 1940, when gas turbine temperatures were low in order to maintain the gas turbine inlet temperature below the permissible temperature, a large quantity of excess air was injected into the combustor and/or part of the heat was released to the outside. In 1950, the method of utilizing the gas turbine exhaust for the Rankine cycle was introduced. So the feed water heating system, exhaust assist burning system, and exhaust reburning system were started in the United States and Europe, as shown in Figure 2.52a–c.

In the 1960s, the supercharger boiler system was introduced in the USSR. The feed water heating system involves preheating the feed water for the steam power plants using the gas-turbine exhaust gas. The exhaust assist burning system involves assist burning in a duct, which leads the exhaust gas from the gas turbine to the exhaust gas boiler. In contrast, the supercharged boiler system involves pressurized burning in a steam boiler attached to the high pressure side of the compressor outlet for the gas turbine and then recovering the exhaust heat in a stacked gas cooler (see Fig. 2.52d). Since the density of the fuel gas increases in a pressurized boiler, the heat transfer coefficient on the gas side increases, leading to a decrease in the surface area and an increase in the compactness of the boiler. However, because the construction cost increases, large-capacity plants, with the exception of the Velox boiler, have not been built. Also, the exhaust gas reburning system uses the exhaust gas as boiler combustion air, including residual oxygen, which comes from excess air; and instead of the conventional boiler fan, use of an air preheater has spread throughout Europe. The main machine is a steam turbine, and the output ratio of the gas turbine to the steam turbine is about one-fifth with a relative increase in thermal efficiency of 5 to 8%.

In contrast, the exhaust recovery system has been widely used in the United States since 1970. In particular, when the gas turbine inlet temperature exceeds 1000°C and the exhaust gas temperature of the high-efficiency gas turbine exceeds 600°C, use of the exhaust heat recovery method, in which the gas turbine is the main turbine, increases. The schematic and *T–S* diagrams are shown in Figure 2.53 (Kurosawa and Inui 1981). Because the exhaust gas side of the gas

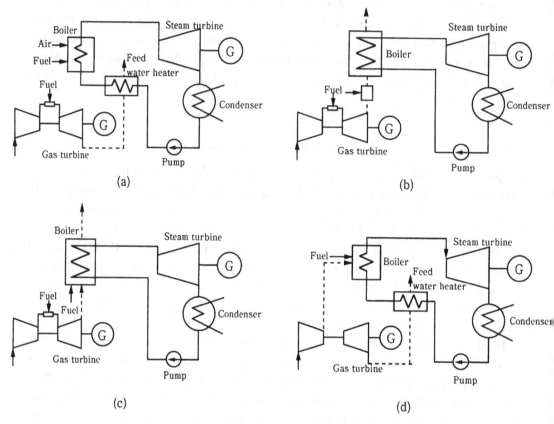

Figure 2.52 Combined cycle. (a) Feed water heater cycle; (b) exhaust assist burning cycle; (c) exhaust reburning cycle; (d) supercharged-boiler cycle.

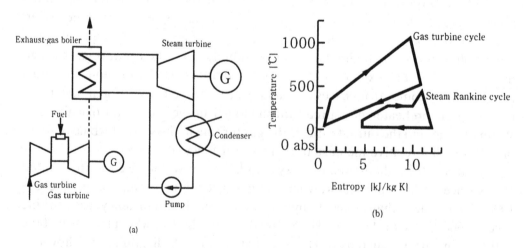

Figure 2.53 Compound combined cycle. (a) Flowsheet; (b) T–S diagram.

turbine can be idealized more, the thermal efficiency is improved. For this system, the output ratio of the gas turbine to the total output is about two-thirds and the plant efficiency is greater than when only the Brayton cycle is used. The heat exchanger for exhaust heat recovery is made up of a compound pressure system such as single and double pressure nonreheat and triple pressure reheat, as shown in Figure 2.54. Regarding the superheat and evaporation of the exhaust heat exchanger and the temperature difference between the gas at the preheating stage and the fluid, the quantity of heat absorbed, the evaporation rate, and the recovery rate of the exhaust heat increase with a decrease in the saturation pressure on the feed water side. Because this involves a decrease in steam pressure, the efficiency of the total cycle decreases. Accordingly, the actual design of the system must be optimized to consider both factors. In Japan, the Sakaide power plant (225 MW, with a firing mixture of coke oven gas and heavy oil and a gas turbine inlet temperature of 788°C), the first exhaust gas reburning system, has been in operation since 1971. As for an exhaust heat recovery system, the Japan National Railroad Kawasaki power plant, 141 MW (100 MW gas turbine and 41 MW steam turbine with a gas turbine inlet temperature of 1065°C), is in operation using kerosene as fuel, and the efficiency at the generator terminal is 40.1% at the level of high calorific value. This is equivalent to or greater than that at supercritical pressure power plants. The system structure is shown in Figure 2.55. With the variety of fuels available in recent years, fuels like clean liquified natural gas (LNG) and liquefied petroleum gas (LPG) are used for large-capacity combined plants. The Tokyo Electric Power Plant, Higashi Niigata Electricity Power Plant No. 3 (capacity 1090 MW), has been in operation since 1984, 12 months of which operation was at the turbine inlet temperature of 1154°C. The plant's efficiency at the electric transmission terminal was recorded at 43.7% at the level of high calorific value, the highest in the world. Since an increase in the gas turbine inlet temperature is currently being emphasized, a small-capacity turbine (11,000 kW) for 1250°C has recently been developed, and investigation of methods for cooling the turbine blades and the use of ceramic materials is in progress (Miwa et al. 1978). Furthermore, coal has gained attention as an alternative fuel to petroleum, and with the development of coal–gas combined power plants as a part of the "Sunshine Plan" for industrial technology under the Ministry of International Trade and Industry, progress has been made in terms of more efficient bed economic methods with low pollution. In particular, in Japan an investigation of the fluidized bed and jet stream is in progress, and under the "Moonlight Plan" for higher efficiency, development toward the goal of 55% total efficiency is in progress. The heat recovery method in combination with the MHD power plant is being considered for the future. In addition, cogeneration systems, which utilize both low temperatures for heat resources and high temperatures for power, have recently become widespread in

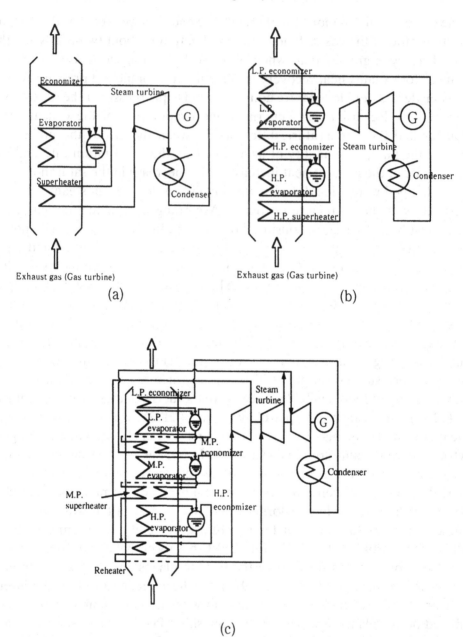

Figure 2.54 Multipressure systems. (a) Single-pressure, nonreheat; (b) double-pressure, nonreheat; (c) triple-pressure, reheat.

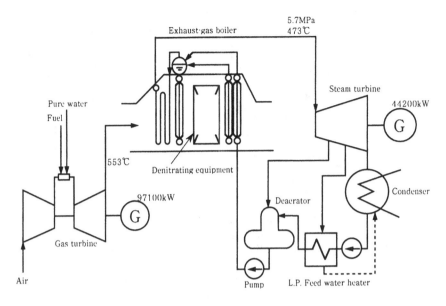

Figure 2.55 Exhaust heat recovery cycle.

hospitals, hotels, airports, and newly developed cities as on-site systems. That is, the heat from the steam can be utilized in air conditioning for the community, as a hot water supply, or for the melting of snow. As a parallel system to the purchase of electricity, conventional large concentrated systems such as thermal and nuclear power plants have the base load while middle and small systems take a share of the peak load, amounting to less loss in the transmission of electricity. In other words, high-temperature and high-quality heat is first exchanged for electricity and/or power by the gas turbine and engine. Next, low-quality energy is utilized to process and feed the hot water, air conditioner, and so on. Energy is utilized from high to low quality (Hirata 1985).

Recently, all possible efforts to utilize heat, like that obtainable by a cascade from higher to lower temperature, for an effective use of heat energy resources are being made by equipping an existing boiler system with a gas turbine and a piston engine. This is called a "repowering system," to which particular attention is being paid in relation to the treatment of waste heat from household and industry. A representative example, operating in Linkoping, Sweden, is shown in Figure 2.56. The hot water boiler, whose fuel consists of household wastes of various sorts (2×10^5 tons/year) and whose design pressure is 20 bar with a normal operating temperature of 175°C, was converted to a steam boiler. The steam is heated further in the waste boilers by the heat from the gas turbine before it is fed to the steam turbine. Both the steam and the gas turbines are built on the same shaft as that of the electric generator. The combination of electricity and heat produc-

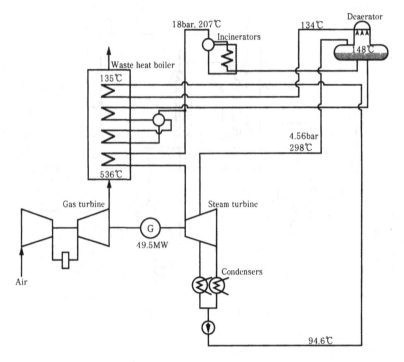

Figure 2.56 Flowsheet of repowering system (Sweden).

tion is very efficient since the fuel energy is used not only for the production of electricity but also for district heating.

2.4.9 Liquified natural gas (electric) power generation

In 1964, LNG was first imported as a clean energy source from Alaska, and it has been used for city gas, fueling electric power generators, food freezing, air separation, low-temperature crushing, ocean fresh water, and so on. LNG is liquefied natural gas in which methane is the main component at −160°C; the exergy of LNG at 1 bar and −162°C is about 62 kJ/kg at the LNG calorific value of about 3100 to 3180 kJ/kg under external conditions of 1 bar and 25°C. The condition of 1 bar and −162°C is shown at point 1 on the *T–S* diagram in Figure 2.57, and the value of 62 kJ/kg is shown by the area delineated by 1-2-3-4-1 (Akagawa and Fujii 1980). An example of the Rankine cycle that effectively utilizes LNG cold energy for power generation is shown in Figure 2.58a–c (Miyahara 1979). Figure 2.58a shows an example of a direct expansion cycle using LNG as the working fluid. The LNG at 1 bar and −162°C is pressurized to a pressure of p_1 (e.g., 5 MPa), heated by its surroundings and then expanded to a pressure of p_2 (e.g., 0.7 MPa)

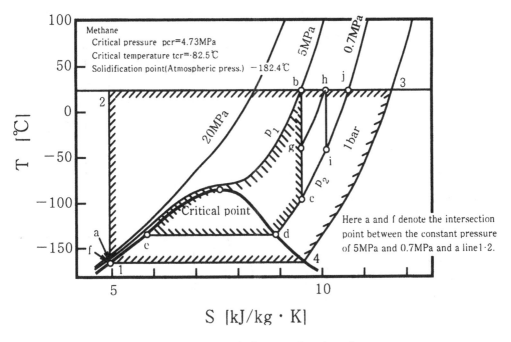

Figure 2.57 *T–S* diagram (methane).

by the LNG turbine. Because the LNG gas temperature at the turbine exit is low, it is further heated by seawater in the heater and used as fuel for electric power generation. The work quality is shown by the area {(a-b-c-d-e-f-a) − (pump work needed at point 1f)} ≅ (a-b-c-d-e-f-a) (Fig. 2.57), and its cold heat energy is utilized. Figure 2.58b shows an example of a direct expansion Rankine cycle of LNG and a mixed cycle using a second medium (a hydrocarbon such as propane, or Freon) as the working fluid. The LNG cold heat of −162°C is used for condensation, and seawater, in its environment, is used for heat. Figure 2.58c shows a mixed media cycle using hydrocarbons such as methane, ethane, propane, and butane. This cycle utilizes the temperature difference of the mixed media in evaporation (condensation), and the temperature change curve of the mixed media approaches the vaporization temperature curve of LNG. Furthermore, by precooling the gas entering the gas turbine as LNG and decreasing the compression power, as shown in the *T–S* diagram in Figure 2.59, an application using a combination of the closed-gas turbine cycle and the direct expansion cycle method in Figure 2.58a can be considered. This method uses an LNG pressure of 7 MPa, a turbine inlet pressure in the gas turbine cycle of 2.8 MPa at a temperature of 720°C, an outlet pressure of 0.4 MPa, and a flow-rate ratio of N_2/LNG equal to 5. The diagonal line in the figure demonstrates the available work generated by cold heat, and the utilization efficiency of cold heat is greatly increased.

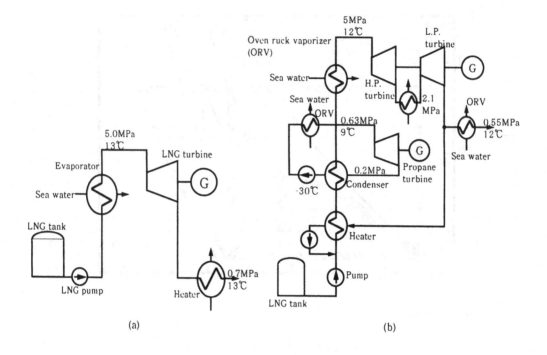

(a)

(b)

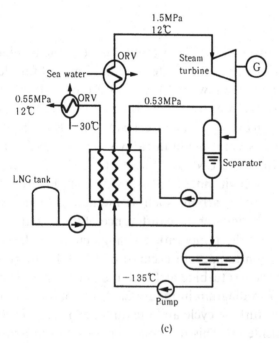

(c)

Figure 2.58 Rankine cycle operation on liquefied natural gas cold heat. (a) Direct expansion cycle; (b) direct expansion two-fluid cycle; (c) mixing-medium cycle.

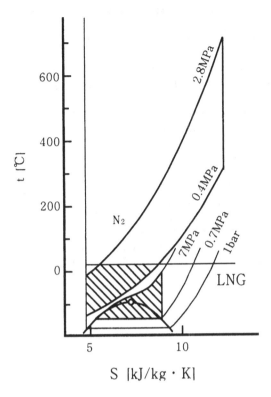

Figure 2.59 Compound cycle.

References

Aikawa, K., and I. Kawaguchi. 1975. *Journal of Thermal and Nuclear Power* 26, no. 9: 967–78.

Akagawa, K., and T. Fujii. 1980. *Journal of the JSME* 83, no. 739: 650–6.

Akagawa, K., T. Fujii, J. Ohta, and S. Takagi. 1986. *Transactions of the JSME*, Series B 52, no. 480: 3052–7.

Church, E.F. 1950. *Steam turbines*, 3rd ed. New York: McGraw-Hill, p. 407.

Hirata, M. 1985. *Journal of Energy and Resources* 6, no. 2: 137–41.

Iida, T. 1975. *Journal of Thermal and Nuclear Power* 26, no. 9: 999–1005.

Ikeda, T., and S. Fukuda. 1980. *Journal of the JSME* 83, no. 745: 1528–34.

Ishigai, S., and S. Nakanishi. 1982. Reports of Special Project Research, Japan, vol. 1, pp. 191–8.

Ito, F., K. Takazawa, and T. Terayama. 1982. *Journal of the JSME* 85, no. 764: 728–33.

Iwanaga, K., T. Nagasaka, T. Sugiura, K. Momoeda, M. Wakabayashi, and N. Komiyama. 1985. *Journal of Thermal and Nuclear Power* 36, no. 12: 1297–307.

Kalina, A.I. 1983. American Society of Mechanical Engineers Paper, 83-JPGC-GT-3.

Kittaka, H., and S. Nakanishi. 1985. *Reviews of the Marine Technical College*, no. 28: 11.

Kretmeier, F., et al. 1979. Design Conference 1979 on Steam Turbines for 1980s, MEP, p. 385.

Kudo, K., H. Taniguchi, et al. 1983. *Thermal and Nuclear Power* 35, no. 1: 31.

Kurosawa, M., and T. Inui. 1981. *Journal of Thermal and Nuclear Power* 32, no. 10: 1151–9.

Laupichler, F. 1926. *Arch. für Wärmewirtshaft und Dampfkesselwesen*, 7 Jahrgang, Heft 5: 139–43.

Mitachi, M., and T. Saito. 1983. *Transactions of the JSME* 49, no. 437: 205–12.

Miwa, K., H. Hirai, and T. Sato. 1978. *Journal of Thermal and Nuclear Power* 29, no. 10: 1027–37.

Miyahara, S. 1979. *Journal of the JSME* 82, no. 358: 358–63.

Nakanishi, S., and H. Kittaka. 1984. *Journal of the Marine Engineering Society of Japan* 19, no. 4: 341.

Nakano, M. 1982. *Journal of Thermal and Nuclear Power* 33, no. 11: 1171–88.

Oberly, W.N. 1950. *Power generation* 54: 55–85.

Rant, Z. 1956. *Forschung.* 22, Heft 1: 36.

Rant, Z. 1960. *Brennstoff Wärmekraft* 12, nr. 1: 1.

Rant, Z. 1961. *Brennstoff Wärmekraft* 13, nr. 11: 297.

Taniguchi, H. 1979. *Journal of the Science of Machines* 31, no. 1: 174–80.

Uehara, H. 1981. *Journal of the Marine Engineering Society of Japan* 16, no. 3: 271–7.

Yoshida, K., and K. Aikawa. 1980. *Journal of Thermal and Nuclear Power* 31, no. 9: 1015–29.

3

General planning of the boiler gas-side heat transfer surface

EIICHI NISHIKAWA
Kobe University of Mercantile Marine

3.1 The gas-side heat transfer surface and its technological problems

This third chapter will describe some technological problems of the gas-side heat transfer surface (HTS). Figure 3.1 shows the general view of the boiler gas-side HTS system and the technological problems concerned. The gas-side HTS can be divided into the radiant heat transfer section and the convective heat transfer section, and the latter can be further divided into the high temperature section of the superheater (SH) and reheater (RH), and the heat recovery section. The radiant section of a large-capacity boiler consists almost entirely of the water wall, which forms the structure of the furnace, that is, the combustion chamber. Thus the boiler gas-side HTS consists of three main components: furnace, superheater and reheater, and heat recovery section.

3.1.1 Heat recovery section

The role of the heat recovery section is to recover the heat of combustion gas as far as possible. There are two obstacles to the performance of this role. One is low-temperature fouling and corrosion, which will be described later in Section 3.6 and is mainly due to sulfuric acid, and the other is that a very large area of heat transfer surface is required due to the small temperature difference between the exhaust gas and the working fluid. An important technological task in heat recovery section design is, therefore, to make the section as compact as possible while preventing fouling and/or corrosion problems.

3.1.2 Superheater and reheater

These sections are subject to very severe temperature conditions. In addition, since the fluid flowing inside SH and RH tubes is not liquid but dry steam, the tube wall temperature reaches a rather higher level than the steam temperature. At a high temperature level such as 600°C, even a very small temperature rise,

113

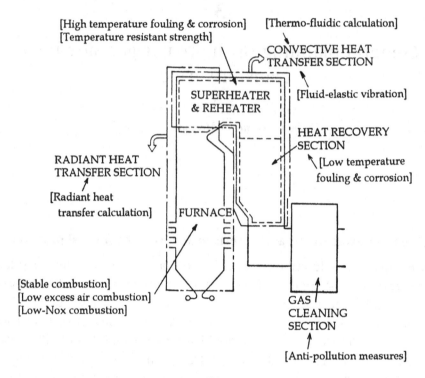

Figure 3.1 Boiler gas-side heat transfer surface system and technological problems.

say, only 5°C, has a great effect on the strength and/or corrosiveness of tube material, so that expensive steel alloys such as austenitic materials are used for the tubes of the SH final stage. Thus the main design task relating to the SH and RH is to consider the optimum HTS arrangement for cost effectiveness, paying careful attention to the limits of material strength and the high-temperature corrosion of both steam- and gas-side tube wall. One conventional strategy for improving power generation efficiency is to adopt a higher grade of steam temperature and pressure conditions, but the extent of improvement would be strongly dependent on the development of materials for SH tubes (see, e.g., Schuster et al. 1994; Weinzierl 1994).

3.1.3 Furnace (combustion chamber)

Most boiler furnaces consist structurally of the combustion space surrounded by water walls. The furnace is designed to perform two functions simultaneously. One is to release the heat of fuel by combustion, and the other is to remove the heat from the furnace to the working fluid inside the water walls. These dual functions can be considered to be the most essential feature of the furnace of a conventional large-capacity boiler.

The first task of combustion technology is to burn the fuel efficiently and steadily, with excess air controlled to be as little as possible, and with the flame shape controlled adequately under the prerequisite that environmental pollutants such as SO_x, NO_x, particulate matter (PM), and noise should be controlled to be as low as possible. The second task is to carry out reliable heat removal so as to avoid the dangerous dry-out of water-side HTS, because heat transfer of the furnace water wall is highly intensive. Further, furnace heat removal is also important for control of the furnace outlet gas temperature in order to adjust the temperature condition of the SH. Furnace heat removal is a very important aspect of boiler safety.

Demand for control of environmental pollutants, especially air pollutants, is increasing year by year at such a rate that clean combustion technology could become the most important one among boiler combustion technologies.

3.2 General planning of the boiler furnace

3.2.1 Performance limits of boiler furnace design

In the last decade, methods of furnace design with regard to heat transfer and/or heat balance have made remarkable progress, mainly owing to the development of computer simulation techniques. A recent simulation method has developed its capability to the extent that a three-dimensional equation system integrating fluid motion, energy transport, and combustion chemical reactions can be calculated numerically so that one can estimate various factors such as temperature, velocity, heat load, and chemical component concentration at any local spot in the furnace (Hirokawa et al. 1992; Endo et al. 1994). This computer design method has greatly contributed to the development of clean combustion technologies controlling the formation of NO_x and/or PM, and to the estimation of heat load distribution in the water wall so as to improve the reliability of heat removal. The computer method has further made it possible to control combustion conditions flexibly, responding to any change in fuel characteristics.

Engineers and researchers have to treat many design variables such as boiler capacity, steam condition, fuel characteristics, burning method, furnace shape, burner arrangement, HTS arrangement, air preheating, air supply system, and exhaust gas recirculation. A computer simulation technique would be very useful indeed as a tool aiding engineers in their design tasks in order to analyze the complicated processes of interaction among these variables. On the other hand, the more deeply an engineer depends on computer techniques apart from his own thinking, the more difficult it would be for him to investigate the essential relations characterizing the furnace performance or to identify what factors dominate the furnace performance. Therefore it would still now be useful as a tool for general planning of the boiler furnace to investigate the basic relations dominating the main characteristics of the furnace. The basic relation is called here the similarity law of the boiler furnace.

Since the furnace occupies the major part of the boiler body for large-capacity boilers, one important task for R&D is to aim for a more compact furnace. What factors influence furnace compactness? Technical parameters indicating furnace performance can be designed within certain limits, being dependent on the following restrictive conditions:

1. *Heat flux limits of furnace HTS.* The maximum allowable heat flux of the water wall is restricted by its water-side burnout (dry-out) heat flux.
2. *Combustion limits.* The lower limit of the furnace volume is dominated by the space required for burning the fuel completely, or to an extent less than the allowable unburned fuel loss.
3. *Furnace outlet gas temperature limits.* The temperature of the gas entering the SH section has to be controlled accurately to be the allowable temperature required to retain the strength of the SH tube material and, in the case of coal firing, to avoid fouling of the SH due to ash slagging.
4. *Slagging limits.* In addition to care over SH fouling, as just mentioned, there should be taken into account the limits of heat release rate in the burner zone to avoid severe fouling of the water wall in that zone.

3.2.2 Effects of performance limits on design factors

3.2.2.1 Critical heat flux of HTS

The heat absorption rate of a furnace can be expressed by the equation

$$Q_f = BH_u \left\{ 1 + \frac{c_f t_f + \mu\phi c_a t_a}{H_u} + \frac{\lambda(\mu\phi + 1)c_{gr}t_{gr}}{H_u} \right.$$
$$\left. - \frac{(1 + \lambda)(\mu\phi + 1)c_{go}t_{go}}{H_u} - \frac{\xi}{100} \right\} = BH_u K_f \qquad (3.1)$$

where Q_f is the furnace heat absorption rate (W), B is the fuel consumption rate (kg/s), H_u is the fuel lower calorific value (J/kg), c_f is the specific heat of fuel (J/kg K), c_a is the constant pressure specific heat of air (J/kg K), c_{gr} is the constant pressure specific heat of recirculated combustion gas (J/kg K), c_{go} is the constant pressure specific heat of combustion gas at furnace outlet (J/kg K), t_f is the fuel temperature (°C), t_a is the air temperature (°C), t_{gr} is the temperature of recirculated combustion gas (°C), t_{go} is the combustion gas temperature at furnace outlet

(°C), λ is the flow ratio of recirculated combustion gas, μ is the excess air ratio, ϕ is the stoichiometric air fuel ratio, and ξ is the ratio of unburned fuel loss (%).

K_f in the above equation expresses the ratio of heat absorbed by the furnace HTS to the total calorific value of the fuel supplied to the furnace. The mean heat flux of the furnace HTS, q_s, can be expressed by the following equation:

$$q_s = K_f \left(\frac{BH_u}{A_e} \right) \qquad (3.2)$$

where A_e is the effective HTS area of furnace inner surface.

The value of q_s is not homogeneous but is distributed over the furnace HTS, and its maximum value is not allowed to exceed the critical heat flux, that is, the water-side burnout heat flux. Taking the distribution of q_s into consideration, a certain value q_s^* less than the burnout heat flux should be adopted as the effective design criterion. Then

$$q_s \leq q_s^* \qquad (3.3)$$

Let us consider the specific furnace heat release rate q_f (W/m^3), which is considered to be the most important one among various design parameters. If the furnace volume is expressed by V_f, then

$$BH_u = q_f V_f \qquad (3.4)$$

and if the inner surface area of the furnace is expressed by A_w and the ratio of effective HTS area by ψ_e, then

$$A_e = \Psi_e A_w \qquad (3.5)$$

If a similar furnace shape and similar arrangement of HTS are supposed,

$$A_e = k_f V_f^{2/3} \qquad (3.6)$$

Combining the above Eqs. (3.3) to (3.6),

$$q_f \leq k_f^{3/2} \left(\frac{\left(q_s^* \right)^{3/2}}{\left(BH_u \right)^{1/2} K_f^{3/2}} \right) \qquad (3.7)$$

Thus, as shown above, the HTS heat flux limits can be reduced practically to the limits of q_f.

3.2.2.2 Combustion limits

To complete the fuel combustion within the furnace space, the fuel injected into the furnace has to reside there for a certain time longer than some critical time t_r^*. The fuel residence time can be estimated by the residence time of the combustion gas produced in the furnace. The average residence time t_r can be calculated by the following equation:

$$t_r = \frac{\varrho_g H_u}{(1 + \lambda)(\mu\phi + 1)q_f} \tag{3.8}$$

The combustion limits can then be expressed as follows:

$$q_f \le \frac{\varrho_g H_u}{(1 + \lambda)(\mu\phi + 1)t_r^*} \tag{3.9}$$

In this manner, the combustion limits can be reduced to the limits of q_f.

Fuel combustion time is mainly dominated by the combustion reaction velocity and the rate at which oxygen is supplied into the reaction zone. The former is dependent on the chemical characteristics of the fuel and, in the cases of oil and coal fuel, also dependent on the physical characteristics of fuel particles. Thus the main technical factors that affect the combustion time are as follows:

- combustion characteristics of the fuel;
- burner techniques such as atomization, pulverization, mixing characteristics of burner flow;
- fluid flow characteristics of the furnace;
- techniques for controlling pollutant formation.

The combustion time of an oil fuel droplet in the boiler furnace atmosphere is estimated to be generally less than 0.1 s. If t_r is assumed to be equal to 0.1 s, then, by Eq. (3.9), q_f is in the order of 5000 kW/m³, which is somewhat larger than the q_f value given by Eq. (3.7) regarding the HTS heat flux limits. In the case of a coal particle, however, its combustion time is much longer. The combustion process of a coal particle can be divided into two steps. The first step is the gaseous combustion of volatile matter, and the second step is the surface combustion of the residual solid matter, called char. Almost all of the combustion time is occupied by the second step of char combustion. When coal particles, the diameters of which range from 10 to 100 μm, are burned in a boiler furnace atmosphere where the unburned fuel ratio is expected to be less than 1%, the combustion time may be estimated generally to be in the order of 1 s. For pulverized-coal-fired boilers, therefore, the combustion limits would greatly contribute to the q_f limits.

3.2.2.3 Furnace outlet gas temperature limits

Since the furnace outlet gas temperature is related directly to the furnace heat absorption ratio K_f, its limits are also reduced to the limits of specific furnace heat release rate q_f by Eq. (3.7).

3.2.2.4 Slagging limits

If the combustion products of coal ash are so severely heated as to become molten, their fouling and corrosive potential is greatly increased. The furnace temperature conditions should therefore be designed not to allow the combustion products to melt, to avoid slagging troubles. For the SH, the slagging limit is estimated from the furnace outlet gas temperature, taking the ash melting point into consideration. For the furnace water wall, the limit is estimated from the specific heat release rate of the furnace section in the burner zone.

3.2.3 General trend of specific furnace heat release rate q_f

In the above discussion on basic design limits regarding furnace heat transfer and fuel combustion, it has been made clear that most of the limits are reduced to the limits of specific furnace heat release rate. It is considered therefore that the specific furnace heat release rate q_f is the most important design index for general planning of the boiler furnace. So let us take a look at the general trend of q_f of various actual boilers. Figure 3.2 shows the relation between q_f of various actual boilers and their equivalent steam generating capacities. In the figure, the mean heat flux of the furnace HTS, q_s, has been calculated for the following conditions:

- For large-capacity boilers, $A_e = 9k_f V_f^{2/3}$, $t_{go} = 1200°C$, $t_a = 300°C$, $\mu = 1.1$, $\lambda = 0$ (i.e., no recirculation of flue gas).
- For small-capacity boilers, $A_e = 5.5 k_f V_f^{2/3}$, furnace heat absorption ratio $K_f = 0.5$.

The important points can be read out from Figure 3.2 as follows:

- Taking a view of the q_f trend for all data, it can be seen that q_f is proportional to (steam generation rate)$^{-1/2}$ as a whole. This means that the HTS heat flux limit q_s^* is dominantly effective. Approximate values of q_s are shown in the figure.
- The q_f of the coal-fired boiler is rather smaller than that of the oil-fired boiler. This difference may be due to the slagging limits. No difference can be observed between the oil- and gas-fired boiler.
- For pulverized-coal-fired boilers, the HTS heat flux limit is the dominant factor for large boilers with a steam generation rate larger than 200–300 t/h, whereas for smaller boilers, q_f does not increase and reaches a ceiling value. The average furnace residence time t_r is shown on the right side of the figure. Judging from t_r, the q_f value at the ceiling corresponds to a residence time of slightly longer than 2 s. It could be considered, therefore, that the dominant factor changes from HTS heat flux limits to combustion limits when the boiler capacity becomes smaller than a 200–300 t/h steam generation rate.
- For very large boilers, data for boilers having a furnace divided by a partition, called a twin furnace, are also plotted. The partition is constructed generally with a water wall or a superheater, so that the partition contributes to making the HTS area relatively wider. This is why twin furnace boilers have achieved a larger q_f than boilers having a normal furnace, as can be read from the figure.

All the various boilers plotted in Figure 3.2 must have been subject to desperate efforts by their engineers to make their furnaces as compact as possible. In

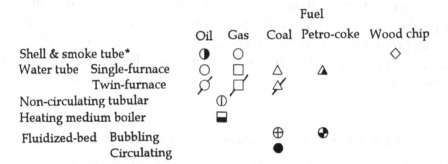

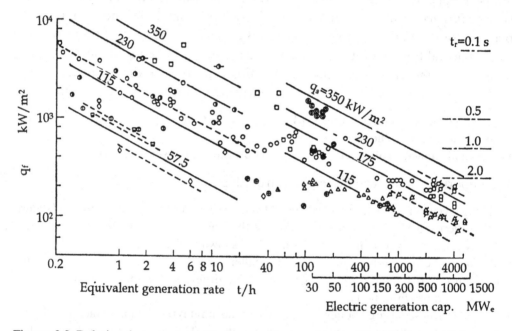

Figure 3.2 Relation between specific furnace heat release rate q_f and boiler capacity (steam generation rate). (q_s) Average heat flux of furnace heat transfer surface estimated approximately; (t_r) residence time of combustion gas.

spite of their efforts, the general characteristics of specific furnace heat release rate cannot get rid of the restrictive relations, as can be seen in the figure. Figure 3.2 should be useful for understanding the characteristics of each type of boiler in comparison with other types.

3.2.4 Similarity law of the boiler furnace

Ishigai (1961) and Yamazaki (1961) developed a useful method for estimating furnace heat transfer, based on the viewpoint of the similarity law. Although the

Ishigai–Yamazaki method was proposed many years ago, it is still used now for general planning of the boiler furnace. As is well known, since boiler heat transfer is dominated by heat radiation, a very troublesome calculation is required for estimating the heat flux of the water wall and/or estimating the furnace outlet gas temperature. By applying the Ishigai–Yamazaki method, however, the main furnace characteristics can be easily estimated by the use of only a calculator instead of a computer. Although the estimated values are approximate, they are accurate enough for general planning.

The Ishigai–Yamazaki method models the furnace as a space filled with a flame ball having a homogeneous temperature, and considers that furnace heat transfer is dominated by radiant heat exchange between the flame ball and the surrounding water wall. That is, the following equation is assumed to be valid:

$$Q_f = A_f C_f \left(T_f^4 - T_w^4 \right) \tag{3.10}$$

where A_f is the surface area of the flame ball (m²), C_f is the effective radiation coefficient of the flame ball (W/m² K⁴), T_f is the flame temperature (K), defined as the arithmetic average of the theoretical combustion temperature and furnace outlet gas temperature, and T_w is the temperature of the cooling surface, that is, the heat absorbing surface (water-wall surface) (K).

From the above equation and Eq. (3.1), the following equation can be derived:

$$\frac{A_e C}{B H_u} = \frac{K_f}{\left(T_{go}/100 \right)^4 - \left(T_w/100 \right)^4} \tag{3.11}$$

where

$$C = C_f \left(\frac{A_f}{A_e} \right) \left(\frac{T_f^4 - T_w^4}{T_{go}^4 - T_w^4} \right) \times 10^8 \tag{3.12}$$

The furnace heat absorption ratio K_f is dependent on various factors, among which the Ishigai–Yamazaki method takes into account only four (μ, t_a, ϕ, and t_{go}) as the main effective design factors and considers the remaining factors as constants. Accordingly, when definite values of those four factors are given, the right-hand side value ($A_e C/BH_u$) of Eq. (3.11) can easily be calculated. Ishigai and Yamazaki have made a nomogram illustrating the relation just mentioned to estimate the effects of those four main factors on the value of ($A_e C/BH_u$). Figure 3.3 represents this nomogram, which has been compiled according to the usual design conditions prevailing in those days, that is, $c_a = 1.03$ kJ/kg, $c_f = t_f = 0$, $\xi = 0.8$, $T_w = 532$ K, and $\lambda = 0$.

The main point of the Ishigai–Yamazaki method is that the value ($A_e C/BH_u$), which is called the Ishigai characteristic value, has been expressed by the simple Eq. (3.11). The component C of this value represents a radiant heat transfer characteristic, such that a very troublesome calculation is required to estimate its

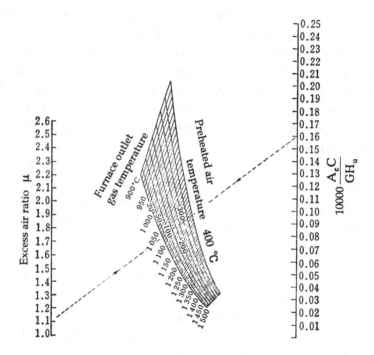

Figure 3.3 Ishigai–Yamazaki's nomogram for general planning of furnace heat transfer.

value directly. By the use of Eq. (3.11), however, it can be estimated easily without any troublesome calculation of radiant heat transfer; further, these four main design factors can be determined by the use of the nomogram. At present, since the computational simulation method has been highly developed, engineers do not worry no matter how troublesome a calculation is required. Boiler furnace heat transfer is dependent on so many parameters that it is not easy to identify the effects of the main parameters even when the computer simulation method is used. As regards these four main design factors, therefore, the Ishigai–Yamazaki nomogram is still now useful for general planning of the boiler furnace.

3.2.5 New boiler concepts breaking through the similarity law of conventional boilers

As seen in Figure 3.2, the HTS heat flux limit is a dominant factor for the characteristics of large capacity boilers having a conventional furnace structure, and the specific furnace heat release rate has to be decreased in proportion to the square root of the boiler capacity (steam generation rate). Thus the furnace volume has to be greatly increased as the capacity increases. In the case of a conventional furnace structure, the fuel supply rate, that is, the heat release rate, can be increased to correspond to the furnace volume. On the other hand, the heat absorption rate can be increased corresponding only to the inner surface area of the furnace. This is the main reason why the HTS heat flux limit has such dominant effects. This similarity law is quite natural, and breaking through it seems to be very difficult. Nevertheless, there have been developed some new concept boilers that aim to circumvent the similarity law of conventional boilers.

3.2.5.1 Pressurized combustion

According to Eq. (3.9), if the combustion gas density ϱ_g is increased by pressurization, the specific furnace heat release rate q_f can be increased without reducing the residence time, and the furnace can be made compact in proportion to the furnace pressure. Two problems would occur because of pressurization. One is the increase in blower power for pressurization, and the other is the elevation of combustion gas temperature.

With regard to the former problem, a gas turbine is generally introduced for recovering the energy required for pressurization.

The latter problem originates from the conflict between two functions, namely, heat release and heat removal, and therefore it is related more fundamentally to the similarity law of conventional boilers. To cope with this problem, the introduction of an additional heat-absorbing device has to be considered. Pursuing the idea of pressurized combustion to its logical conclusion finally leads to the separation of the two functions of the conventional furnace: The furnace is to assume the role only of fuel combustion, while heat transfer is to be carried out by some heat transfer device installed separately from the furnace. This idea opens up the possibility of a new concept of boiler, for instance, a combined cycle of gas turbine and steam power plant, and/or pressurized fluidized bed combustion.

3.2.5.2 Fluidized bed combustion

As is well known, the fundamental characteristic of fluidized bed combustion (FBC) is the ability to keep combustion steady in a fluidized bed at a very low temperature level of 1100–1300 K, so that the FBC can avoid slagging problems and suppress NO_x formation. Further, desulfurization would be possible by the use of suitable fluidized bed material. FBC techniques can be classified mainly into two types: bubbling fluidized bed (BFB) and circulating fluidized bed (CFB). In the case of BFB, generating tube banks are installed within the fluidized bed for heat absorption to keep the bed temperature at a suitably low level. This installation of tube banks is the main device of the BFB for breaking through the restriction of the conventional similarity law because tube banks can be accommodated three-dimensionally in a fluidized bed. But an embedded tube bank is a significant weakness of the BFB because the tube banks are severely attacked by erosion corrosion (see Section 3.6). Although it might be possible to improve the specific heat release rate q_f of the BFB boiler by pressurization (Minchener 1994), this unreliability of embedded tube banks seems, from the author's viewpoint, to be an insuperable obstacle.

In the case of the CFB, the fluidized bed material assumes the role of transporting the heat released by combustion out of the furnace, and the heat is recuperated by the heat exchanger installed separately outside the furnace. Introduction of this separate heat exchanger is the main point of the CFB, because it means the separation of the two functions of the conventional boiler furnace and makes it possible to realize a compact furnace. In Figure 3.2, a few data of actual coal-fired CFB boilers are plotted. As can be seen, their q_f can reach almost the same level as that of the oil-fired boiler. Accordingly, if pressurization is adopted for the CFB, a more compact furnace can be realized.

For the FBC, there is another problem to be noted, of an environmental nature. There is an undesirable possibility of the formation of dinitrogen oxide (N_2O) when combustion proceeds at a low temperature level of 1100 K. N_2O is a very stable air pollutant having both a greenhouse effect and a catalytic destructive effect on the ozone layer in the stratosphere (Klein and Rotzoll 1994). The capability of stable burning of low-grade fuel at a low temperature level is one of the strong merits of FBC, but, on the other hand, the merit itself induces another kind of problem. We have to realize that it is not an easy task to find environmentally sound technologies for fossil-fuel combustion.

3.2.5.3 Jaggy fireball boiler

The jaggy fireball (JAFI) boiler is a new concept of gas-fired boiler recently developed by Ishigai et al. (1992), who have given it that nickname. Figure 3.4 shows the furnace concept of the JAFI boiler. As can be seen, there is no space for the flame zone, and the furnace is almost filled with generating tube banks that are quite different from those of the conventional furnace. Ishigai et al. (1992) point out the merits of the JAFI furnace as follows:

- Karman vortices formed by each tube promote good mixing of fuel gas and air;
- the flame is divided into small segments by tube banks, and therefore the radiant heat flux is decreased so as to keep the heat flux of the generating tube at a safe level lower than the critical heat flux;
- low NO_x combustion can be achieved, owing to the excellent heat removal effect of the generating tube banks;
- the bluff body of each tube can assume the role of effective flame stabilizer.

The JAFI boiler has made it possible to accommodate the heat transfer surface three-dimensionally in the furnace without losing the conventional furnace char-

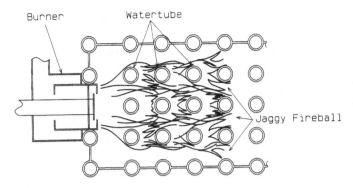

Figure 3.4 Diagrammatic representation of the jaggy fireball concept (courtesy of the Hirakawa Guidom Company).

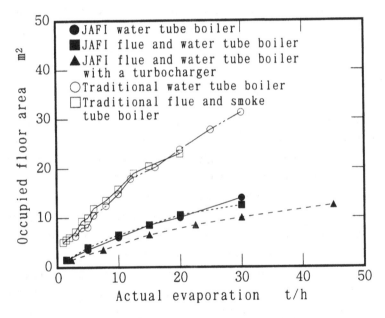

Figure 3.5 Comparison of occupied floor area of JAFI boiler with that of conventional boilers (courtesy of the Hirakawa Guidom Company).

acteristic of having the two functions of combustion and heat transfer. It can be said to be an innovative idea breaking through the restriction of the conventional similarity law. The JAFI boiler has already been placed on the market for actual use, though its steam generation capacity is limited within a range smaller than 20 t/h. Figure 3.5 shows the remarkable effect of the JAFI boiler in making the boiler compact (Ueda et al. 1995).

3.3 Pollutant control techniques

3.3.1 *Environmental problems and fossil-fuel-fired boilers*

Conservation of the environment is the prerequisite for all of our activities. To be able to use energy, there have to exist both a resource-side environment offering energy resources to us and a waste-side environment accepting the various wastes discharged from our energy use processes. We have to devise energy conversion technologies and manage our use of energy under the restriction of environmental conservation.

There are possibilities of environmental deterioration in both resource exploitation and waste disposal. Judging from present phenomena of environmental deterioration, however, it would seem that the main problems have originated in the waste-side environment: local air pollution of NO_x, SO_x, PM, and so on; rain acidification; the threat of climate change due to the "greenhouse" gases such as CO_2, CH_4, and N_2O; and destruction of the ozone layer. Most of these problems have originated from various types of waste discharge.

With regard to air pollution, we are now facing problems on both local and global scales. Local air pollution has been experienced in industrialized countries for many years. In Japan, rapid economic growth in the 1960s led to severe air pollution in some local areas, and public hazards became significant social problems. At first, the main means of coping with air pollution was by diffusion. The solution was to make the stacks higher in expectation of effective diffusion of discharged pollutants, and/or to move the factories to rural areas. These diffusion methods were not as effective as expected, and instead resulted in the spread of the polluted area. As a second step, regulation systems were introduced to control the amount of pollutants discharged. As a result, severe pollution of local areas has been improved, though there are still now some locally polluted areas where effective countermeasures are being demanded. Adding to local pollution, global air pollution has become more serious year by year, and people are also demanding immediate effective measures (Intergovernmental Panel on Climate Change 1996). Any diffusion method is nonsense where global pollution is concerned, and we have to seek effective measures to control and reduce the amount of discharged pollutants.

As is well known, since present main air pollutants such as NO_x, SO_x, PM, volatile organic compounds (VOC), CO_2, and so on are discharged predominantly by the combustion of fossil fuel, the contribution of large fossil-fuel-fired boilers for power generation to both local and global air pollution is not small. For example, large power stations have an undoubted influence on rain acidification in Great Britain (Hill 1994). Figure 3.6, adapted from Ruijgrok (1995), shows the

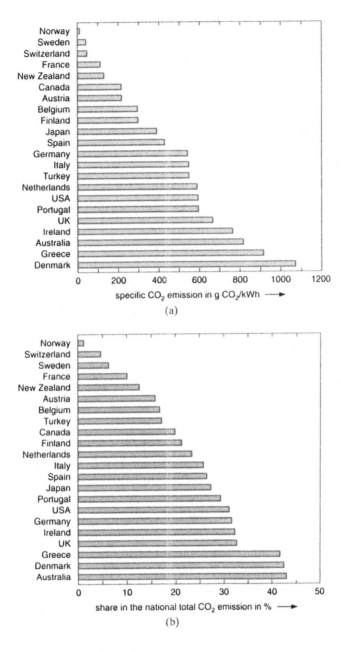

Figure 3.6 (a) Specific CO_2 emissions (in grams of CO_2 per kilowatt-hour) and (b) shares of emissions from the electricity sector in the total national CO_2 emission (data for 1989) (Ruijgrok 1995).

Organization for Economic Cooperation and Development statistics regarding the share of power generation for CO_2 emission.

Judging from the trend of energy demand, especially that of increasing electric energy demand, it should be expected that our demand for energy resources will still depend on fossil fuels for a great while in the future (see, e.g., Bohn 1994). Moreover, taking into consideration the deteriorating level of air pollution year by year, it should also be expected that regulation of pollutant discharge will be tightened rather than relaxed from now on. Control techniques for air pollutants, therefore, will become more and more important for large fossil-fuel-fired boilers, and environmental soundness will be the most important characteristic of a steam power plant (see, e.g., Jager et al. 1994). This section reviews the control techniques for air pollutants.

3.3.2 *General principles of pollution control*

The conceivable pollution control measures are listed in Table 3.1. Diffusion methods aiming to alleviate pollution, as already mentioned, are becoming meaningless due to the globalization of pollution. Many countries – European Union members, the United States, Japan, and so on – have already been enforcing regulations to control the amounts discharged from individual plants, and/or the total amount of pollutant discharge within a certain zone.

Regarding fuel purification, some effects on oil fuel can be expected. For instance, sulfur removal techniques have already been used for many years. In the case of coal, however, fuel purification would hardly be feasible, except in the case of coal gasification if we consider it as one method of fuel purification. As regards fuel change, gas fuel is desirable. Natural gas especially is almost free from sulfur content, and the easiest to control for low-NO_x and low-PM combustion compared with other fossil fuels. Coal is the most troublesome among various fuels for controlling air pollutants.* For instance, Table 3.2 shows the German (Bohn 1994) and Japanese regulation of pollutant discharge for a large capacity boiler. As can be seen, the limits for coal-fired boilers are the most relaxed, responding to the control difficulties of coal combustion. Thus this section describing the techniques of air pollution control focuses on coal combustion.

---

* Substances that have generally been considered as air pollutants among the combustion products of fossil fuel so far are mainly SO_x, NO_x, CO, CO_2, PM, and volatile hydrocarbons. Recently, various trace metals contained in coal are being considered as pollutants. Most of them can be removed from flue gas during the cleaning processes of desulfurization, NO_x removal and PM precipitation, though some vaporous substances such as Hg, F, Cl, and Se are still discharged from the stack (Jones 1994; Meij 1994; Kimura 1996).

Table 3.1. *Pollution control measures of fossil-fuel-fired boilers*

	Pollutants[a]				
Measures	SO_x	NO_x	PM	CO_2	Others
Pollutant control					
Fuel displacement	O	O	O	O	O
Fuel treatment	O	—	O	×	O
Removal from combustion or flue gas	O	O	O	—	O
Burner modification	×	O	O	×	×
Furnace combustion modification	×	O	O	×	×
Flue-gas recirculation	×	O	O	×	×
Steam or water injection	×	O	×	×	×
Pollutant diffusion					
High stack	—	—	—	—	—

[a] (O) effective; (×) ineffective or partially effective; (—) unacceptable or unrealistic.

Table 3.2. *Regulation limits of NO_x and PM for large-capacity boilers*

		Type of boiler		
		Gas-fired	Oil-fired	Coal-fired
NO_x	Japan	60 ppmV $(O_2, 5\%)$	130 ppmV $(O_2, 4\%)$	200 ppmV $(O_2, 6\%)$
	Germany[a]	100 mg/Nm³	150 mg/Nm³	200 mg/Nm³
PM	Japan	30 mg/Nm³ $(O_2, 5\%)$	40 mg/Nm³ $(O_2, 4\%)$	50 mg/Nm³ $(O_2, 6\%)$
	Germany	5 mg/Nm³	50 mg/Nm³	50 mg/Nm³

[a] 100, 150, and 200 mg/Nm³ can be reduced approximately to 50, 75, and 100 ppmV, respectively.

3.3.3 Low-NO_x combustion techniques

3.3.3.1 NO_x formation and its control in the boiler furnace

NO_x is formed during fuel combustion in two ways. One is the thermal fixation of atmospheric nitrogen and the other is the conversion of chemically bound nitrogen in the fuel. NO_x formed in the former way is called thermal NO_x, while that formed in the latter way is called fuel NO_x. NO_x formation mechanisms are explained briefly below (Hashimoto et al. 1993; Schopf et al. 1994).

Thermal NO_x gases are formed through two mechanisms. One is the Zeldovich reaction mechanism through which NO_x is formed in the downstream region of the flame zone; the other is the formation of "prompt" NO_x in the upstream flame region near the burner. The former process of NO_x formation is enhanced with increasing temperature and oxygen concentration and longer residence time in the high-temperature zone. It is hypothesized that prompt NO_x is formed when hydrocarbons in the fuel combine with nitrogen from the air to produce intermediate compounds such as NH, CN, and HCN, and these intermediate compounds are then oxidized further to form NO. It is known that the formation of prompt NO_x is not very much influenced by the temperature conditions.

Fuel NO_x is thought to be formed when nitrogen contained in fuel is thermally decomposed to intermediate compounds such as NH, CN, and HCN and/or atomic nitrogen, and then these compounds are oxidized to form NO. Thus the formation mechanisms of prompt NO_x and fuel NO_x can be said to be similar. It is noticeable that reaction processes forming both prompt and fuel NO_x are reversed in a reducing atmosphere where oxygen concentration is deficient or at a low level, and that NO_x is reduced to N_2 by the effect of hydrocarbons acting as reducers. It is considered therefore that NO_x formed in the upstream flame region is partially reduced to N_2 in the downstream flame zone where O_2 concentration is lacking and some hydrocarbons are still remaining. Figure 3.7 shows an example of experimental results regarding NO_x formation in pulverized coal combustion, obtained by the use of a ceramic test reactor (Hashimoto 1993). The figure compares NO_x formation between two types of combustion atmosphere, $O_2 + Ar$ and $O_2 + N_2$. The difference between the two cases can be considered to originate from thermal NO_x. The following can be read from the figure. When the temperature is below 1400°C, fuel NO_x is dominant. However, when the temperature rises beyond 1400°C, fuel NO_x decreases. On the other hand, thermal NO_x increases with temperature and reaches 25–30% at 1600°C.

The NO_x formation mechanisms outlined above suggest the following basic idea for controlling NO_x formation. In respect of controlling thermal NO_x, the basic measures are

• to lower the combustion temperature,
• to lower the oxygen concentration,
• to shorten the residence time within the high-temperature combustion zone,

among which lowering of the temperature is especially effective. On the other hand, lowering the temperature is not so effective for prompt NO_x and fuel NO_x. For these, if a reducing atmosphere can be formed in the downstream flame zone, it can be expected to reduce the gases to N_2 by use of fuel hydrocarbons. This idea is one principle for controlling NO_x formation and is called the de-NO_x-in-furnace method.

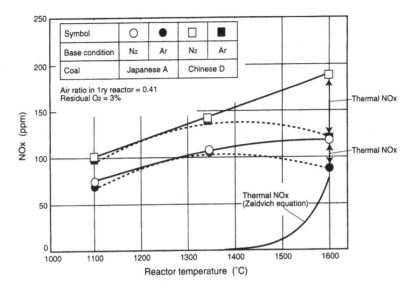

Figure 3.7 Effect of reaction temperature and atmosphere on NO_x formation (Hashimoto 1993).

Pulverized coal combustion proceeds in two steps, the first being volatile combustion and the second char combustion. Volatile combustion is rapidly completed within the burner zone, whereas the latter requires a longer time and its region spreads to the whole furnace space. Therefore the thermal NO_x formation region of the pulverized-coal-fired boiler spreads widely. Since the nitrogen component of coal is contained equally in both volatile and char ingredients, fuel NO_x is also formed in both volatile and char combustion processes. Accordingly, an important requirement for low-NO_x combustion of pulverized coal is to combine appropriately low-NO_x burner techniques with furnace low-NO_x combustion techniques. It should be noted that, since low-NO_x combustion techniques such as lowering the temperature and decreasing the air supply are undesirable conditions for stable and/or complete combustion of coal, low NO_x combustion inevitably induces an increase of unburned fuel loss to some extent. It is therefore necessary to balance the measures for lowering NO_x formation against keeping the unburned loss within some allowable limit. In responding to this requirement, the improvement of coal pulverizing mill techniques to produce better pulverization is important to shorten char combustion time as much as possible (see, e.g., Tamura et al. 1995).

3.3.3.2 NO_x control in the furnace

3.3.3.2.1 Low excess air combustion, lowering of air preheating temperature It can be easily understood, without further explanation, that low excess air com-

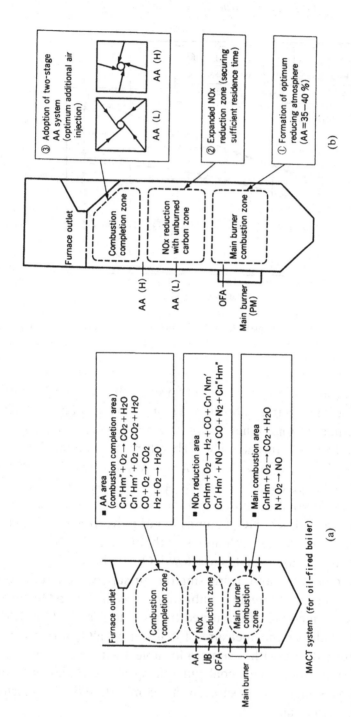

Figure 3.8 An example of in-furnace NO_x reduction system: the (a) MACT and (b) multi-additional air system (courtesy of the Mitsubishi Heavy Industries Company).

bustion and lowering of air preheating temperature are effective for control of NO_x formation.

3.3.3.2.2 Staged combustion In this method, the combustion air is supplied in many stages to lower the combustion temperature. Two-stage combustion was first developed and applied to actual boilers. The air supply to the burner system, which is the first-stage air, is adjusted to be an air–fuel ratio of less than 1. The rest of the air, that is, the second-stage air, is supplied over the burner zone, so is often called the over-fire air. Two-stage combustion is more effective as the first-stage air is decreased, though the combustion becomes unstable. At first, the aim of this method was simply to make combustion slower to lower the combustion temperature. Since then, the staged combustion method has been further improved, both theoretically and technologically, and a more effective method, called de-NO_x-in-furnace combustion, has been developed. This method aims not only to lower the combustion temperature for controlling NO_x formation, but also to apply the NO_x reaction mechanism already mentioned for reducing NO_x. For this purpose, various sophisticated systems of burner and air supply have been developed; examples are shown in Figures 3.8 and 3.9 (Hashimoto et al. 1993). In these systems, staged supply has been introduced not only for air but also for fuel. The burner designated UB in Figure 3.8a and the de-NO_x burner in Figure 3.9 are the second-stage burners, from which fuel is sprayed, and are intended to assume the role of reducer. The system shown in Figure 3.8b has been developed for coal firing. In this system, the formation of a reducing atmosphere in the region over the main burners has been achieved only by appropriate adjustment of the main burner air supply, and the UB burner has been dispensed with. In this example, further improvement has been made to the last-stage air supply, labeled AA (additional air) in Figure 3.9. That is, AA is supplied in two steps to decrease the amount of excess air as much as possible.

3.3.3.2.3 Exhaust gas recirculation In this method, part of the exhaust gas is recirculated to the furnace combustion zone and/or mixed into combustion to slow down the combustion velocity and therefore lower the combustion temperature. Although the effectiveness of exhaust gas recirculation (EGR) increases as the recirculation ratio is increased, combustion becomes unstable. It has been considered, however, that EGR is not as effective for coal firing as for gas firing.

3.3.3.3 Low-NO_x burner

Various types of low-NO_x burners have been developed by boiler makers and burner makers, though their theoretical ideas are not so different from each other. They are based on almost the same principle as that of de-NO_x-in-furnace combustion, and aim to control NO_x formation in the burner zone. The basic tech-

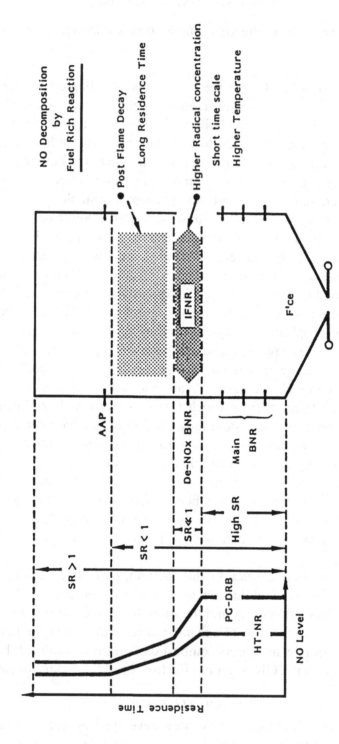

Figure 3.9 Another example of an in-furnace NO_x reduction system, the IFNR system (courtesy of the Babcock-Hitachi Company).

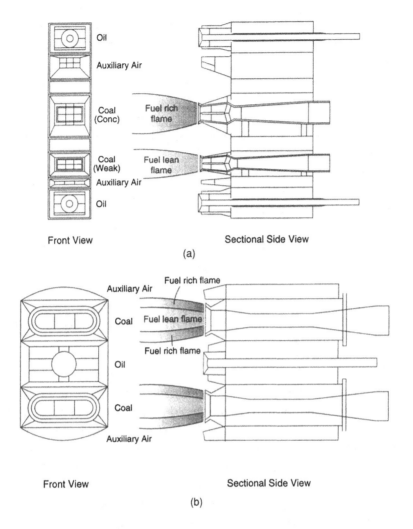

Figure 3.10 An example of low-NO$_x$ burners developed for pulverized-coal firing. (a) PM and (b) A-PM burners (courtesy of the Mitsubishi Heavy Industries Company).

nologies are described here, using some types of low-NO$_x$ burners developed in Japan.

Figure 3.10 shows a method which might be called a rich–lean combustion method. Figure 3.10a shows a burner system in which fuel-rich and fuel-lean burners are arranged alternately (Kaneko et al. 1995). The aims of this method are as follows. First, NO$_x$ formation in the upstream region of the burner flame zone is controlled due to air deficiency in the fuel-rich burner and the lower temperature of the fuel-lean burner. Second, unburned hydrocarbon in the fuel-rich burner functions as reducer and reduces the NO$_x$ formed in the upstream region

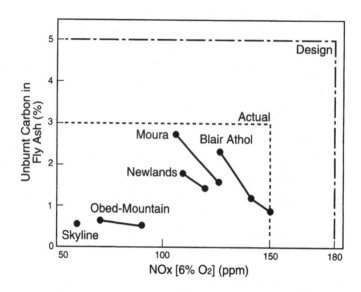

Figure 3.11 Performance test results of the low-NO$_x$ combustion system shown in Figs. 3.8 and 3.10, NO$_x$ vs. unburnt carbon in fly ash for various coals (courtesy of the Mitsubishi Heavy Industries Company).

to N$_2$. This method has been further developed to integrate fuel-rich and fuel-lean burners into a single low-NO$_x$ burner, the A-PM burner (Fig. 3.10b). This burner has been devised to form a rich–lean combustion flame on its own. Integrating this burner, the de-NO$_x$-in-furnace combustion method shown in Figure 3.8, and a high performance pulverizing mill, a low-NO$_x$ coal-firing system has been constructed. The system can achieve the desired low NO$_x$ level, keeping the unburned fuel loss as low as possible. Figure 3.11 shows an example of test results carried out for a large-capacity boiler, where the NO$_x$ concentration could be lowered to 150 ppm (reduced value at 6% O$_2$), keeping the unburned fuel loss less than 3% (Nakamura and Hashimoto 1996).

An example of a low-NO$_x$ burner of ordinary circular type is shown in Figure 3.12 (Tsumura et al. 1993; Ishida et al. 1993). In the top of the figure, the NR burner is the first generation, and the NR2 is the secondary advanced type. The principal ideas demonstrated by these burners for controlling NO$_x$ are shown in the bottom of the figure. One of the structural characteristics of the NR burner is the installation of a stabilizing ring coated with ceramic material. This ring induces a circulating flow that is effective for stabilizing the flame. The ceramic coating is highly heated, and its intense radiation functions effectively for early ignition and for raising the combustion temperature in the upstream region of the burner flame. Owing to these effects of the ring, volatile combustion in zone A proceeds rapidly, and primary air (carrying the pulverized coal) is intended to

Burner type	Hitachi-NR Burner	Hitachi-NR2 Burner
Design concept	Promote NO$_x$ decomposition reaction mechanisms in the flame chemistry by controlling flame structures individually.	The design concept is the same as the Hitachi-NR burner. Increasing combustion efficiency by modifying the structure.
Construction	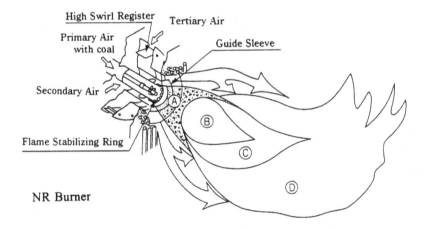	
Features	• To achieve low NO$_x$ with complete combustion	• To achieve extremely low NO$_x$ with complete combustion • Improvement of ignition (decrease of min. boiler load) • Reduction of burner draft loss (decrease of auxiliary power)

Figure 3.12 Another example of a low-NO$_x$ burner, the NR and NR2 burners (courtesy of the Babcock–Hitachi Company). *Bottom*: (A) volatilization zone; (B) reducing species formation zone; (C) NO$_x$ reduction zone; (D) oxidation zone.

be consumed rapidly. Thus an air-deficient atmosphere is formed in the core area of zone A, designated zone B, where the concentration of volatile hydrocarbons becomes high. The other structural characteristic of this burner is the guide sleeve. This sleeve is installed to keep the tertiary air flow, designated "outer air" in the figure, separate from the flame zone until the downstream region is reached. Accordingly, zone C is kept as a reducing atmosphere to reduce the NO_x, using the hydrocarbons from zone B. Moreover, the tertiary air is given an intense swirl. This swirl is helpful for keeping a reducing atmosphere in zone C by suppressing the flow mixing of tertiary air and flame in the border region of the zone. The swirl is also effective in inducing intensive shear flow, which can promote flow mixing in zone D to complete the fuel combustion efficiently. The advanced-type NR2 burner has been further equipped with the structural parts of "PC concentrator" and "space creator" as shown in the top right of the figure. The former device is shaped like a cone on its downstream end, in which the streamline is curved inward to create a pulverized-coal-rich area within the core zone near the coal nozzle. The creation of such a fuel-rich area is intended to promote early ignition and improve rapid combustion as far as possible in zone A. The latter device aims to ensure the separation of secondary and tertiary air and to improve the formation of a reducing atmosphere in zone C. Figure 3.13 compares the combustion test results of NR and NR2 burners (Ishida et al. 1993).

The principal ideas for low-NO_x burners mentioned above can be summarized as follows:

- to consider mixing fuel with primary and secondary air to promote in-flame NO_x reduction;
- to supply tertiary air separately from primary and secondary air to promote staged combustion in the burner zone.

To realize these ideas, various improvements have been made on swirler structure; the supply and adjustment of primary, secondary, and tertiary air; the effective use of bluff bodies*; and so on. The low-NO_x burners shown in Figures 3.10 and 3.12 have been applied to actual large-capacity pulverized-coal-fired boilers for power generation in Japan and in the Netherlands (Kluyver and Gast 1995). Besides the examples mentioned above, there have been developed other types of low-NO_x burner (see, e.g., Hashimoto et al. 1993; Simon 1995).

---

* A bluff body submerged in fluid flow creates shear flow, wake region, and circulation flow. Shear flow contributes effectively to the promotion of flow mixing. Circulation flow and wake are effective for stabilizing the burner flame. So it is a very important task of burner technology to develop the effective use of bluff bodies.

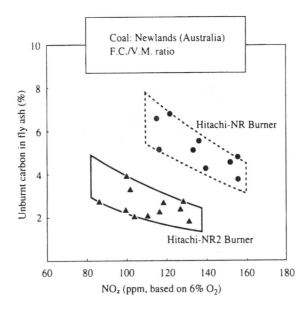

Figure 3.13 Performance test results of the NR2 burner shown in Fig. 3.12.

3.3.4 Coal gasification

3.3.4.1 General introduction

From a long-term perspective, it appears unavoidable that we must, without delay, pursue an innovative strategy that eliminates energy systems based on fossil fuel. However, it is also inevitable that we have to depend on fossil fuel as the main energy resource in the immediate future. If so, we should make every effort to achieve environmentally sound usage of coal fuel, since coal is the most plentiful resource among fossil fuels. As seen in Figure 3.14, coal contributes to CO_2 pollution most significantly, compared with oil and gas, so that efficiency improvement is an important task. Various technologies have been investigated for clean use of coal, among which the main advanced combustion technologies are:

- ultra supercritical (USC) steam power plants for improving cycle efficiency;
- pressurized fluidized bed combustion (PFBC);
- integrated coal gasification combined cycle (IGCC).

Improvement of cycle efficiency has been theoretically discussed in Chapter 2. An important technological issue to be discussed regarding practical USC plants is the balance between the cost of heat-resistant and corrosion-resistant materials required for USC and the efficiency improvement (see, e.g., Bohn 1994). The

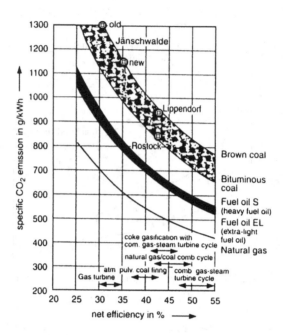

Figure 3.14 Specific CO_2 emission from power stations (Eitz 1995).

corrosion resistance required for materials at such a high temperature level would be so stringent that it would be difficult to expect considerable improvement. Corrosion problems of HTS materials are discussed in Section 3.6.

As for FBC, its fundamental characteristics have already been discussed, though very briefly, in Section 3.2.5. When pressurization is adopted for compactness, a system combined with a gas turbine has to be introduced for achieving satisfactory efficiency. In this case, the low temperature of FBC combustion gas, which is the main characteristic of FBC, changes undesirably to have a reverse effect on the gas turbine because such a low temperature, say at a level of 900°C, is too low to obtain high efficiency of the gas turbine. There is therefore a need to consider an auxiliary device such as a boosting heater. In addition, there are other problems, described in Section 3.2.5 regarding PFBC systems. Thus, if we are going to pursue an efficient and environmentally sound system of PFBC, the final solution, in the author's view, seems to be to use PFBC as a coal gasification technique (Mori and Fujima 1996).

As for coal gasification, it seems to have a more hopeful future compared with USC or PFBC, because it has attracted general interest owing to the following desirable possibilities:

• possibility of considerable fuel conservation and CO_2 control by the introduction of the combined cycle of gas turbine and steam turbine (IGCC);

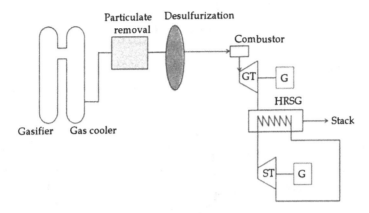

Figure 3.15 Basic system of integrated coal gasification combined cycle (Ministry of International Trade and Industry, Resource and Energy Agency 1996).

- possibility of efficient desulfurization and PM removal directly from fuel gas;
- possibility of wide application, not only for IGCC but also for various technologies such as fuel cells and steel making;
- possibility of usage of not only coal but also other low-grade fuels.

This section will describe coal gasification technologies and IGCC in more detail.

Figure 3.15 shows the basic layout of the IGCC system, of which the main components are gasifier, gas cleaning equipment, and electric generation plant. The gas cleaning equipment is for the removal of sulfur and PM from coal gas. This equipment is required not only for controlling environmental pollution but also to protect the blades of the gas turbine from erosion corrosion. The power-generating plant consists of gas turbine (GT), heat recovery steam generator (HRSG), steam turbine (ST), and electric generator. The HRSG consists wholly of convective HTS, so it is important to consider the optimum design and reliability of convective HTS (see, e.g., Buschmann et al. 1994; Jones 1996). Design problems of convective HTS are described below in Section 3.4 and the following pages.

3.3.4.2 Gasification technology

The key technology of IGCC is of course coal gasification technology. Coal gasification means the process of oxidizing coal partially, using oxygen under a pressurized atmosphere of 2–7 MPa, and producing fuel gas consisting mainly of CO and H_2. Coal resources have already been used for many years to produce fuel gas and material gas for the chemical industry. Various types of coal gasifier have therefore been developed and used in practice. If we focus on the method by which the gasification agent comes into contact with the coal, they can be

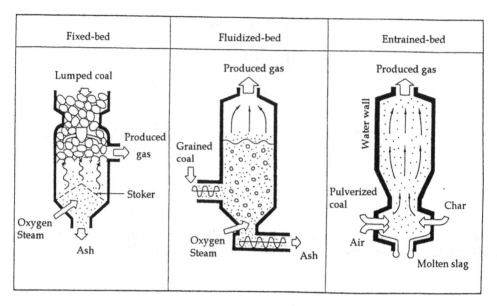

Figure 3.16 Three typical types of coal gasifier (Ministry of International Trade and Industry, Resource and Energy Agency 1996).

classified into three main types as shown in Figure 3.16 (Kaiho 1996; Ministry of International Trade and Industry, Resource and Energy Agency 1996). A different size of coal is used corresponding to the type of gasifier; that is, lumped coal for the fixed-bed type, grained coal for the fluidized-bed type, and pulverized coal for the entrained-bed type. A typical gasifier of the fixed-bed type is the well-known Lurgi gasifier, which has been used widely for many years.

According to Kaiho (1996), the gasification process can be explained basically as follows. If the composition of coal is represented by CH_mO_n, then the amount of oxygen theoretically required for combustion is calculated to be $[1 + 0.25m - 0.5n]$. Since the values of m and n are in the ranges of 0.80–0.85 and 0.12–0.20, respectively, for the kinds of coal being used for gasification such as brown coal and sub-bituminous coal, the amount of oxygen, $0.5(1 - n)$, required for converting all the C content of coal to CO can be calculated to correspond to an oxygen ratio of about 0.4; this ratio is defined as the ratio of oxygen supplied to oxygen theoretically required for combustion. Figure 3.17 shows the data of coal conversion rate and oxygen ratio of various actual gasifiers (Kaiho 1996). The higher the oxygen ratio, the higher the gasification temperature and reaction velocity, with resultant increase of specific gasifying capacity and improvement in the coal conversion rate. To improve gasification productivity, therefore, it is important to increase the oxygen ratio as near as possible to 0.4.

For gasifiers of fixed-bed and/or fluidized-bed types, when oxygen is increased

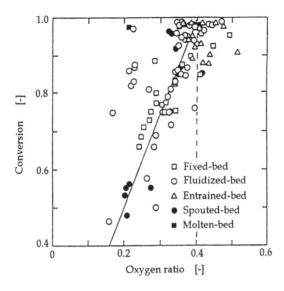

Figure 3.17 Relation between oxygen ratio and conversion ratio (Kaiho 1996).

and the working temperature has risen, local overheating becomes more liable to occur; as a result, operational troubles would be induced due to ash melting and slagging. To cope with this problem, the oxygen of the gasifying agent is diluted with steam to make the combustion slower and to avoid local overheating. The steam supply should be carefully controlled, corresponding to coal characteristics, because supplying steam decreases the temperature and the specific gasifying capacity and, further, induces the problem of tar formation.

On the other hand, the entrained-bed gasifier is operated at a high enough temperature to keep coal ash in a molten state so as to make ash slag flow down the wall and exit the gasifier. Owing to its high working temperature, the entrained-bed gasifier can achieve a high specific gasifying capacity and has an increased production capacity compared with other types. These are the reasons why, among the three types, the entrained-bed type is adopted as the gasifier of IGCC. For this type, it is possible to use air as the gasifying agent. When air is used, energy for heating up all the air including nitrogen is required; thus, some gasifiers are designed to operate at an oxygen ratio of more than 0.4.

Figure 3.18 shows examples of gas composition and coal conversion efficiency for three types of gasifier (Koyama 1996). In the entrained-bed type, the gas consists mainly of CO and H_2. The calorific value achieved for converted gas has amounted to 80–84% of that of coal, a value for which almost the theoretical limit has been reached. This is the conversion efficiency, also called the cold-gas efficiency. Taking the sensible heat of converted gas into account, the total thermal efficiency attained is about 95% provided that effective usage of the sen-

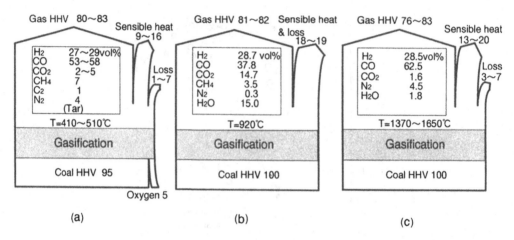

Figure 3.18 Gas composition and heat balance of oxygen-blown coal gasifiers (Koyama 1996). (a) Fixed-bed; (b) fluidized-bed; (c) entrained-bed. Adapted from Sharman et al. (1980), Adlhoch et al. (1993), and Mahagaokar et al. (1991), respectively.

sible heat is realized. When air is used as the gasifying agent for the entrained-bed type, the sensible heat ratio is increased, so that its recovery is important. In this case, IGCC is suitable because sensible heat can be effectively recovered by HRSG.

3.3.4.3 Gas cleaning

Since converted fuel gas contains various pollutants such as sulfuric materials, PM, and trace elements, as does the flue gas of a coal-fired boiler, gas cleaning is an important process of IGCC. Desulfurization and PM removal should be carried out in advance directly from fuel gas, at high pressure and temperature, upstream of the GT. As for NO_x, control can be carried out by low-NO_x combustion in the GT combustor and/or de-NO_x treatment of flue gas in the downstream region of HRSG.

Sulfur is contained in the form not of SO_x but of other compounds such as H_2S and COS. It is therefore necessary to adopt an appropriate type of desulfurizer for those sulfuric compounds. The wet-type desulfurizer, which consists of two processes, scrubbing by water and desulfurizing by absorber, can remove simultaneously some of the other pollutants such as trace elements and ammonia. Although the wet-type desulfurizer is desirable for cleaning, the sensible heat is undesirably wasted without being used. To avoid sensible heat loss, a dry-type desulfurization system has also been developed, consisting of a PM remover using a ceramic filter and a dry-type desulfurizer. When air is used as the gasifying agent, the sensible heat of the converted gas reaches a large ratio, so that the dry type would be suitable. There is not much NO_x formation in the gasifier, but intermediate nitrogen compounds such as ammonia and HCN are formed and are contained in the converted fuel gas. Since these intermediate compounds are easily oxidized to NO_x in the combustion process, a GT combustor has to be considered to control NO_x formation.

3.3.5 Removal of pollutants from flue gas

3.3.5.1 NO_x removal

The NO_x removal technique that has been used most widely so far is the selective catalytic reduction (SCR) method, which reduces NO_x to N_2 using some form of reducer and a suitable catalyst. Most SCRs use ammonia as reducer, though urea is used in some cases because ammonia is toxic and explosive. SCR proceeds by the reducing reactions described in Eq. (3.13) below with the help of a catalyst. The activity of the catalyst is greatly dependent on the temperature, and therefore an appropriate catalyst has to be selected corresponding to the temperature level of the flue gas to be treated. In the case of large-capacity boilers, the SCR is generally installed just after the economizer, where the catalyst adopted is one whose activity is most effective at a temperature level of 300–400°C. Figure 3.19 shows a typical example of catalyst structure, which is composed of many clusters of catalyst elements (Inoue 1996). The element shown in the figure has been made to form a passage for flue gas in the shape of a square mesh. Other shapes have been developed, such as a plate-type element for the regenerative exchanger. The mesh size is small, in the range of 3.7–7 mm, to accommodate as wide a contact area as possible. For the flue gas of a coal-fired boiler, a coarse mesh of 7 mm is applied to avoid it becoming plugged due to fouling such as PM deposition. The element body is made of ceramic materials such as TiO_2 and Al_2O_3, and catalyst metals such as V_2O_5 or WO_3 are coated on the element surface.

The reducer (ammonia) is added into the flue upstream of the SCR, as shown in Figure 3.20. The addition rate is almost the same, in terms of molecular weight, as that of the NO_x, as predicted by the following equations:

$$\left.\begin{array}{llr}\text{Reduction reactions:} & 4NO + 4NH_3 + O_2 \rightarrow 4N_2 + 6H_2O & \text{(a)} \\ & 6NO_2 + 8NH_3 \rightarrow 7N_2 + 12H_2O & \text{(b)} \\ \text{Oxidation reactions:} & 4NH_3 + 5O_2 \rightarrow 4NO + 6H_2O & \text{(c)} \\ & 4NH_3 + 3O_2 \rightarrow 2N_2 + 6H_2O & \text{(d)} \end{array}\right\} \quad (3.13)$$

When the appropriate kind of catalyst has been decided, the reduction performance is mainly dependent on the following factors:

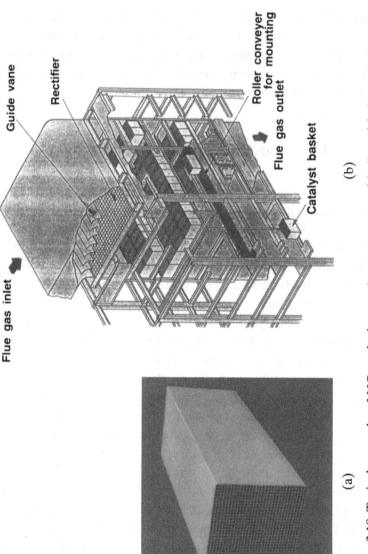

Flue gas inlet

Guide vane

Rectifier

Roller conveyer
for mounting

Flue gas outlet

Catalyst basket

(a)

(b)

Figure 3.19 Typical example of NO$_x$-reducing catalyst for coal-fired boilers. (a) Catalyst element; (b) construction of de-NO$_x$ reactor for coal-fired boiler (courtesy of the Ishikawajima-Harima Heavy Industries Company).

- flue gas temperature;
- addition rate of ammonia reducer;
- catalyst contact area with flue gas;
- flue gas residence time within SCR, that is, gas velocity;
- catalyst activity.

If the flue gas temperature is above or below the suitable range for the catalyst, the reduction performance is lowered. For instance, the performance of an SCR using a vanadium/titanium catalyst is lowered due to the activation of the oxidation reactions described in Eq. (3.13) when the temperature becomes too high. As for contact area, surface fouling by the deposition of fly ash, calcium sulfate, and so on is a serious problem for SCR of coal-fired gas. The catalyst activity also deteriorates over operational time due to the contaminant effects of alkali metals, arsenic, and so on.

Figure 3.20 shows an example of an NO_x removal system for a coal-fired boiler (Hashimoto et al. 1993). One problem to which very careful attention should be paid is the "ammonia slip" of an SCR system using ammonia, or the amount of ammonia that does not react with NO_x and remains in the outgoing flue gas. Since ammonia not only smells bad but is toxic, ammonia slip has to be controlled to as small an amount as possible. Moreover, excess ammonia reacts with SO_3 and forms various ammonium compounds such as ammonium sulfate, sulfurous ammonium, and ammonium hydrogensulfate at certain temperature conditions below 320°C. These ammonium compounds deposit onto the catalyst element and downstream equipment such as the air heater, and cause serious fouling problems. Even when the total addition rate can be controlled accurately, local ammonia slip will occur if addition to the flue gas flow is badly distributed. To decrease ammonia slip, it is very important to pay careful attention to the total addition rate and to uniform distribution. As seen in Figure 3.20, duct design for uniform gas flow and design of ammonia addition equipment are important. Not only design but also monitoring and maintenance are important because ammonia slip will increase with operational time due to the deterioration of the catalyst for various reasons mentioned above (Cho 1994; Gutberlet 1994). Figure 3.20a shows an example of an SCR system installed upstream of AH. Other systems have been developed, such as SCR installed after removal of PM by an electrostatic precipitator (ESP), as shown in Figure 3.20b and c.

3.3.5.2 *Flue-gas desulfurization*

Most fuel sulfur is contained in the form of SO_2; a small part, at a level of a few percent, is in the form of SO_3. Various principles have been applied in the construction of flue-gas desulfurization (FGD) plants. Among them, the most widely

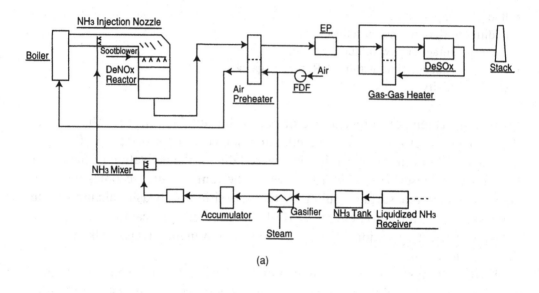

(a)

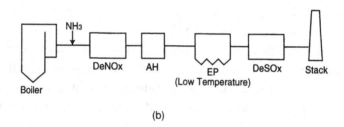

(b)

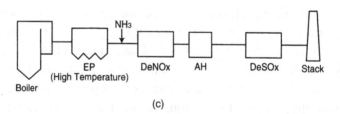

(c)

Figure 3.20 Examples of flue-gas de-NO$_x$ system. (a) Flow sheet of a flue-gas de-NO$_x$ system; (b, c) integrated flue-gas cleaning system for (b) high and (c) low dust concentration (courtesy of the Mitsubishi Heavy Industries Company).

applied one is the principle that converts SO_2 to sulfates by the use of absorbent Ca and/or Mg compounds and then removes them from the flue gas. Various systems based on this principle have been developed. As for the technological characteristics and/or cost performance of each practical system, the reader is referred to Makansi (1993) and Sage and Ford (1996).

The various FGDs can be classified mainly into three types with respect to the method of bringing the absorbent into contact with the flue gas:

- direct injection into furnace (furnace desulfurization);
- semi-dry system;
- wet system.

The wet-type FGD using lime as absorbent has been the one most generally used for actual boilers, especially for large-capacity boilers. This type of FGD converts sulfate further to gypsum, which is a valuable and useful material. Figure 3.21 shows an example of an advanced FGD plant for an actual power-generating coal-fired boiler of 700 MW (Tamaru et al. 1996). This plant uses lime as an absorbent and recovers sulfate in the form of gypsum. The absorber of this plant carries out two processes of SO_2 absorption and oxidative reaction, producing gypsum only in one tower. The flue gas temperature is lowered by the absorber, due to its wet method, and this lowers the buoyancy of the flue gas, likely to be visible as a white cloud due to steam condensation at the stack outlet. The gas-to-gas heater (GGH) shown in Figure 3.21, which is installed downstream of the ESP, is for reheating the flue gas to the temperature level of 90–100°C. As seen, the GGH is devised cleverly to use the flue gas heat so as to have both effects: fuel saving, and adjusting the flue gas temperature for the absorber. When a regenerative-type heat exchanger is used, there occurs the undesirable possibility of flue gas leaking into the air and detracting from the cleaning effect. To eliminate such a possibility, a nonleak type such as the tubular heat exchanger is often used. The nonleak type is sometimes convenient in its ability to be arranged in many ways for installation. Another type uses heat pipe technology. Since the wet-type FGD produces a large amount of waste water, waste water treatment is necessary for conservation of the water environment.

The wet-type FGD is especially desirable for coal-fired boilers because its scrubbing effect can also remove other pollutants such as various trace elements, chlorides, and fluorides contained in flue gas. However, these contaminants are mixed with the waste water, which should therefore be treated very carefully. As the practical use of FGD spreads, many advanced FGD systems have been developed incorporating various improvements in cost saving, compactness, simplifying of waste water treatment, decreased refuse, and so on. For instance, a hybrid system combining furnace injection and a semi-dry system has been

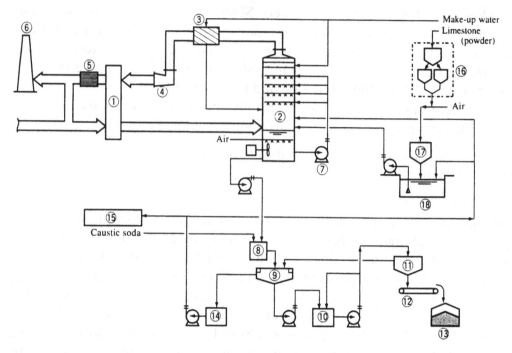

Figure 3.21 Example of flue-gas desulfurization system. (1) Gas–gas heater; (2) absorber; (3) mist eliminator; (4) FGD boost-up fan; (5) silencer; (6) stack; (7) absorber recirculating pump; (8) neutralization tank; (9) gypsum thickener; (10) centrifuge feed tank; (11) gypsum separator; (12) gypsum conveyor; (13) gypsum silo; (14) mother liquor tank; (15) waste-water treatment unit; (16) pneumatic conveyor; (17) limestone powder storage silo; (18) limestone slurry pit (courtesy of the Ishikawajima-Harima Heavy Industry Company).

developed to save the high cost of the wet-type FGD (Yoshii et al. 1992), and a dry-type FGD has also been developed that uses activated carbon as removal agent.

Since flue gas cleaning plants are, of course, designed to prevent environmental pollution, they are required to have high reliability. One of the significant factors influencing the reliability of such plants is the fouling of equipment, piping, ducting, and so on. The fouling problem due to ammonia slip has already been discussed. FGD plants have to deal with very corrosive fluids, so equipment and piping are protected by rubber lining or are made of corrosion-resistant materials. More information on practical techniques and concrete measures coping with fouling problems can be found in Thermal and Nuclear Power Engineering Society (1990), Landwehr (1993), Makansi (1993), Gutberlet (1994), Kuron (1994), Nonhoff and Lux (1995), and other authors.

3.4 Flow and heat transfer of convective HTS

3.4.1 Heat transfer of cross-flow tube banks

The bank of circular tubes with axes normal to the outer flow, that is, the cross-flow tube bank, has been used very widely for various heat exchangers including boiler convective HTSs. In the case of boilers, the shape of the furnace and/or gas flue is not simple, so that tube banks are often equipped with tube axes inclined to the gas flow. These inclined-flow tube banks are designed on the basis of the characteristics of the cross-flow tube bank. It is fundamental, therefore, to study the gas-side heat transfer characteristics such as the heat transfer coefficient, pressure drop of cross-flow tube banks.

3.4.1.1 Characteristics of outer flow

The pressure drop (flow resistance) of the outer flow through cross-flow tube banks and the heat transfer coefficient, respectively, are defined as follows:

$$\Delta P = \zeta_m (\varrho/2) U_m^2 n \tag{3.14}$$

$$\mathrm{Nu} = \frac{hD}{\lambda} = c \mathrm{Re}^m \mathrm{Pr}^{1/3} \tag{3.15}$$

where ΔP is the pressure drop (flow resistance), ζ_m is the flow resistance coefficient, ϱ is the fluid density, U_m is the velocity of outer flow at the minimum flow passage area, n is the number of rows of tube bank, Nu is the Nusselt number, h is the mean heat transfer coefficient, D is the tube outer diameter, λ is the thermal conductivity of fluid, Re is the Reynolds number $U_m D/\nu$, ν is the kinematic viscosity, Pr is the Prandtl number, and c and m are correlation constants.

The coefficients ζ_m, c, and m in the above equations are dependent in a very complicated way on tube pitch, tube arrangement, and/or Reynolds number, so that it would be a very substantial research task to develop an accurate estimation method for these coefficients. There has been a long history of research into cross-flow tube banks, in which various correlations have been proposed so far to predict those coefficients. Most of these proposals, however, have been developed without attention being paid to the effect of flow structure on the heat transfer characteristics and/or to the relation between flow structure and tube arrangement. But the investigation of flow structure is essential, not only for estimating the heat transfer characteristics but also for dealing with other problems such as flow-induced vibration.

So, first we shall start by studying the flow structure of the tube bank. Since the outer flow of tube banks is very complicated, we study the flow structure, paying

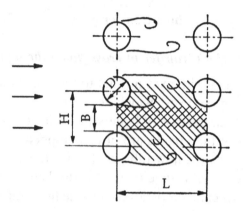

Figure 3.22 Outer flow through cross-flow tube bank.

attention to the following two points. As is well known, Karman vortices are formed when a circular tube is set with its axis normal to the flow, and they have strong effects on the characteristics of the tube wake.* In the case of tube banks, each tube, except for tubes in the first row, would be situated in the wake of its upstream tube, as shown in Figure 3.22. The first factor to be considered is the behavior of these Karman vortices when investigating the flow structure of the cross-flow tube bank. The second point for our attention is the behavior of the gap flow between adjacent tubes. The gap flow, which is shown in Figure 3.22 as the flow in the cross-hatched area, has a very noticeable characteristic like a two-dimensional jet flowing into a limited space – the hatched region in Figure 3.22. The flow between tube columns, therefore, can be seen to be formed by a row of two-dimensional jets with distance H between them.

Seeing that the flow structure of the cross-flow tube bank is predominantly characterized by the two points just mentioned, that is, the behavior of Karman vortices and of the two-dimensional jets, the flow pattern can be classified in the form of the maps shown in Figure 3.23 (Nishikawa et al. 1977). Each domain A, B, and C in the figure corresponds, respectively, to the domain of pattern A, B, and C illustrated in Figure 3.24. For in-line arrangements, the most general flow pattern is pattern A in which Karman vortices are formed by all tubes except those of the first row. When the longitudinal distance ratio L/D decreases to less than 2, pattern C changes to pattern B, and Karman vortices are not formed by any tube because the wake area of each tube becomes too narrow to form one. Fluid transport between bulk and wake region becomes worse since there is no

---

* Any cylindrical body, not only a circular cylinder, the section of which is not streamlined, can shed Karman vortices unless the Reynolds number is extremely small, say, less than about 40, though its shedding frequency is dependent on section shape, Reynolds number, and so on.

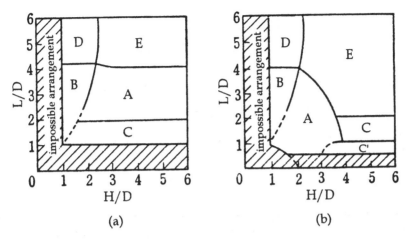

Figure 3.23 Flow pattern map of cross-flow tube bank. (a) In-line and (b) staggered arrangement. Domains A, B, and C correspond to patterns A, B, and C shown in Fig. 3.24, respectively. Domain C′ of (b) has the same pattern with zone C of (a). In domain D, the flow deflection of gap flow occurs due to the Coanda effect. In domain E, each tube can be considered to be a single tube.

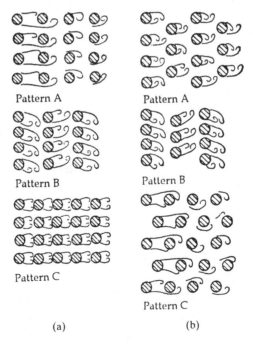

Figure 3.24 Flow pattern of outer flow through cross-flow tube bank. (a) In-line and (b) staggered arrangement.

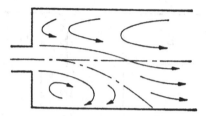

Figure 3.25 Definition of two-dimensional jet flowing into a semi-closed space due to Coanda effect.

formation of a Karman vortex in domain C, so that heat transfer also worsens. For staggered arrangements, flow pattern A is the most common, and all tubes shed Karman vortices. When H is fairly large compared with L, pattern A changes to pattern C, which is almost the same as pattern A of in-line arrangements. In domain E, the tube spacing is very large and each tube in the tube bank behaves as a single tube.

The change of flow pattern from one to another is not gradual but rather abrupt on the borderline shown in Figure 3.23. The borderlines in the figure should not be considered rigorously fixed. They are slightly dependent on Reynolds number, properties of the equipment, and so on, though the dependence would be very little within the ordinary range of Reynolds number over which boiler tube banks are used.

Pattern B, which is observed in both in-line and staggered arrangements, is very noticeable. Gap flows are swinging in a switching motion synchronized with Karman vortex shedding. This interesting flow motion is considered to be induced by the Coanda effect. As is well known, a two-dimensional jet discharging into a limited space does not flow straight forward but deflects to one side of the flow passage, as shown in Figure 3.25. This phenomenon is known as the Coanda effect. Domain D, where the L to H ratio is very large, is distinguished from other domains since the gap flow deflection due to the Coanda effect is so strong that the whole flow in the bulk area just behind the tube bank deflects to one side of the flow passage and the Karman vortex formation becomes unsteady. In the flow region within the tube banks, the tube column just downstream prevents the Coanda effect on the gap flow when the L to H ratio is not large. Figure 3.26 shows the critical value of L/D and/or L/H below which the preventive effect of the downstream tube row appears. Borderlines dividing domains D and E, and domains B and A, are determined according to those critical values, as shown in Figure 3.26.

Figure 3.27 shows the distribution of mean surface pressure of each tube in the 1st, 2nd, 3rd, and 5th rows, respectively. In the figure, the surface pressure is shown

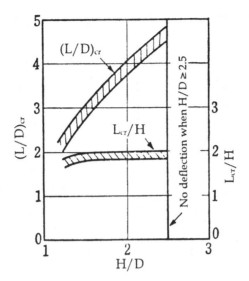

Figure 3.26 Critical tube arrangement $(L/D)_{cr}$ and/or L_{cr}/H of gap flow deflection. (Beyond $(L/D)_{cr}$ and/or L_{cr}/H, gap flow defects due to the Coanda effect. When $H/D \geq 2.5$, the deflection does not occur at any value of L/D.)

by the relative value of $P_{n\theta}/P_{10}$. $P_{n\theta}$ and P_{10} represent the surface pressure at an angle θ of a tube in the nth row and at an angle $0°$ of a tube in the 1st row, respectively, where n denotes the row order counted from upstream, and the angle is the circumferential angle measured from the upstream stagnation point. Figure 3.27a1–3 correspond to the patterns A, B, C, respectively, whereas Figure 3.27b1 and 2 correspond to pattern A and Figure 3.27b3 corresponds to pattern B. In the case that Karman vortices are not formed, such as a 2nd row tube of Figure 3.27a1 and/or all tubes except the 1st row of Figure 3.27a3, the pressure distribution is quite different from the case where Karman vortices are formed. Judging from the way the distribution tendency varies with row number, the flow structure of the cross-flow tube bank is considered to have grown at the 3rd row, at which point the distribution therefore represents a typical example of individual tube arrangement.

3.4.1.2 *Method of estimating flow resistance and heat transfer coefficient*

The above discussion reveals that the flow structure of the cross-flow tube bank is essentially determined by two flow phenomena, that is, the gap flow as a two-dimensional jet and the Karman vortex flow. With the behavior of these two flows in mind, we will try to analyze the flow resistance and heat transfer coefficient of cross-flow tube banks.

(a) *Flow resistance*: The characteristic length that dominates the behavior of

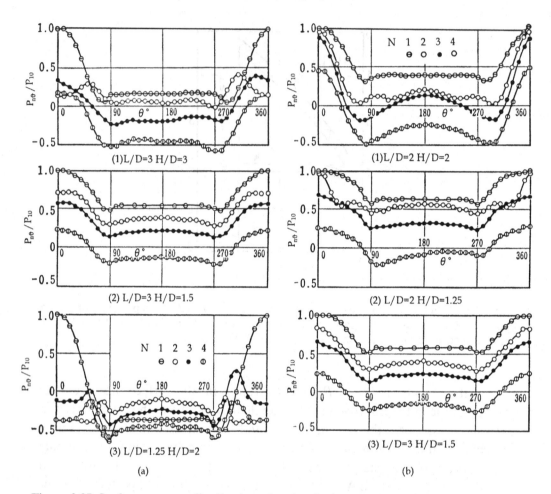

Figure 3.27 Surface pressure distribution of each tube in cross-flow tube bank. (a) In-line and (b) staggered arrangement. (N) Row number counted from the upstream; (P_{10}) pressure at the upstream stagnation point ($\theta = 0$); ($P_{n\theta}$) pressure at θ degrees of a tube in Nth row.

two-dimensional jet flow is the non-dimensional passage length L/B. Figure 3.28 shows how the flow resistance coefficient ζ_m defined by Eq. (3.14) is dependent on L/B. These data, plotted in the figure, are taken from Pierson (1937) and Zhukauskas et al. (1968). It is obvious from the figure that L/B is a dominant factor determining the flow resistance, especially in the case of in-line arrangements (Fig. 3.28a). Taking a careful look at this figure, we see that the gradient of the correlation line between ζ_m and L/B varies, corresponding to each flow pattern, so that it would be possible to deduce a simple method of estimating the flow resistance from the figure.

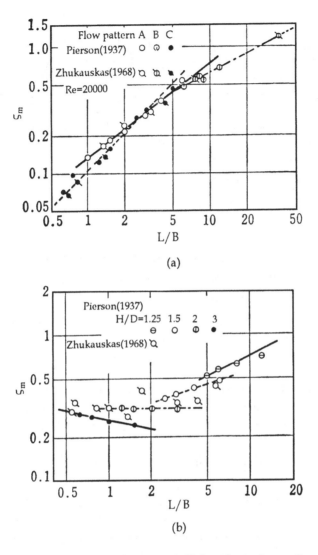

Figure 3.28 Relation between flow resistance coefficient ζ_m and non-dimensional passage length regarding gap flow L/B. (a) In-line and (b) staggered arrangement.

In the case of staggered arrangements, the relation seems to be more complicated than for in-line arrangements. This difference can be explained by the flow passage model shown in Figure 3.29. The point of difference in the staggered arrangement model compared with that of the in-line arrangement is that a tube is standing in a gap of the flow from the row just upstream, and forms Karman vortices. The model suggests that the influence of Karman vortex flow on flow resistance becomes very important, in addition to the influence of gap flow. In the

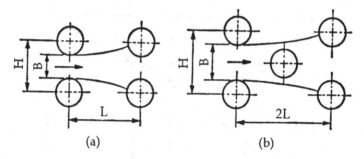

Figure 3.29 Flow model of outer flow of cross-flow tube bank. (a) In-line and (b) staggered arrangement.

case of a staggered arrangement, therefore, these two influences should be taken into account in analyzing ζ_m. According to the model, ζ_m may be divided into two components. One, ζ_j, is the flow resistance due to the gap flow itself, and the other, ζ_p, is due to the flow passing over the tube standing in the gap flow. Then ζ_m is expressed as follows:

$$\zeta_m = \left(\zeta_j + \zeta_p\right)/2 \tag{3.16}$$

where the denominator 2 is due to the definition of ζ_m, which corresponds to the flow resistance per single tube row. It can be judged from the model that the smaller L/B is, the stronger the relative influence of Karman vortex flow becomes.

Consequently the analysis and model shown in Figures 3.28 and 3.29 give very simple equations, listed in Table 3.3, for estimating the flow resistance coefficient of cross-flow tube banks. These equations are applicable for the regions of Re number and tube pitch shown in Table 3.3. As can be seen, the equations are very simple and make it possible to calculate easily and with far better accuracy than with the conventional methods developed so far.

(b) *Heat transfer coefficient*: Figure 3.30a and b shows the influence of L/B on the correlation constants c and m defined by Eq. (3.15). As with in-line arrangements, correlations between these constants and L/B are different, obviously corresponding to whether Karman vortices are formed or not. In the case of arrangements where Karman vortices are not formed, the heat transfer coefficient becomes rather small compared with arrangements where Karman vortices are formed. It is obvious from the figures that the Karman vortex flow has a great effect on the heat transfer characteristics of cross-flow tube banks. Fluid transport between the bulk region and the wake region will be weak when Karman vortices are not formed. This might be the reason why the heat transfer coefficient becomes worse. The correlation apparent in the figures is informative

Table 3.3. *Equations for estimation of flow resistance and heat transfer coefficient of cross-flow tube banks (region of each flow pattern is shown in Fig. 3.23)*

		In-line arrangement		Staggered arrangement
Flow resistance	$\Delta P = \varsigma_m \dfrac{\varrho}{2} U_m^2 n$ $n \geq 10$	$\varsigma_m = 0.135(L/B)^{0.75}$ $\varsigma_m = (0.225 - 1.35\mathrm{Re} \times 10^{-6})$ $\times (L/B)^{0.5}$ $\varsigma_m = 0.105(L/B)$	Region A Region B Region C	$\varsigma_m = 0.33\mathrm{Re}^{-0.15}$ $\times \{(H/D + 1)$ $\times (D/L)$ $+ 1.5(L/B)^{0.75}\}$
	Applicable conditions	$2000 \leq \mathrm{Re} \leq 40{,}000$ $1.1 \leq L/D \leq 3, 1.05 \leq H/D \leq 3$		$2000 \leq \mathrm{Re} \leq 40{,}000$ $1 \leq L/D \leq 3.9,$ $1.25 \leq H/D \leq 3$
Heat transfer	$\mathrm{Nu} = c\mathrm{Re}^m\mathrm{Pr}^{1/3}$ $\mathrm{Re} = U_m D/v$ $\mathrm{Nu} = hD/\lambda$	$c = 0.19(L/B)^{0.35}$ $m = 0.65(L/B)^{-0.06}$ $c = 0.085(L/B)^{0.65}$ $m = 0.72(L/B)^{-0.09}$	Regions A, B Region C	$c = \{0.32 + 0.02 (H/D$ $- L/D)\}$ $m = 0.6$
	Applicable conditions	$2000 \leq \mathrm{Re} \leq 50{,}000$ $1.25 \leq L/D \leq 3, 1.25 \leq H/D \leq 3$		$2000 \leq \mathrm{Re} \leq 50{,}000$ $1.25 \leq L/D \leq 3,$ $1.25 \leq H/D \leq 3$

for designing effective tube arrangements from the viewpoint of heat transfer. For instance, it may be said, referring to Figure 3.23, that those arrangements in which L/D is designed to be smaller than 2 are not effective. When L/D is limited to a very small value due to a strong demand for very compact heat exchangers, such as a steam generator of a nuclear power plant, it would be better to make H/D small, corresponding to the value of L/D, so as to keep L/B to within a certain large range.

Although many efforts have been made so far to develop a method of estimating the heat transfer coefficient, the behavior of constants c and m of the in-line arrangement is so complicated that it has been considered to be impossibly difficult to find a useful method. Judging from Figure 3.30a, however, correlation of c and m is not so complicated and is dominantly dependent on L/B, so that a simple estimation method can be derived.

As for staggered arrangements, the variation of c and m with L/D and/or H/D is not very large, as can be seen in Figure 3.30b. Almost all the tube arrangements actually used for boilers belong to pattern A where Karman vortices are formed. This is the reason why the variation of c and m is relatively small. According to the figures, m can be considered to be approximately constant at 0.6, and c can be simply related to the tube arrangement.

These new estimation methods, which are the equations shown in Table 3.3, are derived from Figure 3.30. They are very simple and far more accurate than other

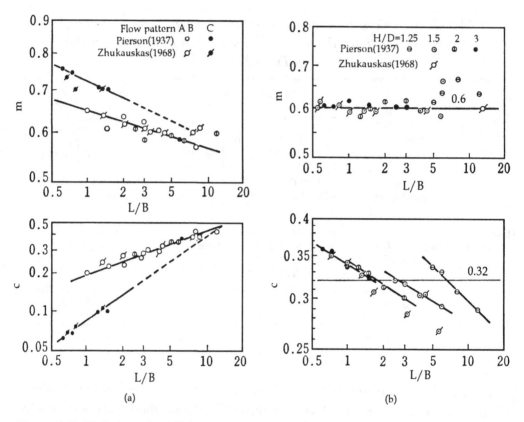

Figure 3.30 Variation of coefficients c and m of Eq. (3.15) with non-dimensional gap flow length L/B. (a) In-line and (b) staggered arrangement.

methods developed so far. Figure 3.31 shows the accuracy of these equations when applied to actual results.

3.4.2 Extended HTSs

HTSs for heat recovery, used in devices such as economizers, air heaters, and various exhaust boilers, usually require a very wide heat transfer area due to their low heat flux. An extended HTS is therefore often adopted to design these heat exchangers to be as compact as possible. Many kinds of extended surfaces for various heat exchangers have been devised, among which those usually used are some kind of finned tube, as shown in Figure 3.32. When an extended HTS is planned, there are many design parameters of the fin such as shape, pitch, height, thickness, and tube arrangement. There are an infinite number of combinations, and heat transfer characteristics vary with each combination. Even for bare tube banks, estimation of heat transfer characteristics is not easy, as discussed previ-

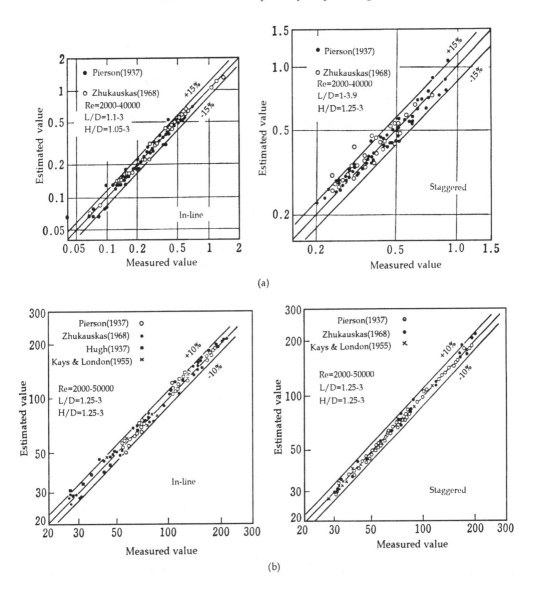

Figure 3.31 Accuracy of estimation methods proposed in Table 3.3. (a) Flow resistance coefficient; (b) heat transfer coefficient.

ously. The flow structure of finned tube banks is even more complicated, and so far there has not been developed any generally adaptable method that can predict the heat transfer coefficient and/or flow resistance of finned tube banks. When searching for an optimum finned HTS among many combinations of design parameters, therefore, it would be convenient to use some appropriate index for comparing the performance of heat exchangers.

Let us compare the finned and bare tube bank with regard to some character-

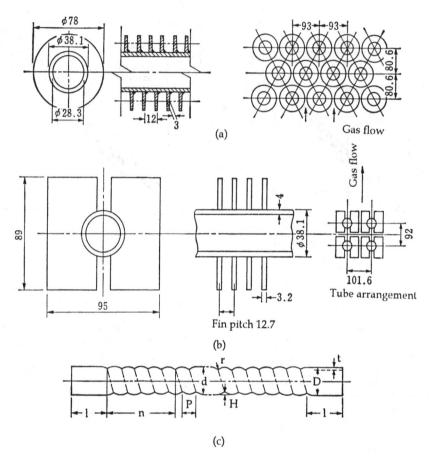

Figure 3.32 Examples of extended heat transfer surface for boilers. (a) Spiral fin; (b) square fin; (c) smoke tube ($D = d = 38.1, t = 2.9, P = 23, H = 3, r = 4.0, l = 80$) (courtesy of the Ishikawajima-Harima Heavy Industries, Osaka Boiler, and Kawasaki Heavy Industries companies).

istic items of heat exchanger to look for some useful index. The required HTS area S can be expressed by the equation

$$S = \frac{Q}{K\Delta_T} \tag{3.17}$$

where Q is the heat rate to be exchanged, K is the overall heat transfer coefficient, and Δ_T is the logarithmic mean temperature difference.

The HTS compactness rate is defined here as the HTS area accommodated within a unit volume of heat exchanger. This can be calculated by the equation

$$S_v = \frac{S}{HL} \tag{3.18}$$

where S_v is the HTS compactness rate, s is the HTS per unit length of heat transfer tube, L is the tube pitch in the flow direction, and H is the tube pitch perpendicular to the flow direction.

The whole volume V and whole weight W of the heat exchanger can then be equated as follows:

$$V = \frac{S}{S_v} \tag{3.19}$$

$$W = \frac{wS}{s} \tag{3.20}$$

where w is the weight per unit length of the heat transfer tube.

When the heat flow Q is equal to that of other exchangers of the finned tube and bare tube type, the ratios of V and W can be expressed as follows:

$$\alpha_v = \frac{V_f}{V_b} = \frac{S_f}{S_b} \cdot \frac{S_{vb}}{S_{vf}} = \frac{K_b S_b}{K_f S_f} \cdot \frac{(HL)_f}{(HL)_b} \tag{3.21}$$

$$\alpha_w = \frac{W_f}{W_b} = \frac{S_f}{S_b} \cdot \frac{w_f}{w_b} \cdot \frac{S_b}{S_f} = \frac{K_b S_b}{K_f S_f} \cdot \frac{w_f}{w_b} \tag{3.22}$$

where suffixes f and b mean finned tube and bare tube heat exchanger, respectively. The whole weight, including fluid contained in heat transfer tubes, is also an important index of heat exchanger performance. The weight ratio of contained fluid is shown in the following equation:

$$\alpha_f = \frac{(\pi/4)d_{if}^2(S_f/s_f)}{(\pi/4)d_{ib}^2(S_b/s_b)} = \frac{S_f}{S_b} \cdot \frac{s_b}{s_f} \cdot \frac{d_{if}^2}{d_{ib}^2} = \frac{K_b S_b}{K_f S_f}\left(\frac{d_{if}}{d_{ib}}\right)^2 \tag{3.23}$$

where d_i is the inner diameter of heat transfer tube.

What is the ratio of flow resistance of the finned type to the bare? If flow passage area F and flow velocity U_m are assumed to be equal to those of other exchangers of the finned tube and bare tube type, then the ratio of the row number of tube banks is as follows:

$$\frac{n_f}{n_b} = \frac{\{S_f/(Fs_f/H_f)\}}{\{S_b/(Fs_b/H_b)\}} = \frac{S_f}{S_b} \cdot \frac{s_b}{s_f} \cdot \frac{H_f}{H_b}$$

So, according to Eq. (3.14), the flow resistance ratio α_p can be expressed by the equation

$$\alpha_p = \frac{\Delta P_f}{\Delta P_b} = \frac{\zeta_{mf}U_m^2 n_f}{\zeta_{mb}U_m^2 n_b} = \frac{S_f}{S_b} \cdot \frac{s_b}{s_f} \cdot \frac{H_f}{H_b} \cdot \frac{\zeta_{mf}}{\zeta_{mb}} = \frac{K_b S_b}{K_f S_f} \cdot \frac{H_f}{H_b} \cdot \frac{\zeta_{mf}}{\zeta_{mb}} \tag{3.24}$$

The four ratios mentioned above can be useful as indicators for estimating the effect of a finned tube bank. Among the four, α_v, α_w, and α_f are for estimation of compactness, and α_p is for estimation of heat transfer enhancement. If the ratio is smaller than 1, the effect of finned tube adoption is positive; the smaller the ratio, the larger the effect. The factor Ks appears in all equations of the four ratios, so that it is considered the most important factor of α. The dimension of the factor is [W/(Km)], which denotes heat transfer performance per unit length of heat transfer tube. As with the effect of heat transfer enhancement, the ratio of the above factor to the flow resistance coefficient Ks/ζ_m is important.

One problem in finned tubes is the deterioration of heat transfer performance due to fouling of the tubes. Since the gap between neighboring fins is 10 mm or so, as in the examples shown in Figure 3.32, the flow passing through the gap is very sensitive to fouling. Figure 3.33 shows that flow resistance is greatly influenced by the fouling of the fin surface. For example, in the case of a heat recovery boiler where the exhaust gas contains sulfuric acid and PM, there occurs such extensive fouling in certain operating conditions that the gap is filled up with deposited materials. The influence of fouling should be carefully taken into account when finned tube banks are adopted. Too much account, however, might cancel out the effect of the finned tube bank.

In view of the above, a practical and correct estimation of fouling influence is a key task for the effective use of extended HTSs. Maintenance work would also be an important consideration for extended HTSs in service. This is one of the reasons why the fin shape actually used is limited to certain types as shown in Figure 3.32. Demand for effective recovery or use of exhaust heat energy and/or low-temperature heat will become intensified in the future, as will demand for extended HTSs. So R&D on extended HTSs will be just as important in the future as it is now.

3.5 Vibration induced by gas flow

Flow-induced vibration of tube banks is a troublesome problem for the designer of convective heat transfer sections. There are two cases of vibration induced by gas flow:

- resonant vibration of a gas column formed within tube banks or the ducting system;
- vibration of tubes in tube banks.

In the former case, there occurs an extremely strong acoustic vibration that may have psychological and/or physiological impacts on operators and neighbors, and might induce vibration in the casing of tube banks. In the latter case,

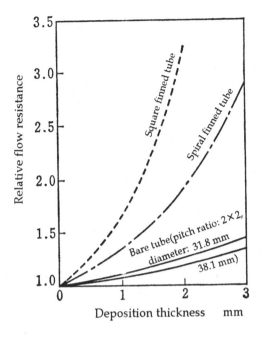

Figure 3.33 Increase of flow resistance due to outer fouling.

tubes are set into oscillatory motion at their natural frequency and this may sometimes result in breakage of tubes due to fatigue. For both cases, it is not easy to estimate exactly the probability of occurrence at the design stage, so the flow-induced vibration of tube banks is still an important factor that should be investigated. Here the mechanism of flow-induced vibration in cross-flow tube banks is explained and the measures for preventing it are described briefly.

3.5.1 Gas column resonance

Resonant vibration of a gas column is almost always originated by Karman vortices formed in tube wakes, and generally the characteristic length of the gas column is assumed to be equal to the width of the flow passage perpendicular to the flow direction. Resonant noise would not necessarily reach an unbearably loud level unless the following two conditions are satisfied simultaneously:

i. the shedding frequency of Karman vortices is equal or nearly equal to the natural frequency of the gas column;
ii. the excitation energy originating resonance is at a certain level higher than the dissipative energy of the resonating gas column, that is, the resonator.

Regarding (i), the vortex shedding frequency f_v varies in a complex way with the tube arrangements.* Although gas column vibration is a resonant phenomenon, as just mentioned, there exists an interaction between gas column vibration and vortex formation. When gas column vibration occurs, it has a feedback effect on vortex formation, and the shedding frequency is forced to keep a constant value equal to the natural frequency of the gas column within a certain velocity range. This feedback effect is called a locking-in phenomenon. The first task is to obtain data on f_v to estimate the possibility of gas column vibration. Factors such as high sensitivity to tube arrangement and the locking-in phenomenon make it difficult to get correct data for f_v in cross-flow tube banks, and there are some discrepancies among the measurement data for f_v published so far. Chen (1968) and Fitz-Hugh (1973) have published data, which had been obtained, however, without any attention being paid to the flow pattern shown in Figures 3.23 and 3.24. The data for f_v shown in Figure 3.34 were measured by Ishigai et al. (1975).

In Figure 3.34, f_v is expressed by the non-dimensional Strouhal number St, defined by $St = f_v D / U_m$. It should be noted that St data are plotted in domain C of the in-line arrangement where Karman vortices are not formed, as shown in Figure 3.23. These data do not relate to the Karman vortex shedding frequency but to the dominant frequency of gas velocity fluctuation that was measured in the wake region of each tube of the tube banks. It should also be noted that the gap flow is strongly deflected due to the Coanda effect, and that neither Karman vortex formation nor the dominant frequency of velocity fluctuation are observed in domain D.

The natural frequency f_c of the gas column, regarding condition (i), can be easily calculated on the basis of sound velocity of the gas. Attention should be paid to the fact that the sonic velocity changes a little when some obstacles such as tubes, especially smaller sized obstacles compared with the sonic wavelength, exist in the gas column. Their influence can be estimated approximately by the following equation:

$$\frac{\left(\text{sound velocity of the gas column containing tubes}\right)}{\left(\text{sound velocity of the gas}\right)} = \frac{1}{\left(1 + \sigma\right)^{1/2}} \quad (3.25)$$

where σ is the ratio of the volume of obstacles contained in the gas column, that is, the volume of tubes, to the volume of the gas column. A more exact estima-

---

* As seen in Figs. 3.22 and 3.24, Karman vortices are generated and grow at the free shear layers elongated from both separation points of the tube while they are moving downstream. When a vortex on the shear layer on one side has grown to sufficient strength to break down the shear layer on the other side, the vortex on the other side is forced to stop developing and is shed from the tube. The wake of the tube, up to the point where Karman vortices stop developing, is called the vortex formation region. The size and/or shape of the vortex formation region varies with tube arrangement. This is why the shedding frequency varies in such a complicated manner.

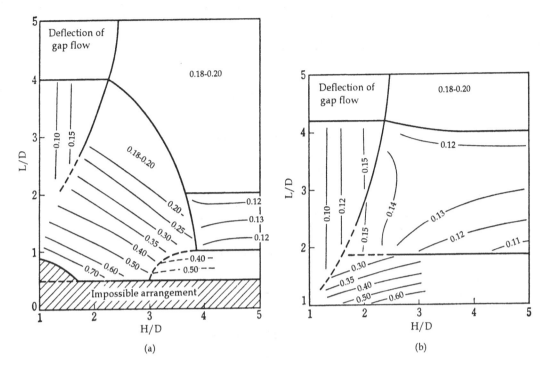

Figure 3.34 Strouhal number St of cross-flow tube bank (St = $f_v D/U_m$). (a) Staggered and (b) in-line arrangement.

tion of sound velocity of the tube bank gas column can be obtained from Planchard (1980).

The meaning of condition (ii) is interpreted, for an example, as follows. When the flow velocity is not high enough to supply the excitation energy even if f_v coincides with f_c of the first mode resonance of the gas column, the resonant oscillation would not be induced to a problematic level. But unbearable resonant sound might occur at a higher mode, say the second or third, when the flow velocity is high enough to supply the excitation energy for inducing strong resonance (Kobayashi et al. 1979). There has been found so far, however, no definite theory that can estimate quantitatively condition (ii). Chen (1974) proposed a discriminant based on Figure 3.35, in which the data regarding gas column resonance obtained for many actual tube banks, though limited to in-line arrangements, are plotted. That is,

$$\psi = \frac{\mathrm{Re}}{\mathrm{St}} \left(1 - \frac{D}{L}\right)^2 \left(\frac{D}{H}\right) > 2000 \tag{3.26}$$

Funakawa (1978) carried out experiments on gas column resonance for both in-line and staggered tube banks, in which the tube pitches were in the ranges of

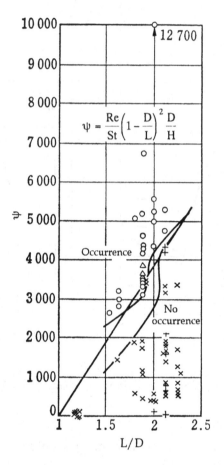

Figure 3.35 Estimation correlation of gas column resonance proposed by Chen (1974).

$1.47 \le H/D \le 2.68$, and $1.60 \le L/D \le 2.11$, and for various casings of tube banks of which sound absorption coefficients were changed. He summarized the experimental results as shown in Figure 3.36 and proposed that resonant vibration occurs when the approaching gas velocity and/or dynamic pressure exceeds a certain critical velocity determined by the tube arrangements and the logarithmic decrement of the sound resonant system. Figure 3.36 shows that, in the case of staggered arrangement, the critical velocity becomes lower as H/D becomes smaller, and when H/D is kept constant, the critical velocity becomes lower as L/D becomes larger.

A general measure for preventing gas column resonance is to insert one or more baffle plates into the tube bank section to change the gas column natural frequency to a higher value. Baffle plates are inserted beforehand in the manufacturing process when there is estimated to be a high possibility of resonance occurring, judging from the design conditions and effects such as those shown in

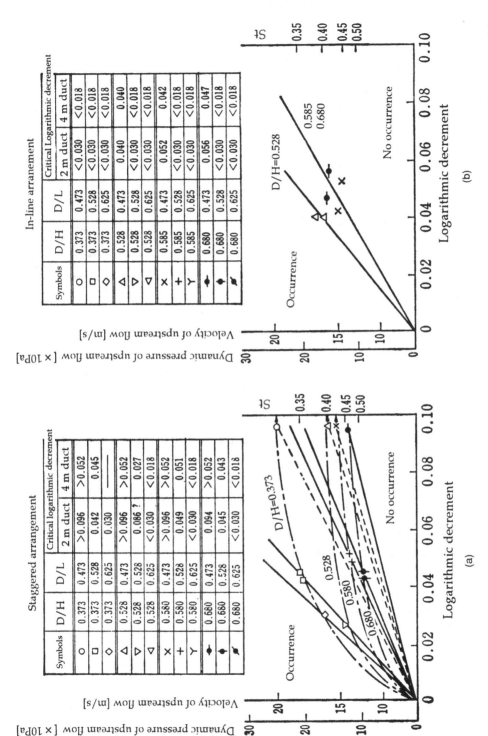

Figure 3.36 Measured results of occurrence limits of gas column resonance by the use of experimental tube banks. (a) Staggered and (b) in-line arrangement (Funakawa 1978).

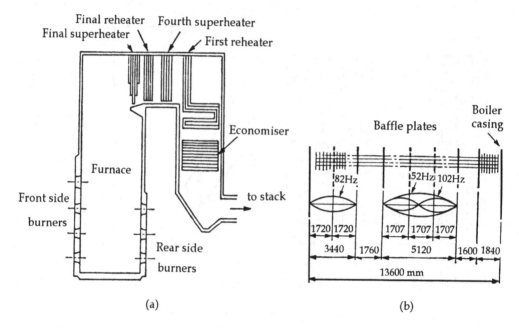

Figure 3.37 Actual example of gas-column oscillation and the baffle plate method for a 340 MW power generation boiler. (a) Sketch of boiler; (b) resonant mode, frequency, and baffle plates (Kobayashi 1979).

Figures 3.34–3.36. The positions of insertion are chosen to be asymmetrical in the gas column, and sometimes more than one plate is inserted to avoid higher mode resonances. Figure 3.37 shows an example of an actual boiler, one designed for power generation of 340 MW. The noise began to get louder when power load exceeded 200 MW, and, further, the loudness reached 100 phon beyond two-thirds of the design load. The boiler had had four baffle plates inserted in its tube banks beforehand, the positions of which are shown in Figure 3.37b by solid lines, but those were not effective. The noise measurements indicated that resonance of the vibration mode shown in Figure 3.37b was probably occurring, so three more plates were added at the positions shown by broken lines. This measure allowed the noise to be suppressed satisfactorily.

3.5.2 Tube vibration

There are three mechanisms that can be considered to induce tube vibration in cross-flow tube banks:

  i. vibration induced by Karman vortices;
 ii. vibration induced by the buffeting effect of flow turbulence;
iii. fluid-elastic vibration.

Since the circulation around the tube changes, corresponding to the periodic shedding of Karman vortices, an alternating lift force acting on the tube is produced and changes periodically, corresponding to the vortex shedding. Mechanism (i) is induced by this alternating lift force. It may be generally said in the case of gas flow that vibration (i) does not grow to a very serious level unless resonant vibration occurs, when the vortex shedding frequency f_v happens to be equal to the tube natural frequency f_n.

Mechanism (ii) is a kind of forced vibration excited by the buffeting effect of fluid flow, which is explained as follows. Even when Karman vortices are not formed, the fluid flow through the tube banks produces fluctuating forces having various frequency modes that act on tubes and force them to oscillate. These tubes oscillate in the flowing fluid in a selective forced oscillation mode at their natural frequency. This phenomenon is called buffeting vibration.

Mechanism (iii) is a kind of self-excited vibration due to interaction between tube vibration and fluid flow, such that the vibration may grow to a very danger-ous, destructive level once this type of vibration has occurred. As mentioned in Section 3.4.1.1, the fluid flow through tube banks is very sensitive to tube arrange-ments, especially when the tube pitch becomes small. When tubes are induced to oscillate by some mechanism such as the buffeting effect, the oscillation of these tubes may possibly have a feedback effect on fluid flow, which itself may begin to fluctuate periodically. There is a possibility that the feedback effect is positive, and if so, the induced flow fluctuation produces forces in the fluid that amplify the tube oscillation further. In this way, the positive feedback process could cause tube vibration to grow dangerously. This mechanism is called fluid-elastic vibration.

Figure 3.38 shows a typical example of how these three vibration phenomena occur in a single row tube bank in a gas flow (Ishigai et al. 1973). The vibration amplitude due to the buffeting effect increases gradually as the flow velocity goes up. If f_v coincides with f_n at a certain velocity, then a resonant vibration would appear. The peak due to mechanism (i) disappears when the velocity increases further, while the vibration due to (ii) gradually increases. Increasing the veloc-ity further, to a point where the velocity reaches some critical velocity represented as $U_{m.cr}$, leads to the occurrence of fluid-elastic vibration at which the amplitude would increase dangerously.

(a) *Resonant vibration induced by Karman vortices*: Mechanism (i) was explained simply as a resonance phenomenon. However, the actual mechanism is a bit more complicated. There occurs some interactive effect between tube vibration and Karman vortex shedding; that is, the flow separation points on the tube surface move fore and aft corresponding to tube vibration, so that the mag-nitude of the periodic exciting fluid force and its phase vary with tube vibration. Regarding the estimation of vibration magnitude, so far several empirical for-

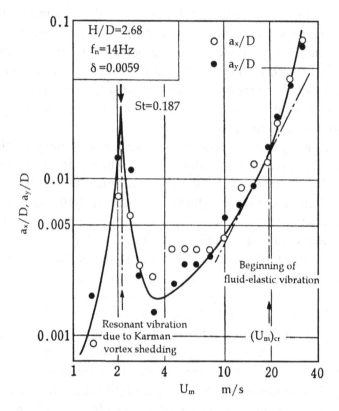

Figure 3.38 Measured example of flow-induced vibration of elastically supported tube in single row tube bank (*H/D*: 2.68; δ: logarithmic decrement; a_x, a_y: vibration amplitude in the drag and lift directions).

mulae have been proposed; these were summarized by Paidoussis (1982) as follows.

Hartlen and Currie:
$$\frac{a_y}{D} = \frac{0.0505}{\left(3.36 + 4\delta_r^2\right)^{1/2} \cdot \text{St}^2}$$

Griffin, Skop, and Ramberg:
$$\frac{a_y}{D} = \frac{1.29k}{\left\{1 + 0.43\left(4\pi\text{St}^2 \cdot \delta_r\right)\right\}^{3.35}}$$

Iwan and Blevins:
$$\frac{a_y}{D} = \frac{0.07k}{\left(2\delta_r + 1.9\right)\text{St}^2}\left\{0.3 + \frac{0.72}{\left(2\delta_r + 1.9\right)\text{St}}\right\}^{1/2}$$

Sarpkaya:
$$\frac{a_y}{D} = \frac{0.32}{\left\{0.06 + \left(4\pi\text{St}^2 \cdot \delta_r^2\right)\right\}^{1/2}}$$

(3.27)

where a_y is the amplitude perpendicular to the flow direction, δ_r is the non-dimensional vibration parameter* equal to $(m\delta)/(\varrho D^2)$, δ is the logarithmic decrement, m is the tube mass per unit length, ϱ is the fluid density, and k is the constant dependent on the tube support condition, equal to 1 when both ends are fixed, 1.305 when one end is fixed and the other is free, and 1.155 when both ends are jointed.

The magnitudes estimated by the above formulae are almost coincident with each other over quite a wide range of non-dimensional vibration parameters, though Griffin et al.'s estimation is small compared with other estimations in the range of large δ_r. The δ_r of a boiler tube in gas flow is on the order of 10^1 to 10^2, and therefore the amplitude ratio is not very large, being on the order of 10^{-2} to 10^{-1}. In the case of water flow, however, δ_r is small, on the order of 10^{-2} to 10^{-1}, so the amplitude ratio would reach beyond the order of 1.

(b) *Fluid-elastic vibration*: A typical example of this vibration can be seen in the single row tube bank, where a self-excited vibration is induced by the "jet switching mechanism." As mentioned in Section 3.4.1.1, the gap flow between adjacent tubes behaves as a two-dimensional jet in which the flow direction is sensitive, due to the Coanda effect, especially when the gap becomes narrower. When tubes go into oscillation, this can affect the gap flow and make its direction change periodically. This periodic fluctuation of gap flow induced by tube vibration was called the jet switching phenomenon by Connors (1970). When the gap flow changes its direction periodically, the fluid resistance, that is, drag, of each tube varies also, corresponding to tube vibration, as shown in Figure 3.39. If jet switching occurs, it will have a positive feedback effect on tube vibration and induce a dangerous self-oscillating state.

The previous example shown in Figure 3.38 was a single-row tube bank of tube pitch $H/D = 2.68$, tube diameter equal to 28 mm, tube natural frequency $f_n = 14$ Hz, and $\delta_r = 10$. In this example, it can be read from the figure that fluid-elastic vibration began at a velocity U_m of about 20 m/s and its amplitude increased rapidly. If the critical velocity $U_{m.cr}$, beyond which fluid-elastic vibration occurs, is

---

* The physical meaning of the non-dimensional vibration parameter δ_r is explained briefly as follows. Let us suppose a very simple vibration system, that is, (1) a tube is vibrating in a harmonic mode; (2) the exciting fluid force F can be expressed as $F = \tilde{C}_D (\varrho/2) U^2 Dl$, where l is the tube length; and (3) the fluctuating fluid force coefficient $\tilde{C}_D$ is constant and independent of tube vibration. Then the exciting energy supplied from fluid flow during one cycle is $E = \oint F \dot{y}\, dt$, where y indicates displacement, and the decremental energy consumed during one cycle is $W = \oint 2\delta f_n ml\, \dot{y}^2\, dt$. When the tube is vibrating in an equilibrium state at its amplitude of a_y, E must be balanced with W. Accordingly, the amplitude ratio is expressed as

$$\frac{a_y}{D} = \frac{1}{8\pi}\tilde{C}_D\left(\frac{1}{St^2}\right)\left(\frac{1}{\delta_r}\right)$$

As explained above, δ_r is thus an important characteristic number determining the amplitude of a vibration system. The property of an actual vibration system is not as simple as mentioned above, however, with the result that the relation between amplitude and δ_r is as shown in Eqs. (3.27).

expressed by a non-dimensional velocity $[U_{m.cr}/f_n D]$, then it can be generally correlated with δ_r, as in the following equation:

$$\frac{U_{m.cr}}{f_n D} \cong K_{cr}\left(\frac{m\delta}{\varrho D^2}\right)^{\beta} \equiv K_{cr}\delta_r^{\beta}$$

(3.28)

where K_{cr} and β are constants that are dependent on tube arrangements, δ_r, and so on. As with single-row tube banks, the constant β is considered to be almost equal to 0.5 when δ_r is within the range of 10^0 to 10^2, and only K_{cr} varies, dependent on tube pitch. K_{cr} values for single-row tube banks, obtained from experiments, are shown in Figure 3.40.

Thus, as explained above by a typical example of single-row tube banks, fluid-elastic vibration is induced by the interaction between tube vibration and fluid flow. Turning to the case of ordinal tube banks, which comprise many rows and many columns, how is fluid-elastic vibration induced? In the following, a theory proposed by Tanaka (1980b, 1980c, 1981), Tanaka and Takahara (1981), and Tanaka et al. (1983) is introduced briefly.

Let us focus our attention on one tube, say tube O, in the tube bank model shown in Figure 3.41. In the case of multirow and multicolumn tube banks, tube O's vibration might be affected by various fluctuating fluid forces induced by the vibrations of many tubes surrounding tube O such as tube 1, 2, 3, 4, U, D, L, R, and so on. Tanaka's model, however, takes into account, fluid forces due only to tube O itself and four neighboring tubes U, D, L, and R and neglects the effects of other tubes since they are considered to be relatively small. The fluctuating fluid force F of each tube is assumed to be proportional to its vibration amplitude, so that F can be expressed by the following equation:

$$F_{XLY} = \frac{1}{2}\varrho U_m^2 \tilde{C}_{XLY} Y_L$$

(3.29)

where F is the fluctuating fluid force, $\tilde{C}$ is the fluctuating fluid force coefficient, Y is the vibrational displacement of tube L in the direction perpendicular to fluid flow, and the suffix combination XLY means the fluid force along the x-axis (the first suffix) induced by tube L's (center suffix) displacement along the y-axis. Then total fluctuating fluid force induced by all five tubes can be expressed by the following equations:

$$\left. \begin{aligned} F_X &= \frac{1}{2}\varrho U_m^2 \sum_1^5 \left\{\tilde{C}_{XjX} X_j + \tilde{C}_{XjY} Y_j\right\} \\ F_Y &= \frac{1}{2}\varrho U_m^2 \sum_1^5 \left\{\tilde{C}_{YjX} X_j + \tilde{C}_{YjY} Y_j\right\} \end{aligned} \right\}$$

(3.30)

where j indicates the five tubes O, U, D, L, and R. There is generally some phase difference between displacement and exciting fluid force, so the coefficient $\tilde{C}$ is a

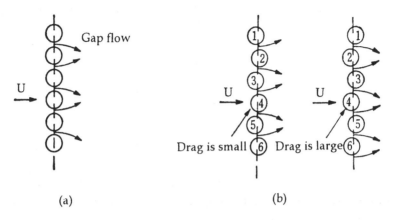

Figure 3.39 Interaction between gap flow deflection and tube vibration in the case of single-row tube bank. (a) Deflection of gap flow due to the Coanda effect; (b) switching of gap flow deflection induced by tube vibration in the flow direction (Connors 1970).

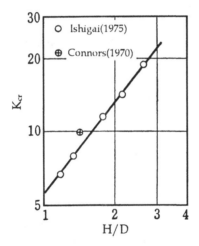

Figure 3.40 Variation of coefficient K_{cr} in Eq. (3.28) with H/D for single-row tube bank (Ishigai 1975).

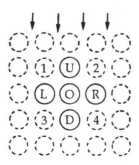

Figure 3.41 Modeling of cross-flow tube bank (Tanaka 1980b).

complex number. The oscillatory motion of tube O due to the above fluid forces can be expressed by a Lagrangian equation, as follows:

$$\left.\begin{array}{c} m_e\left(\ddot{X} + \dfrac{\delta}{\pi}\omega_n\dot{X} + \omega_n^2 X\right) = F_X \\[2mm] m_e\left(\ddot{Y} + \dfrac{\delta}{\pi}\omega_n\dot{Y} + \omega_n^2 Y\right) = F_Y \end{array}\right\} \qquad (3.31)$$

where m_e is the equivalent mass of the tube and ω_n is the circular natural frequency of tube equal to $2\pi f_n$.

The following is a brief procedure for solving Eq. (3.31). The displacements X and Y are expressed summarily by one symbol Z. The number of variable Z is therefore double the number of tubes composing the tube bank. If the damping coefficient is expressed by C, and, since fluid force F_Z is a linear function of Z, the spring constant can be expressed summarily by K, then the vibration equation can be rearranged for all tubes by the use of coefficient matrices, that is,

$$\left[m_e\right]\{\ddot{Z}\} + \left[C\right]\{\dot{Z}\} + \left[K\right]\{Z\} = 0 \qquad (3.32)$$

If the oscillation mode is assumed to be a harmonic mode, then Z can be expressed as

$$Z = Ze^{\lambda t}$$

and, therefore, introducing unit matrix I, the vibration equation can be finally rearranged as follows:

$$\begin{bmatrix} m_e\lambda + C & K \\ I & -\lambda I \end{bmatrix} \begin{Bmatrix} \dot{Z} \\ Z \end{Bmatrix} \qquad (3.33)$$

Equation (3.33) is a set of linear equations in which the number of elements is four times as many as the number of tubes. The solution of the equation is λ, which is obtained by making the coefficient matrix equal to zero. $\lambda = \lambda_R + i\lambda_I$, where λ_I indicates circular natural frequency and λ_R indicates increase or decrease in amplitude depending on whether its sign is plus or minus. Therefore, looking at the maximum λ_R, the stability, or whether fluid-elastic vibration would occur or not, can be judged from its sign.

The above is a brief explanation of Tanaka's model and its analysis. To carry out the analysis with regard to some actual tube banks, concrete data of the fluctuating fluid force coefficient $\tilde{C}_{XjX}$, $\tilde{C}_{XjY}$ ($j = $ O, U, D, L, R) and phase relation have to be known. Those data can be obtained only by experimental investigation on tube banks.

Tanaka et al. (1983) carried out experiments on several in-line tube bank models, including one with a tube pitch ratio of $L/D = H/D = 1.33$. Experiments were also done using water as the fluid, in which case the non-dimensional vibra-

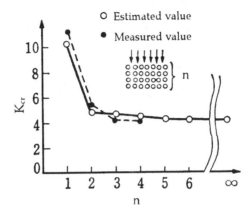

Figure 3.42 Variation of K_{cr} with row number for the in-line cross-flow tube banks of $H/D = L/D = 1.33$ (Tanaka et al. 1981).

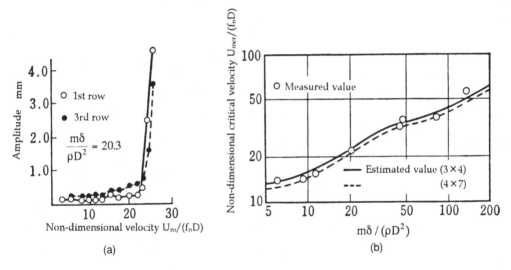

(a)

(b)

Figure 3.43 Fluid-elastic vibration characteristics of in-line cross-flow tube banks of $H/D = L/D = 1.33$ (Tanaka 1980b). (a) Variation of amplitude with velocity; (b) relation between non-dimensional decrement parameter $m\delta/(\varrho D^2)$ and critical velocity $U_{m.cr}/(f_n D)$.

tion parameter δ_r is small. Experimental results for a tube bank model of which the tube pitch was $H/D = L/D = 1.33$ are shown in Figure 3.42, where the variation of K_{cr} in Eq. (3.28) with the number of tube rows is plotted when the exponent β is equal to 0.5. Figure 3.43 shows the results for the tube banks when the number of tube rows is four. The relation between vibration amplitude and flow velocity is shown in Figure 3.43a, where the amplitude increases abruptly at a certain velocity so that the critical velocity can be detected clearly. The critical velocity thus detected was non-dimensionalized and plotted in Figure 3.43b with a non-dimensional vibration parameter. Calculated results of the theory described above are also plotted in Figures 3.42 and 3.43. Close agreement can be seen between theoretical and experimental results. Tanaka's experiments have revealed further that the variation of the exponent β to values different from 0.5 is not negligible when δ_r changes widely, and the following experimental equations have been obtained:

$$\begin{aligned}
\text{For } H/D = L/D = 1.33, \quad & \delta_r \geq 4.5, & U_{m.cr}/f_n\,D &= 5.8\delta_r^{0.46} \\
& 4.5 > \delta_r \geq 0.4, & U_{m.cr}/f_n\,D &= 3.7\delta_r^{0.37} \\
& 0.4 > \delta_r \geq 0.01, & U_{m.cr}/f_n\,D &= 3.4\delta_r^{0.26} \\
\text{For } H/D = L/D = 2, \quad & \delta_r > 7, & U_{m.cr}/f_n\,D &= 3.0\delta_r^{0.75}
\end{aligned} \qquad (3.34)$$

Thus, as described above, the mechanism of fluid-elastic vibration of cross-flow tube banks has been almost cleared up by Tanaka's theory. To detect concrete values of K_{cr} and β of Eq. (3.28), however, data for the fluctuating fluid flow coefficient $\tilde{C}$ have to be obtained by laborious experiments on individual tube banks.

Paidoussis (1982) has summarized various data reported so far regarding the non-dimensional critical velocity for cross-flow tube banks, as shown in Figure 3.44. Although the effects of tube pitch and/or arrangement cannot be read out from the figure, it is still convenient for a rough estimation of the possibility of fluid-elastic vibration. Since the non-dimensional parameter δ_r is on the order of 10^1 to 10^2 in the case of gas flow and the non-dimensional critical velocity would be on the order of 10^1, fluid-elastic vibration would be small for boiler tube banks within the range of ordinary tube pitches. When the flow velocity is very high and tubes are arranged very compactly, such as in the steam generator of a nuclear power plant, and/or the fluid is liquid in which case δ_r becomes small, careful attention should be paid to fluid-elastic vibration. If the vibration is possible, it would be prudent to increase the natural frequency of the tube by strengthening the tube support (Sato et al. 1980).

Discussion has been focused here on the flow-induced vibration of cross-flow tube banks. Parallel-flow tube banks, used for heat exchangers, for instance, in a nuclear reactor or steam generator, are also sometimes troubled by flow-induced vibration. Paidoussis (1982) is a useful reference for such problems.

3.6 Gas-side fouling and corrosion

3.6.1 The struggle against corrosion

Fuel cost is determined by two factors, specific fuel consumption and fuel price. The former corresponds to thermal efficiency and the latter corresponds to fuel

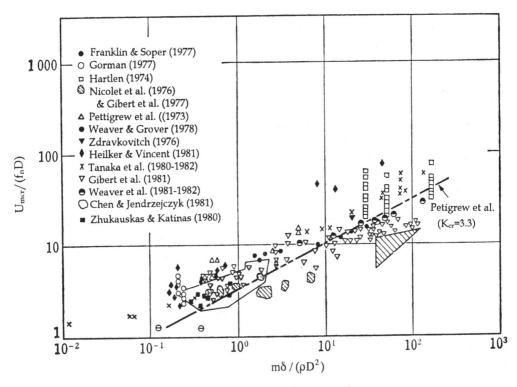

Figure 3.44 Accumulation of experimental data of fluid-elastic vibration limit for cross-flow tube banks (Paidoussis 1982).

quality. It may be said that one of the most important tasks to be overcome for improving fuel cost has been the struggle against corrosion problems, especially gas-side corrosion of (HTSs). A key way to achieve a higher temperature and higher pressure system for improving the thermal efficiency of steam power plants is to develop heat- and corrosion-resistant materials for superheaters and reheaters. As with the improvement of boiler efficiency, it is important to develop countermeasures against low-temperature fouling and corrosion of heat recovery sections such as air heaters. It goes without saying, also, that the use of low-grade fuel is largely dependent on the ability to overcome corrosion problems.

The struggle against corrosion is critical not only in steam plants but also in diesel plants and gas turbines that also use fossil fuels. By the mid-1950s, two epoch-making technologies had been developed for the marine diesel engine. One was to put alkalized lubricant into service for cylinder lubrication, and the other was the practical application of chrome-plating technology for protecting the cylinder wall. Both were very effective against corrosive attack from combustion products of heavy fuel oil. The diesel engine was inherently superior to the steam turbine with respect to specific fuel consumption. Adding to this superiority, the diesel engine was now able to use low-grade fuel oil as a result of this technology. Since then, the diesel has gained a definite advantage over the steam turbine for marine engines. In the case of the gas turbine, since its blades are exposed directly to high temperature and high-speed gas flow, their corrosion liability is so severe that fuel quality has to be controlled very carefully. In the 1970s, desperate efforts were made to develop a gas turbine that could use heavy fuel oil, with the intention of applying the gas turbine to marine engines. But the struggle against high-temperature corrosion was too difficult and this idea could not end in success.

The struggle against corrosion will still continue in the future in various technological fields such as materials for use in power plants. It should be kept in mind, however, that this outlook is based on the premise that the heat engine continue to occupy its dominant position and that fossil fuels be used as energy sources. There remains the possibility of drastic change in the fuel and/or energy situation according to the future trend of environmental problems.

The general process by which the gas side of an HTS becomes fouled and corroded is shown in Figure 3.45. Environmental pollution due to various products of fuel combustion is increasingly becoming a very serious problem. These products also give various types of trouble with boiler equipment before being discharged to the environment. The characteristics of combustion products are dependent on individual fuel quality; similarly, the characteristics of fouling and corrosion are also dependent on the individual fuel. Fouling and corrosion problems in the cases of oil and coal combustion are discussed here briefly according to the process shown in Figure 3.45.

3.6.2 Fouling and corrosion of oil-fired boilers

3.6.2.1 General introduction

Table 3.4 shows the average chemical composition of deposits collected from various points of boiler HTSs. The following can be read from the table.

- Almost all deposits are composed predominantly of compounds of sodium (Na), sulfur (S), and vanadium (V). Although iron constituents are found in some deposits, these are considered not to be combustion products but corrosion products of the tube material. Thus the main impurities of heavy fuel oil having a strong effect on the fouling of boiler HTSs are sodium, sulfur, and vanadium.
- Although the sodium and vanadium content is much less, on the relative order of 10^{-3}, than that of sulfur content, the deposited amount of sodium and vanadium on HTSs is nevertheless on the same order as that of sulfur.
- Deposits on high-temperature HTSs contain rather high proportions of sodium and vanadium, but the content of these elements is much lower in deposits on low-temperature HTSs.

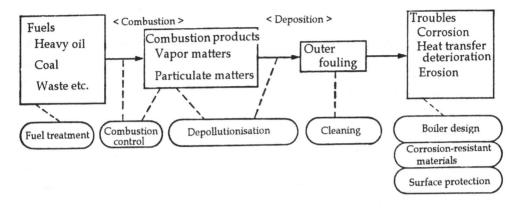

Figure 3.45 General view of outer fouling and problems.

Table 3.4. *Typical examples of chemical composition of oil-fired boiler deposits[a]*

Component		① Furnace bottom	② Upper wall of furnace	③ Plate type SH	④ First SH	⑤ Hanging type RH	⑥ Hanging type 2nd RH	⑦ Horizontal RH	⑧ Econo-mizer	⑨ Air heater
pH (1 g/100 ml H_2O)		6.6	7.1	4.2	3.3	3.4	3.4	2.4	2.4	2.0
Water soluble	(%)	99.3	68.8	68.9	57.6	41.6	52.2	57.6	62.5	71.8
Total sodium as Na_2O	(%)	40.5	26.7	26.4	21.0	17.8	18.9	14.0	8.3	3.1
Total sulfur as SO_3	(%)	54.3	35.6	37.7	32.9	22.9	27.5	33.3	34.1	35.2
Total vanadium as V_2O_5	(%)	1.4	20.7	28.5	21.9	45.2	31.4	23.5	31.6	10.5
Total iron as Fe_2O_3	(%)	0.1	4.3	2.4	11.7	9.2	10.0	24.3	7.7	6.1
Others	(%)	3.7	12.7	5.0	12.5	4.9	12.2	4.9	17.5	54.9

[a] Fuel impurities: S, 1.5–2.0%; Na, 10–30 ppm; V, 40–50 ppm.

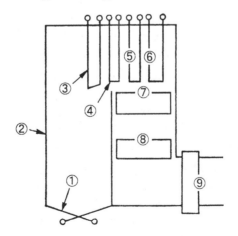

- As for the amounts deposited on HTSs, in general the fouling of HTSs in the intermediate temperature region is less than in the high- or low-temperature region. The
intermediate temperature region designated here refers to temperatures of HTSs
ranging roughly from 200 to 350°C, such as in the low-temperature section of SH and
RH, a generator situated just downstream of the SH and/or RH, the high-temperature
section of the economizer, and so on.

Summarizing the above, the characteristics of fouling and corrosion are strongly
dependent on the temperature: the deposits in the high-temperature region
contain high proportions of sodium and vanadium; the deposits in the low-
temperature region contain predominantly the sulfur component; and, at intermediate temperatures, the fouling is generally small. With regard to these ten-
dencies, fouling phenomena of the oil-fired boiler are characterized as
high-temperature fouling and low-temperature fouling; or, viewed from the
standpoint of corrosion, high- and low-temperature corrosion are called vanadium attack and sulfuric corrosion, respectively.

3.6.2.2 Fuel impurities and combustion products

Experiments using test furnace and chemical analysis of deposits have shown that
the main substances depositing onto HTSs from combustion gas are sulfuric acid
in the low-temperature region and sodium sulfate (Na_2SO_4) and vanadium oxide
(V_2O_5) in the high-temperature region. These compounds are therefore considered to be the combustion products involved in the gas-side fouling of HTSs.

Low-grade heavy fuel oil generally contains sodium components amounting to
several tens of parts per million, mainly in the form of NaCl, and vanadium and
sulfur components in the form of hydrocarbonaceous compounds. The vanadium
content ranges on the order of ten to hundreds of parts per million, and that of
sulfur is on the order of a few percent. Those impurities are converted to other
compounds through reactions in the combustion furnace as follows. Sulfur is oxidized to SO_2, a small part of which is further oxidized to SO_3. The conversion
ratio of SO_2 to SO_3, which is on the order of several percent, is influenced by combustion conditions, air excess ratio, existence of catalysts, and so on. The concentration of SO_3 in the flue gas is on the order of 10 ppm. Sodium is converted first
to its oxide Na_2O, which is further converted to Na_2SO_4 through the reaction with
SO_3. Since the Na_2O concentration is ordinarily much less than that of SO_3, the
concentration of Na_2SO_4 is determined by the quantity of sodium. Vanadium is
oxidized to V_2O_5. When the air excess ratio is very small, say, less than 1.05, the
oxidation does not proceed to the final stage but stops at the lower stage, producing V_2O_4 and so on. The general combustion products and substances depositing onto HTSs of oil-fired boilers are summarized briefly in Table 3.5, where
special combinations of impurities are also shown.

Table 3.5. *Main combustion products including outer fouling of heavy oil-fired boiler (Nishikawa 1979)*

Fuel impurities	Combustion products	Deposition substance	
		High-temperature zone	Low-temperature zone
Na	Na_2CO_3, $NaCl^a$	Na_2CO_3, NaCl	No deposition
S	SO_3^b	No deposition	SO_3
V	V_2O_5	V_2O_5	No deposition
Na, S	Na_2SO_4, SO_3^c	Na_2SO_4	SO_3
S, V	Mainly SO_3, V_2O_5	V_2O_5	SO_3
Na, V	$NaVO_3$, NaCl, Na_2CO_3, $V_2O_3^d$	$NaVO_3$, NaCl, Na_2CO_3, V_2O_3	No deposition
Na, S, V	SO_3, Na_2SO_4, V_2O_5	Na_2SO_4, V_2O_5	SO_3

[a] It is supposed that heavy oil contains Na in the form of NaCl.
[b] Most sulfur oxides are in the form of SO_2 and only 1–10% are in the form of SO_3.
[c] It is supposed that the SO_3 content is large enough to convert all of the Na_2O into Na_2SO_4.
[d] V_2O_5 would not be produced when the Na content is high enough, and, contrary to that, when the V content is sufficient, NaCl and Na_2CO_3 would not be found.

3.6.2.3 The deposition mechanism of combustion products

How do Na_2SO_4 and V_2O_5 deposit onto high-temperature HTSs, and SO_3 onto low-temperature HTSs? The following is considered to be a possible mechanism causing deposition (Nishikawa 1979):

vaporous substance → vapor diffusion and condensation

$$\text{particulate substance} \rightarrow \begin{cases} \text{inertial collision} \\ \text{particle diffusion} \begin{cases} \text{concentration diffusion} \\ \text{thermal diffusion} \end{cases} \end{cases}$$

Among the above, vapor diffusion and condensation are considered to be the dominant mechanisms for oil-fired boilers.

Inertial collision means that particles suspended in the flowing gas collide with the HTS and attach to it due to their inertial force. When particles get into the boundary layer of the HTS, their velocity is decelerated and their inertial force is decreased, so that there exists a certain critical diameter of particle below which the inertial force is too small to allow it to collide with the HTS. The critical diameter is affected by gas velocity, shape of collision surface, particle density, and so on. For instance, in the case that the gas velocity is 10 m/s and the HTS is a tube of diameter 38.1 mm, the critical diameter is on the order of $10\mu m$. The size of particulate substances suspended in the combustion gas of an oil-fired boiler is generally smaller in diameter than $10\mu m$, whereas the combustion gas of a diesel engine contains soot particles of larger diameter, and therefore careful attention should be paid to fouling by inertial collision.

When vapor diffusion is the dominant mechanism, the deposition rate $\dot{m}$ can be estimated approximately on the basis of mass transfer theory (Giedt 1957). That is,

$$\dot{m} \cong \frac{\text{Nu}}{\text{Pr}^{1/3}\text{Sc}^{2/3}} \cdot \frac{\eta}{D} \cdot \left(M_G - M_S\right) \qquad (3.35)$$

where Pr is the Prandtl number of the combustion gas, Sc is the Schmidt number relating to the depositing vaporous substance, η is the viscosity coefficient of the combustion gas (Pa s), Nu is the Nusselt number, D is the outer diameter of the heat transfer tube (m), M_G is the mass fraction of the depositing vaporous substance in the combustion gas (if M_G is larger than the saturated mass fraction M_G^* corresponding to the combustion gas temperature, then $M_G = M_G^*$, and M_S is the saturated mass fraction of the depositing vaporous substance at the temperature of the deposition surface, that is, the heat transfer tube surface.

The Nusselt number of the tube concerned can be estimated by Eq. (3.15). To obtain Sc, the diffusion coefficient of the vapor of the depositing substance is required. When it is unobtainable, an approximate value can be estimated by the use of the following empirical relation, proposed by Sherwood and Pigford (1952), regarding a small concentration of vapor mixed with air. That is,

$$\text{Sc} = 0.145 \, M^{0.556} \qquad (3.36)$$

where M is the molecular mass of the vaporous substance.

Figure 3.46 shows the saturated vapor pressure curves of Na_2SO_4 and V_2O_5. The curves are different for individual investigators due to the difficulties of measurement. Figure 3.47 shows the data for SO_3 in the form of a relation between concentration and dewpoint (Nakashima 1985). As can be seen in both figures, the saturated vapor pressure is very sensitive to atmospheric temperature and decreases rapidly with temperature. When the temperature of the deposition surface decreases slightly, therefore, the value of M_G^* in the thermal boundary layer decreases rapidly, and as a result the deposition rate decreases rapidly.

As suggested by Eq. (3.35), when vapor diffusion is dominant, the deposition rate becomes zero when the temperature of the HTS is above the saturation temperature corresponding to the vapor pressure of the vaporous substance contained in the combustion gas. Figure 3.48 shows some experimental results for the

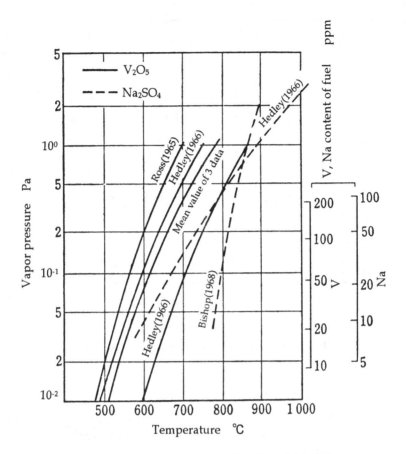

Figure 3.46 Vapor pressure of V_2O_5 and Na_2SO_4.

relation between deposition rate and deposition surface temperature obtained with test furnaces simulating oil-fired boilers. These figures confirm that the dominant mechanism of deposition in the case of the oil-fired boiler is vapor diffusion and condensation.

Judging from the saturated vapor pressure curves shown in Figures 3.46 and 3.47 and the general concentration levels of V_2O_5, Na_2SO_4, and SO_3 vapors contained in the combustion gas of heavy fuel oil, the saturation temperature of each vapor can be estimated to be on the order of 600, 700–800, and 150°C, respectively. The former two temperatures are on the same level as the surface temperature of the SH and RH, and the latter is on the same level as that of the low-temperature HTSs. This is why the fouling of oil-fired boilers is characterized by the designations of high-temperature fouling and low-temperature fouling, and why fouling in the intermediate temperature region is not as serious as high- or low-temperature fouling.

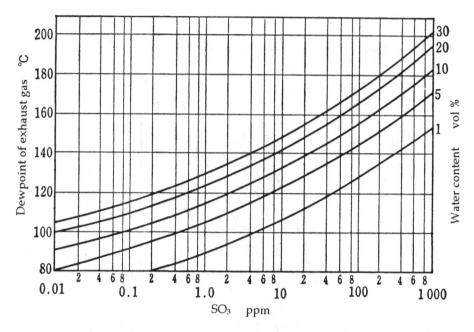

Figure 3.47 Variation of dewpoint with SO_3 and water contents in exhaust gas.

3.6.2.4 High-temperature corrosion by V_2O_5 and Na_2SO_4

The corrosion of low-temperature HTSs is mainly due to attack by sulfuric acid. The mechanism is not very complicated and is already well known. So let us proceed to discuss high-temperature corrosion.

(a) *Vanadium attack*: When V_2O_5 and Na_2SO_4 coexist, they interact and form various sodium vanadates, the type of which is dependent on the conditions of temperature and the mass ratio of the original substances. Table 3.6 shows what kinds of compounds are formed by the Na_2SO_4–V_2O_5 system corresponding to the conditions just mentioned. Table 3.7 shows the melting points of these compounds. Their melting points are not so high that they would easily attain the molten state on SH and/or RH tubes. These sodium–vanadium compounds are strongly corrosive, for the following reasons:

i. sodium–vanadium compounds absorb oxygen from the atmosphere and supply it to the surface of the tube metal;
ii. sodium–vanadium compounds dissolve oxides of the tube metal, especially iron oxide, so that the iron oxide layer cannot grow thick enough to be a protective coating against corrosion, and accelerative corrosion is induced.

Figure 3.49 shows some experimental results on the effect of the mixing ratio of Na_2SO_4 and V_2O_5 on the corrosion rate and oxygen absorption rate. As is seen, the change of corrosion rate is almost the same as that of the oxygen absorption

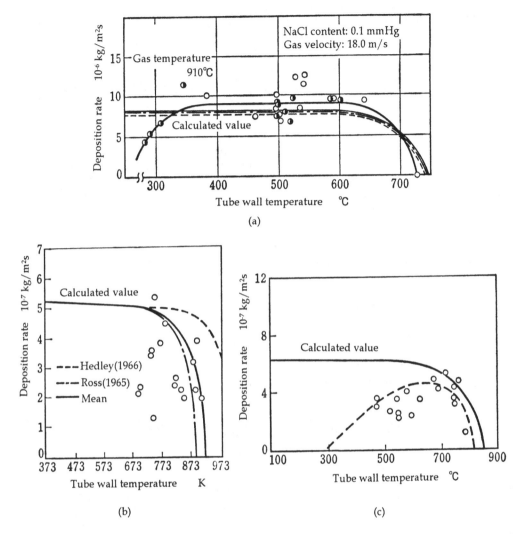

Figure 3.48 Deposition rate of some typical substances of high-temperature fouling of oil-fired boilers (vapor pressure data from Fig. 3.46). (a) NaCl; (b) V_2O_5; (c) Na_2SO_4.

rate. The mixing ratio showing the strongest corrosiveness is read from the figure as being around 20%, at which ratio the compound "NV_6" ($Na_2O \cdot V_2O_4 \cdot 5V_2O_5$) is formed, as shown in Table 3.6. This is why an artificial mixture of 20 wt% Na_2SO_4 and 80 wt% V_2O_5 is often used in experimental models simulating vanadium attack.

It is not easy to estimate quantitatively the corrosion rate of actual boiler tubes. Figure 3.50 shows an example of measurements carried out in an actual boiler of a power station (Combustion Engineering Marine Power Systems 1976). In the figure, the corrosion rate is shown as the rate of decrease of tube wall thickness.

Table 3.6. *Various compounds formed in* Na_2SO_4–V_2O_5 *system (Pollmann 1965)*

(a) Effect of mixing ratio

| | Mixing ratio (mole% of Na_2SO_4) | | |
Compound[a]	Equivalent ratio	Range of actual formation	Range of pure compound formation
V_2O_5	0	0–10.5	0–0.5
NV_6	14.3	0.5–29	10.5–15
N_5V_{12}	29.4	15–46	29–31
$NaVO_3$	50	31–95	46–55
Na_2SO_4	100	55–100	—

(b) Effect of temperature

Temperature (°C)	Compound
≤500	No reaction
550–700	$NV_6 + Na_2SO_4$
800–1000	$NV_6 + N_5V_{12} + Na_2SO_4$
1000≤	$NaVO_3 + Na_2SO_4$

[a] NV_6 and N_5V_{12} express $Na_2O \cdot V_2O_4 \cdot 5V_2O_5$ and $5Na_2O \cdot V_2O_4 \cdot 11V_2O_5$, respectively.

Table 3.7. *Melting point of various compounds formed in oil-fired boilers*
(Harada et al. 1980)

Compound	Melting point (°C)	Compound	Melting point (°C)
V_2O_4	1970	$3Na_2O \cdot V_2O_5$	850
V_2O_5	690	$10Na_2O \cdot 7V_2O_5$	573
$Na_2O \cdot V_2O_5$	630	$Na_2O \cdot V_2O_4 \cdot 5V_2O_5$	625
$Na_2O \cdot 3V_2O_5$	560	$5Na_2O \cdot V_2O_4 \cdot 11V_2O_5$	535
$2Na_2O \cdot V_2O_5$	640	$NaVO_3$	560
$2Na_2O \cdot 3V_2O_5$	565	Na_2SO_4	884

The figure shows also the influences of vanadium content of heavy fuel oil and air excess ratio, and shows further the anti-corrosion effect of magnesium additive.

(b) *Sulfide attack by* Na_2SO_4: Na_2SO_4 is dissociated by some reducing agents such as elemental metals, and liberated elemental sulfur reacts with the tube metal to form sulfide. The liberated sodium oxide reacts with carbon

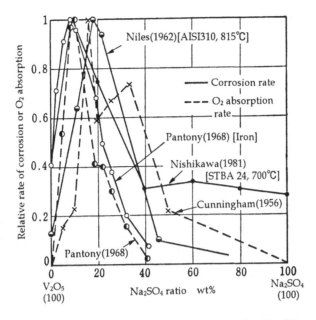

Figure 3.49 Corrosivity and O_2 absorptivity of V_2O_5–Na_2SO_4 mixture.

dioxide and forms sodium carbonate (Na_2CO_3). When the sulfide and tube metal coexist, they interact and form a eutectic mixture having a low melting point. Thus, if such a eutectic mixture is melted onto the tube surface, accelerative corrosion will take place. The process of sulfide attack just mentioned has long been seen to be important for gas turbine blades, for which temperature conditions are very severe. Harada et al. (1980) have pointed out that sulfide attack also damages high-temperature HTSs in boilers. As can be judged from Figures 3.46 and 3.48, Na_2SO_4 deposits on higher-temperature HTSs compared with V_2O_5, so that attention should be paid to sulfide attack as the tube temperature rises.

3.6.3 Fouling and corrosion of coal-fired boilers

3.6.3.1 Impurities and combustion products

Coal contains various impurities, called ash components, the characteristics and amounts of which are dependent on the location of the coal mine. The main chemical elements are as follows.

Sulfur (S): Sulfur is contained mainly in the form of iron sulfides, chiefly FeS_2, organic compounds combined with the original coal substances, and sulfates such as $CaSO_4$ and $FeSO_4$. FeS_2 is usually decomposed into iron oxides and sulfur oxide

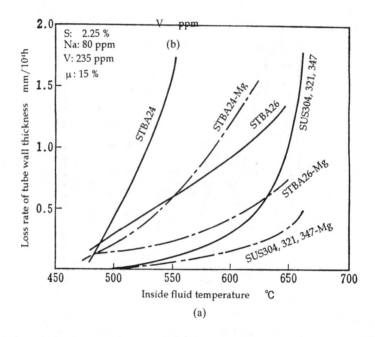

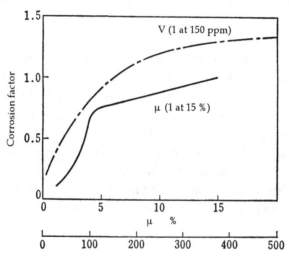

Figure 3.50 Proposal for estimation of high-temperature corrosion of oil-fired boilers. (a) Influence of temperature (parameter "Mg" means the results when the additive MgO was used by mole ratio 2 to V content); (b) corrosion factor for the loss rate of (a) (CE Marine Power Systems 1976).

in the combustion process. When it is burned in a reducing atmosphere due to a deficient air supply, however, the decomposition does not proceed completely and FeS is produced. FeS and iron interact and form a eutectic mixture having a low melting point, so that slagging often occurs and causes problems in the fire bed in the case of stoker firing. If the air deficiency is too much, there is an undesirable possibility that FeS will hardly react at all and deposit directly on the water wall tubes, inducing sulfide corrosion. Such a possibility should be kept in mind, for instance, when two-stage combustion is adopted for preventing NO_x production. In general, though, the sulfur component of FeS is released completely and reacts to form SO_2, which is further oxidized in part to SO_3 and induces low-temperature fouling and corrosion.

Alkali metal elements (Na, K): These are the most important impurities with regard to fouling. They react with SO_3, as mentioned before, and produce sulfate compounds. In the case of coal containing very little sulfur, they form carbonate compounds, as already shown in Table 3.5, which also deposit onto HTSs. Sodium is contained almost entirely in the form of NaCl, so that NaCl deposits directly onto HTSs when the SO_3 content is too low to change all of the NaCl to sodium sulfate.

Phosphorus (P): Most of the phosphorus contained in coal is believed to exist as the fluoride of calcium phosphate, $Ca_5F(PO_4)_3$. This changes to the stable calcium phosphate in an oxidizing atmosphere. But in a reducing atmosphere, such as in stoker firing, it changes to a phosphorus oxide having a low melting point, which induces slagging troubles in some cases. Phosphorus would contribute, therefore, very little to fouling in the case of pulverized coal firing.

Chlorine (Cl): Almost all chlorine is contained as NaCl. Some British coals have high contents of chlorine. Crossley (1946) investigated the fouling characteristics of many power generation boilers in the United Kingdom, and made clear that there was a strong relation between fouling and chlorine content. Since then, chlorine content has been adopted as one of the indicators for estimating the extent of fouling. As discussed above in this text, however, the important element taking the role of high-temperature fouling and corrosion is sodium rather than chlorine, so that NaCl contained in coal should be recognized as a supplier of sodium. It could be said, in reality, that the effect of chlorine pointed out by Crossley suggested indirectly the effect of sodium on fouling.

Other elements (Si, Al, Ti, Fe, Ca, Mg, etc.): Most of the impurities in coal, that is, the ash component, consist of the elements silicon, aluminum, titanium, iron, calcium, magnesium, and so on, and therefore fly ash and slag, which are main ingredients of combustion products, are also composed of them. Characteristics such as melting point and viscosity of slag and ash are dependent on the composition of these elements.

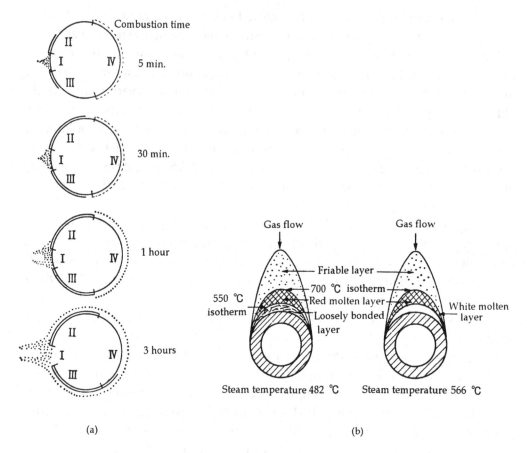

Figure 3.51 Growing process and general structure of high-temperature fouling of pulverized coal-fired boilers. (a) Initial growing process of fouling observed by the model furnace test (Bishop 1968) (tube wall temperature 550°C, gas flows from left). (b) Sketch of fouling structure of SH tube (Borio et al. 1978).

3.6.3.2 Deposition mechanism of combustion products onto HTSs

The general structure of high-temperature fouling on HTSs of pulverized-coal-fired boilers is as shown in Figure 3.51. Bishop (1968) made a detailed observation of the very beginning of the process of fouling growth by the use of a model furnace. Figure 3.51a shows a sketch of his observation, and Table 3.8 shows the chemical composition of individual layers of deposits. The first layer, counted from the tube surface, consists mainly of Na_2SO_4 and K_2SO_4, the depositing mechanism of which is vapor diffusion, as with oil-fired boilers. The second layer consists of these sulfates and fly ash. If Na_2SO_4 and K_2SO_4 coexist they form a eutectic mixture having a low melting temperature. The eutectic mixture becomes molten when the layer thickness increases and the temperature of the outer surface of the layer reaches the melting point.

Table 3.8. *Chemical composition of deposits of Fig. 3.51a (Bishop 1968)*[a]

Deposit	Chemicals (wt%)											
	SiO_2	Al_2O_3	Fe_2O_3	CaO	MgO	Na_2O	K_2O	TiO_2	Mn_3O_4	P_2O_5	SO_3	Cl
Upstream side 1st layer (Zones II, III)	6.6	4.9	7.0			28.0	5.5				41.8	—
Upstream side 2nd layer (Zones II, III)	16.8	10.7	18.5	4.2	1.3	15.9	3.6	0.4	0.4	—	24.2	—
Downstream side (Zone IV)	16.3	12.1	11.9	4.0	1.0	18.1	4.5	0.5	0.2	—	27.1	—
Ash content	31.9–38.0	22.8–25.4	12.8–17.7	6.0–7.0	1.2–1.7	6.6–7.3	1.0–1.1	1.0–1.3	0.1–0.2	0.4–0.9	4.6–8.2	—

[a] Coal properties: Ash, 4.4–5.0%; S, 0.9%; Cl, 0.86–0.9%.

Fly-ash particles collide with HTSs by their inertial force. According to Moza et al. (1980), these colliding fly-ash particles can adhere and deposit on HTSs when one or more of the following conditions are satisfied, that is:

• the ash particle is in a molten state;
• the HTS, that is, the deposition surface, is covered by some molten salt;
• the deposition surface is covered with some substance that becomes molten the instant an ash particle collides with the surface.

Judging from these conditions, the first and second layers formed by deposition of Na_2SO_4 and/or K_2SO_4 in the molten or nearly molten state would contribute to catching colliding particles. Similarly, sulfates of sodium and potassium and fly ash deposit onto the HTS and develop the second layer further. As the second layer increases in thickness and its surface temperature rises beyond the saturation temperature of Na_2SO_4 and K_2SO_4, these sulfates would stop condensing onto the HTS. After that, only fly ash would be deposited and would grow a third layer.

Figure 3.51b shows a typical example of actual boiler fouling (Borio et al. 1978). Looking at the structure of deposited layers, it can be seen that the fouling of actual boilers develops along the lines of the process described above. As stated, alkali metal contents (sodium and potassium) play a definite role in the fouling of coal-fired boilers.

3.6.3.3 Effects of ash composition on fouling

As pointed out by Moza et al. (1980), serious fouling will be developed if the fly ash carried by the gas flow is in a molten state. The combustion gas temperature at the inlet of the SH and/or RH section should therefore be designed with careful attention to the melting characteristics of fly ash. Methods or indicators have been proposed to estimate the fouling characteristics of coal ash (e.g., ASTM Standards; Attig and Duzy 1969; Albrecht and Pollmann 1980).

3.6.3.4 Corrosion of coal-fired boilers

Low-temperature corrosion is mainly caused by sulfuric acid under the same mechanism as in oil-fired boilers. Therefore only high-temperature corrosion will be described here. There are two main mechanisms: sulfidation attack due to alkali sulfates of sodium and potassium, and corrosion caused by complex compounds of the sulfates $Na_3Fe(SO_4)_3$, and $K_3Fe(SO_4)_3$, which are formed by alkali sulfates and Fe_2O_3 contained in fly ash. For coal-fired boilers, the latter mechanism is more important than the former.

Na_2SO_4 and/or K_2SO_4 and Fe_2O_3 deposited onto HTSs react with SO_3 contained in combustion gas and form complex compounds, as described by the following equation:

$$3Na_2SO_4 + Fe_2O_3 + 3SO_3 \rightleftharpoons 2Na_3Fe(SO_4)_3 \qquad (3.37)$$

These complex compounds are apt to become molten due to their low melting points and attack the iron material of HTSs, that is,

$$10Fe + 2Na_3Fe(SO_4)_3 \rightleftharpoons 3Fe_3O_4 + 3FeS + 3Na_2SO_4 \qquad (3.38)$$

FeS in the right-hand side of the above equation is reduced to Fe_3O_4 and SO_3 by further reactions, and Na_2SO_4 repeats the reaction process of Eq. (3.37). Thus complex compounds of $Na_3Fe(SO_4)_3$, $K_3Fe(SO_4)_3$, and FeS advance the above corrosion process cyclically. Figure 3.52 shows an example of some corrosion test results, by which the corrosion mechanism described above can be confirmed. The complex compounds are not formed in the high-temperature region beyond 1300°F (704°C). That is why the corrosion rate decreases abruptly when the temperature reaches this level, as shown in the figure.

3.6.3.5 Erosion

Erosion due to fly ash is one of the gas-side problems troubling coal-fired boilers. Solid or liquid particles in the gas flow collide with the HTS and/or structural members and erode or injure those surfaces mechanically. Quantitative estimation of erosion is not easy since it is difficult to clarify the mechanism of erosion. A general relation has been derived empirically as follows (Hutchings 1979):

$$E = k \cdot V^n \cdot f(\theta) \qquad (3.39)$$

where E is the mass ratio of eroded material to the total mass of colliding particles and V is the velocity of a colliding particle. The constant k, which is almost independent of material properties, takes a value ranging from 1 to 10, and the

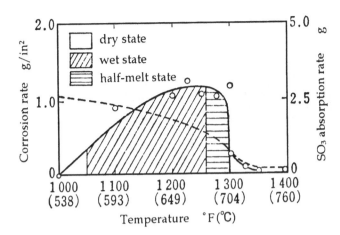

Figure 3.52 Temperature effect on corrosivity of $Na_3Fe(So_4)_3$ sample. Test specimen: TP-321. (Solid line) Corrosion rate; (dashed line) absorption rate of SO_3 (Chain and Nelson 1961).

exponent n takes a value ranging from 1 to 2 when V is measured in m/s. $f(\theta)$ is an non-dimensional factor expressing the effect of the collision angle θ, with a maximum value of 1. $f(\theta)$ becomes maximum at about $\theta = 20°$ for ductile materials and 90° for brittle materials.

The design criterion for actual boilers is estimated from the gas velocity, as shown for instance in Figure 3.53. This figure shows the change in design gas velocity adopted by the Babcock and Wilcox Company (Isoda et al. 1982). The design velocity has become lower with time, possibly with a view to improving reliability against erosion. The gas flow passage of a boiler is very complicated, so there may exist some spots where local velocity and/or particle content are higher than the average level. Attention should be paid to those spots to avoid erosion. Figure 3.54 shows some local areas that are likely to be attacked by erosion (Shiroma 1962).

There is a possibility that a sootblower induces erosion troubles in coal-fired boilers since its strong jet entrains and accelerates the surrounding gas containing fly ash (Tanaka 1980a).

3.6.4 Impact of fouling on heat transfer

There can be found only a few papers and/or reported data so far regarding how gas-side fouling affects the performance of HTSs. According to the measurement results of a report referred to in Reid (1971), which measures how the heat transfer characteristics of a water wall of a coal-fired boiler change gradually from the beginning when the water wall surface is in a very clean state, the heat flux of the

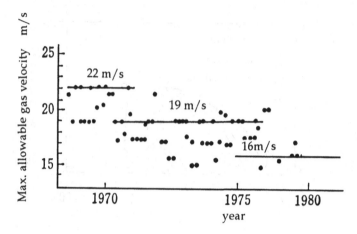

Figure 3.53 Example of actual design gas velocity by Babcox and Wilcox Company (Isoda et al. 1982).

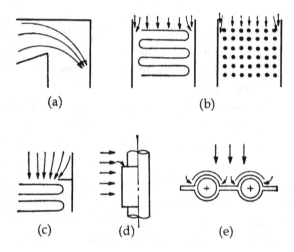

Figure 3.54 Some typical areas apt to be attacked by erosion. (a) Particles concentrating due to centrifugal force; (b) velocity increasing due to low flow resistance; (c) flow intensifying due to baffle plate; (d) particles concentrating at protector end; (e) particles concentrating at open edge of protecting cover (Shiroma 1962).

water wall deteriorates rapidly at first after starting operation, and then the deterioration rate slows down gradually and approaches final equilibrium 6 hours or so after starting up. The heat flux attained at equilibrium was measured at about 60% of that in the clean state. Mulcathy et al. (1966) carried out detailed measurements of thermal properties such as radiant emissivity and heat conductivity of gas-side deposits of a boiler fired with pulverized Australian coal. They carried out further the estimation of change of heat flux and furnace outlet gas temperature by the use of a homogeneous simulation model, which they themselves had developed, and obtained data on the thermal characteristics of the deposits. According to their simulation results, the heat flux deteriorates rapidly due to a very small amount of deposit, and finally decreases to about 60% of the value in a clean state.

3.6.5 Countermeasures against gas-side fouling and corrosion

Various measures have been developed and devised corresponding to each step of fouling and/or the corrosion process, as shown in Figure 3.45.

3.6.5.1 Excess air control

Low-temperature fouling and corrosion, and also Na_2SO_4 production, are mainly dependent on the content of SO_3 that is produced in the combustion area. To inhibit SO_3 production by low excess air combustion, therefore, would be an effective means of preventing gas-side fouling and corrosion. Low excess air combustion is also effective against V_2O_5 production. Figures 3.55 and 3.56 show the effects of low excess air combustion on the formation of SO_3 and V_2O_5 formation, respectively. Looking at these figures, it is clear that valuable effects cannot be expected unless the excess air ratio is lowered to smaller than 1.05, which corresponds to an excess O_2 amount of 1%. This requirement of 1.05 can also be read out from Figure 3.50b. Very strict combustion control would be required regarding various items such as air distribution to each burner and fuel supply control, to keep the combustion steady at such a very low excess air level, while at the same time preventing soot formation.

3.6.5.2 Additives

Various additives have been tried and/or used practically for many years to change harmful combustion products to other, chemically harmless compounds, and to prevent fouling and corrosion. Among them, magnesium chemicals such as MgO and $Mg(OH)_2$ have been recognized both practically and theoretically as being the most effective, especially for high-temperature fouling and corrosion. When magnesium chemicals are added to the combustion zone of the furnace, they react with vanadium oxides and form magnesium vanadates such as $3MgO \cdot V_2O_5$, which has a very high melting point of 1191°C. Magnesium chemicals also react with SO_3 and convert it to harmless magnesium sulfate, $MgSO_4$.

To attain the desired effects of magnesium additives, however, careful attention should be paid to the selection of an appropriate method of injection into the combustion zone. It is recommended to spray magnesium additives into the

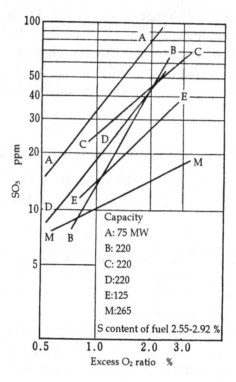

Figure 3.55 Effect of excess oxygen control on SO_3 formation (some data of actual boilers; Harada et al. 1980).

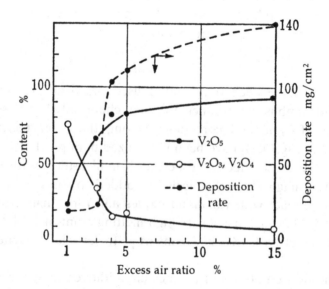

Figure 3.56 Effect of excess air ratio on vanadium oxidation and deposition rate (Harada et al. 1980).

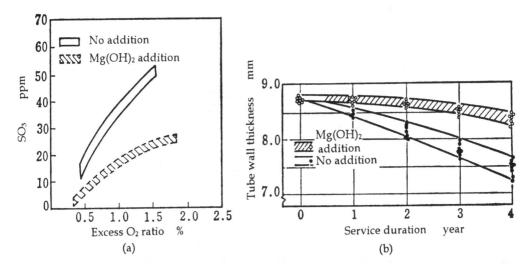

Figure 3.57 Effect of Mg(OH)$_2$ additive. Actual results for oil-fired boilers of power generation. (a) Decrease of SO$_3$ formation; (b) decrease of corrosion of SH tube made of SUS321 (Harada 1986).

combustion zone in the form of fine particles. The amount required is about three times the molar ratio of magnesium to vanadium (Nishikawa 1983). Harada (1986) developed a magnesium hydroxide slurry method and applied it successfully to many large oil-fired boilers for power generation. The method is that a slurry of Mg(OH)$_2$ is added into the fuel line just before the burners so as to atomize the slurry into the combustion zone together with fuel. Figure 3.57 shows the effect of this method. It should be noted, however, that magnesium additives are effective mainly for oil-fired boilers but not as effective for coal-fired boilers, especially in the temperature range beyond 700°C where the corrosion caused by alkali sulfate becomes dominant. No effective additive has yet been developed for coal-fired boilers.

3.6.5.3 Sootblowers

The sootblower is a well-known device, intended to blow off the deposits on HTSs by steam or air jets. The performance of a sootblower is affected by various factors in a complicated manner. Generally the performance is proportional to the square of the jet nozzle diameter, proportional to jet pressure, and inversely proportional to the nth power of the distance between the jet nozzle and the HTS. The nth is in the range from first to second. As with the sootblower for tube banks, it is critically important whether or not the sootblower can blow the deposits off the first row of tubes, because particulate deposits blown off the first row collide against tubes in successive rows and clean up the deposits on them.

In the case of coal-fired boilers, ash deposition is very severe, and powerful sootblowers are therefore used.

3.6.5.4 Corrosion-resistant materials for HTS protection

(a) *Chromium*: Chromium is an excellent resistant metal against gas-side corrosion, especially vanadium attack. The higher the chrome content, the more resistant the material is, so the best measure would be to use chrome steel for high-temperature HTSs.

(b) *Coating the HTS*: The most successful application of HTS coating has been the enamel coating of the HTS of a regenerative gas air heater, which may have improved the boiler efficiency. For high-temperature HTSs, thermal spraying techniques to build a resistant layer have been tried by coating the HTS with a ceramic substance such as aluminum oxide or chromium oxide. Covering the upstream side of the high-temperature tube with a metal sheet made of highly resistant alloy has also been tried. The sheet is intended to work as a protector.

References

Adlhoch, W. 1993. Rheinbraun's HTW process to gasify coal in a fluidized bed. 2nd World Coal Institute Conference, pp. 1–24.

Albrecht, W., and S. Pollmann, 1980: *VGB Kraftwerstechnik* 60, no. 2: 90–7.

Attig, R.C., and A.F. Duzy. 1969. Proceedings of American Power Conference, p. 290.

Bishop, R.J. 1968. *Journal of the Institute of Fuel* 41: 51–65.

Bohn, H. 1994. *VGB Kraftwerkstechnik* 74, no. 3: 151–64.

Borio, R.W., A.L. Plumley, and W.R. Sylvester. 1978. *Ash deposits and corrosion due to impurities in combustion gases*, ed. R.W. Bryers. New York: Hemisphere, pp. 163–83.

Buschmann, M., R. Bischler, H. Kowatsch, et al. 1994. *VGB Kraftwerkstechnik* 74, no. 12: 880–3.

Combustion Engineering Marine Power Systems. 1976. Improved Marine Boiler Reliability TASK I.

Chain, C., Jr., and W. Nelson. 1961. *Transactions of the American Society of Mechanical Engineers, Journal of Engineering for Power* 83: 468–74.

Chen, Y.N. 1968. *Transactions of the American Society of Mechanical Engineers*, Ser. B 90, no. 1: 134–46.

Chen, Y.N. 1974. *Transactions of the American Society of Mechanical Engineers*, Ser. B 96, no. 3: 1072–5.

Cho, S.M. 1994. *Chemical Engineering Progress* 90, no. 1: 39–45

Connors, H.J., Jr. 1970. Proceedings of Flow-Induced Vibration in Heat Exchangers, American Society of Mechanical Engineers, 42–56.

Crossley, H.E. 1946. Research Work at the Fuel Research Station, Department of Scientific and Industrial Research on Boilers, paper no. 4.

Eitz, A.W. 1995. *VGB Kraftwerkstechnik* 75, no. 7: 540–5.

Endo, Y., S. Miyamae, and Y. Ando. 1994. *Ishikawajima-Harima Engineering Review* 34, no. 2: 93–100.

Fitz-Hugh, J.S. 1973. International Symposium on Vibration Problems in Industry, Keswick, England, paper no. 427.

Funakawa, M. 1978. Preprint of the JSME, no. 784/5: 101–4

Giedt, W.H. 1957. *Principles of engineering heat transfer.* D. van Nostrand.

Gutberlet, H. 1994. *VGB Kraftwerkstechnik* 74, no. 1: 54–9

Harada, Y., S. Naito, T. Tsuchiya, et al. 1980: *Mitsubishi Juko Giho* 17, no. 6: 875–84.

Harada, Y. 1986: *Boshoku Gijutu*, 35, no. 12: 718–31.

Hashimoto, T., N. Kobayashi, T. Kiga, et al. 1993. *Thermal and Nuclear Power* 44, no. 10: 1115–37.

Hill, T.A. 1994. *VGB Kraftwerkstechnik* 74, no. 5: 389–91.

Hirokawa, M., K. Inoue, K. Roko, et al. 1992. *Kawasaki Technical Review*, no. 115: 44–9.

Hugh, E.C. 1937. Experimental investigation of effects of equipment size on convection heat transfer and flow resistance in cross flow of gases over tube banks. *Transactions of the American Society of Mechanical Engineers* 59, paper PRO-59-7, pp. 573–81.

Hutchings, I.M. 1979. Proceedings of Corrosion/Erosion of Coal Conversion Systems Material Conference, ed. A.W. Levy. Berkeley, California, pp. 393–428.

Inoue, H. 1996. *Journal of the JSME* 99, no. 930: 389–91.

Intergovernmental Panel on Climate Change 1996. Climate Change 1995, Second Assessment Report of IPCC, Cambridge University Press.

Ishida, T., S. Morita, and H. Kaneda. 1993. *Hitachi Review* 42, no. 1: 25–30.

Ishigai, S. 1961. *Boira-Yoron (Essentials of Boilers).* Tokyo: Corona.

Ishigai, S., H. Kobayashi, Y. Ueda, et al. 1992. HTD-199, Heat Transfer in Fire and Combustion Systems, 28th National Heat Transfer Conference of American Society of Mechanical Engineers, ed. A.M. Kanury, pp. 189–95.

Ishigai, S., E. Nishikawa, and E. Yagi. 1973. Proceedings of International Symposium on Marine Engineering, Tokyo, pp. 1/5/23–33.

Ishigai, S., E. Nishikawa, and E. Yagi. 1975. *Journal of the Marine Engineering Society of Japan* 10, no. 1: 66–73.

Isoda, K., and F. Koda. 1982. *Hitachi Review* 64, no. 10: 713–9.

Iwashita, S., M. Kashiwazaki, K. Kondo, et al. 1979. *Thermal and Nuclear Power* 30, no. 1: 71–8.

Jager, G.W., D.H. Kallmeyer, and M. Wagner. 1994. *VGB Kraftwerkstechnik* 74, no. 4: 297–9.

Jones, C. 1994. *Power* 138, no. 10: 51–9.

Jones, C. 1996. *Power* 140, no. 2: 33–7.

Kaiho, M. 1996. *Shigen to Kankyo (Resources and Environment)* 5, no. 6: 359–68.

Kaneko, S., Y. Gengo, T. Hashimoto, et al. 1995. *Mitsubishi Juko Giho* 32, no. 1: 23–6.

Kimura, T. 1996. *Bulletin of Applied Coal Structure Association*, no. 15: 14–30.

Klein, M.R., and G. Rotzoll. 1994. *VGB Kraftswerkstechnik* 74, no. 12: 912–9.

Kluyver, J.P., and C.H. Gast. 1995. *VGB Kraftwerkstechnik* 75, no. 7: 534–9.

Kobayashi, T., and M. Funakawa. 1979. *Journal of the JSME* 82, no. 728: 720–6

Koyama, S. 1996. *Journal of the JSME* 99, no. 930: 379–81.

Kuron, D. 1994. *VGB Kraftwerkstechnik* 74, no. 4: 337–46

Landwehr, J.B. 1993. *Power* 137: 62.

Mahagaokar, V. 1991. Shell's SGPI-1 test program – Final overall results. 10th EPRI Conference on Coal Gasification Power Plants, pp. 13/1–15.

Makansi, J. 1993. Controlling SO2 emissions, special report, *Power* 137: 23–66.

Meij, R. 1994. *Fuel Process Technology* 39: 199–217.

Minchener, A.J. 1994. *Proceedings of Institution of Mechanical Engineers*, Part A, *Journal of Power and Energy*, 209: 9–17.

Ministry of International Trade and Industry, Resource and Energy Agency. 1996. Advanced coal-fired generation technologies towards the 21st century. Governmental report of Resource and Energy Agency, Tokyo.

Mori, S., and Y. Fujima. 1996. *Journal of the JSME* 99, no. 930: 371–4.

Moza, A.K., L.G. Austin, and R.E. Tressler. 1980. *Transactions of the American Society of Mechanical Engineers, Journal of Engineering for Power* 102: 679–83.

Mulcathy, M.F.R., J. Boow, and P.R.C. Goard. 1966. *Journal of the Institute of Fuel* 39: 385–98.

Nakamura, K., and T. Hashimoto. 1996. *Thermal and Nuclear Power* 47, no. 8: 831–9.

Nakashima, T. 1985. *Boira-Kenkyu (Journal of the Japan Boiler Association)* no. 212: 9–17.

Niles, W.D., and H.R. Saunders. 1962. *Transactions of the American Society of Mechanical Engineers, Journal of Engineering for Power* 84: 178–85.

Nishikawa, E., and S. Ishigai. 1977. *Transactions of the JSME,* Ser. B 43, no. 373: 3310–9.

Nishikawa, E. 1979. *Journal of the Marine Engineering Society of Japan* 14, no. 9: 754–62.

Nishikawa, E., S. Ishigai, M. Kaiji, et al. 1981. *Transactions of the JSME,* Ser. B 47, no. 419: 1371–9.

Nishikawa, E. 1983. *Journal of the Marine Engineering Society of Japan* 18, no. 4: 296–302; no. 5: 370–9; no. 7: 503–10; no. 8: 589–94.

Nonhoff, G., and R. Lux. 1995. *VGB Kraftwerkstechnik* 75, no. 3: 171–8.

Paidoussis, M.P. 1982. Mechanical Engineering Research Laboratories, Report 82–1, McGill University.

Pantony, D.A., et al. 1968. *Journal of Inorganic and Nuclear Chemistry* 3: 755.

Pierson, O.L. 1937. *Transactions of the American Society of Mechanical Engineers* 59: 563–72.

Planchard, J. 1980. *Computer Methods in Applied Mechanics and Engineering* 24: 125–35.

Pollmann, S. 1965. *VGB Kraftwerkstechnik* 45, no. 1: 1–18.

Reid, W.T. 1971. *External corrosion and deposits.* Amsterdam: Elsevier.

Ruijgrok, W. 1995. *VGB Kraftwersktechnik* 75, no. 3: 225–30.

Sage, P.W., and N.W.J. Ford. 1996. *Proceedings of the Institution of Mechanical Engineers*, Part A, *Journal of Power and Energy* 210: 183–90.

Sato, T., Y. Mukai, K. Otake, et al. 1980. *Thermal and Nuclear Power* 31, no. 11: 1195–202.

Schopf, N., W. Peters, B. Konig, et al. 1994. *VGB Kraftwerstechnik* 74, no. 12: 884–8.

Schuster, H., D. Bose, M. Klein, et al. 1994. *VGB Kraftwerstechnik* 74, no. 1: 30–5.

Sharman, R.B. 1980. The British Gas/Lurgi slagging gasifier and its relevance to power generation, EPRI Conference on Synthetic Fuels, pp. 1–29.

Sherwood, T.K., and R.L. Pigford. 1952. *Absorption and extraction.* New York: McGraw-Hill.

Shiroma, A. 1962. *Netsu-kanri (Thermal Power)* 14, no. 10: 24–33.

Simon, E., D. Lasthaus, and H. Schuster. 1995. *VGB Kraftwerkstechnik* 75, no. 8: 646–50.

Suzuki, T., N. Iiyama, and A. Yukimura. 1993. *Ishikawajima-Harima Engineering Review* 33, no. 5: 298–301.

Tamura, M., N. Kiga, N. Endo, et al. 1995. *Ishikawajima-Harima Engineering Review* 35, no. 5: 301–5.

Tamaru, T., H. Inoue, and T. Abe. 1996. *Ishikawajima-Harima Engineering Review* 36, no. 5: 395–403.

Tanaka, H. 1980a. *Thermal and Nuclear Power* 31, no. 3: 275–80.

Tanaka, H. 1980b. *Transactions of the JSME,* Ser. B 46, no. 404: 600–9.

Tanaka, H. 1980c. *Transactions of the JSME,* Ser. B 46, no. 408: 1398–407.

Tanaka, H., and S. Takahara. 1981. *Transactions of the JSME,* Ser. B 47, no. 417: 778–85.

Tanaka, H. 1981. *Transactions of the JSME,* Ser. B 47, no. 419: 1224–33.

Tanaka, H., S. Takahara, and K. Ohta 1983. *Transactions of the JSME,* Ser. B 49, no. 444: 1668–75.

Thermal and Nuclear Power Engineering Society. 1990. *Thermal and Nuclear Power* 41, no. 7: 911–28.

Tsumura, T., S. Morita, K. Kiyama, et al. 1993. Proceedings of the JSME and the American Society of Mechanical Engineers (ASME), International Conference on Power Engineering, Tokyo, 2: 325–30.

Ueda, Y., S. Ishigai, K. Toh, et al. 1995. Proceedings of JSME–ASME International Conference of Power Engineering, Shanghai, pp. 357–62.

Weinzierl, K. 1994. *VGB Kraftwerstechnik* 74, no. 2: 102–6.

Yamazaki, Y. 1961. Radiant heat transfer of boiler combustion chamber, Japan Central Institute of Electric Power, Report no. 60009.

Yoshii, S., I. Ichinose, A. Hashimoto, et al. 1992. *Mitsubishi Juko Giho* 29, no. 1: 60–3.

Zukauskas, A., B. Makaryavichyus, and A. Slanchyaskas. 1968. *Teplootdacha Pychkov Tryb v Poperechnoi Potoke Zhidkosti*. Mintis/Vilnius.

4

Thermal and hydraulic design of steam-generating systems

KOJI AKAGAWA
Professor Emeritus, Kobe University

4.1 Two-phase flow in steam-generating tubes

The two-phase steam–water mixture encountered in boilers, steam generators, and boiling water reactors plays quite an important role in the heat transfer process and pressure drop in these systems. The primary parameter of such a two-phase flow is the flow structure of the two-phase mixture, that is, the phase distribution, which is referred to as the flow pattern. This flow pattern is of prime importance in the planning and design of steam generating systems, because it strongly affects the flow stability and heat transfer characteristics. For instance, when the mass flow rate is very low in a horizontal steam generating tube, phase stratification takes place and steam flows in the upper part of the tube and water in the lower part. This flow pattern is referred as "separated flow." Consequently, the upper part of the tube wall may possibly become overheated or burned out owing to the low heat transfer coefficient, or the tube may be bent by the temperature difference between the upper and lower parts of the tube wall. To eliminate such troubles or to avoid such a separated flow pattern, quantitative information on the flow pattern, such as the operating conditions and geometrical configuration, should be available at the design stage of the plant.

The importance of flow pattern is not limited to the phase distribution problem, but is also exemplified in flow stability and flow distribution in parallel channels. Thus, this section begins with a discussion on flow patterns and heat transfer with regard to the initial planning and design of boiler evaporating tubes and cooling systems in nuclear reactors.

4.1.1 Flow pattern in vertical and horizontal tubes

The flow configuration in a gas–liquid two-phase flow depends on the properties of each phase (density, viscosity, surface tension), the geometrical dimensions of the channels (geometry of cross-section, dimensions, tube inclination angle, flow

direction relative to gravity), and the flow rates. The flow patterns can be estimated by flow regime maps that are constructed on reasonably selected coordinate systems. A typical flow pattern map for vertical tubes with coordinates of
mass flux G (kg/m^{2}s) and actual quality (thermodynamic nonequilibrium quality)
x_p is shown in Figure 4.1, which is based on data from Bennet et al. (1965),
Weisman (1979), and Weisman and Kang (1981).

The transition of flow pattern along the tube length is represented in Figure
4.1 in the case of $G = 1000$ kg/m^{2}s and $p = 7$ MPa. In the range of the flow quality
$x_p < 0.01$, the flow is "bubbly flow" (small bubbles are dispersed in the continuous liquid phase), whereas in the range of $0.02 < x_p < 0.055$, the flow pattern is
"slug flow" (the flow is composed of bullet-shaped large gas bubbles followed by
liquid slugs entrained by small bubbles) and "froth flow" or "churn flow" (the
length of the large bubble becomes rather long and the liquid slug is severely
agitated). In the latter two flow patterns, the large bubbles and liquid slugs are
arranged alternately along the tube, and the flow exhibits oscillatory behavior
even in the steady state. These flow patterns are therefore sometimes referred to
as "intermittent flow." In the region of $0.12 < x_p < 0.65$, further down the tube,
the flow becomes "annular flow" (a thin liquid film flows on the tube wall
around the vapor core including entrained liquid droplets), and in the range of
$x_p > 0.65$ the flow becomes "droplet flow" or "mist flow" (droplets are dispersed
in the vapor flow).

The hatched regions in Figure 4.1 represent the transition regions between
these flow patterns. The boundary lines between flow patterns shift toward lower
quality with increase in the mass flux. For instance, the transition from intermittent flow to annular flow at $G = 2000$ kg/m^{2}s shifts to $x_p = 0.03$–0.05 from the
boundary $x_p = 0.05$–0.12 at $G = 1000$ kg/m^{2}s, and the transition to the mist flow
shifts to $x_p = 0.45$ from 0.65.

The fluid in steam-generating tubes is not necessarily in thermodynamic equilibrium across the tube cross-section; liquid superheat is, for example, an essential feature in the boiling process. Even if the liquid in the core is not at the
saturation temperature, boiling may occur near the wall owing to liquid superheat ("surface boiling"). This means that the actual quality may have a positive
value, even when the thermodynamic equilibrium quality in the tube cross-section
is zero or negative. Thus the abscissa in the flow pattern maps of Figures 4.1–4.4
should be considered the actual quality, but not the thermodynamic equilibrium
quality. The thermodynamic equilibrium quality at the initiation of bubbly flow
is approximately $x_e = -0.01$, and that of intermittent flow is -0.005 to $+0.003$
according to Hewitt's experiments. The method of estimating the actual quality
is described in Section 4.2.

The flow pattern map for vertical steam-generating tubes is shown in Figure
4.2. The transition quality from bubbly to intermittent flow, and that from inter-

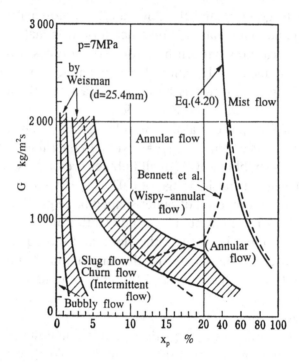

Figure 4.1 Flow pattern map in vertical tubes.

mittent to annular flow, increases with an increase in the pressure. The transition quality from annular to mist flow increases in the range of $p < 5\,\text{MPa}$ and decreases in the range of $p > 5\,\text{MPa}$ with increase in the pressure.

The flow pattern maps for horizontal steam-generating tubes are shown in Figures 4.3 and 4.4. The trend of the flow pattern transition is similar to that in vertical tubes, as shown in Figures 4.1 and 4.2, except for the existence of "stratified flow" (where phase separation with smooth interface takes place under gravity, as mentioned above) and "wavy flow" (where the interface is wavy in form).

Weisman's flow regime maps (Weisman 1979; Weisman and Kang 1981), which are a basis for drawing Figures 4.1–4.4, have been constructed with regard to a large number of experimental data including the effects of density, viscosity, surface tension, tube diameter, and mass flux, and the boundaries of the flow pattern transition are given by empirical formulae. The coordinate system is characterized by dimensionless parameters including the above-mentioned factors. In the saturation condition, the properties of both phases are functions of only the pressure. The total flow rate is constant along the tube length owing to the continuity relationship, and the flow rates of both phases are functions of quality. Thus the flow regime map can be expressed as a function of the total mass flux

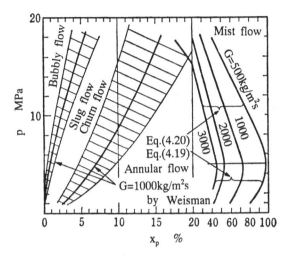

Figure 4.2 Flow pattern map in vertical steam-generating tubes.

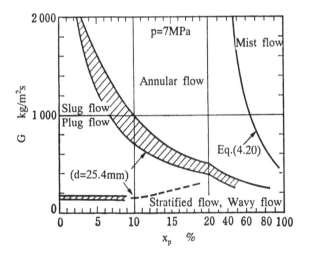

Figure 4.3 Flow pattern map in horizontal steam-generating tubes.

G, the pressure p, and the quality x. This is how the coordinate system for the flow pattern map was selected.

4.1.2 Prediction method of stratified flow pattern

Stratified flow may cause the burnout of horizontal steam-generating tubes under high heat flux conditions, and therefore the flow rate should be large enough beyond a certain limit to avoid phase stratification. An example of the relation-

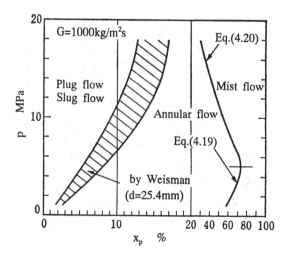

Figure 4.4 Flow pattern map in horizontal tubes.

ship between the heat flux and the inlet velocity (volumetric flux) j_L in horizontal tubes heated by an upper wall is shown in Figure 4.5 (Ishigai et al. 1969a). The stratified flow at the given flow rate (inlet velocity) exists over a certain limited range of heat flux. This can be explained phenomenologically as follows. When the heat flux is low enough, any small bubbles generate flow along the upper wall without coalescence, which does not lead to phase stratification. On the other hand, at high heat flux, the high velocity of the two-phase mixture and the disturbance induced by the high generation rate of the vapor prevent phase stratification but induce intermittent flow. At moderate heat flux, phase stratification occurs owing to the lower velocity and fairly high steam flow rate. The existence of a minimum mass flux (minimum inlet velocity) to avoid phase stratification is indicated in Figure 4.5, and the effect of heat flux is not observed explicitly. The criterion is approximately determined by the Froude number (Fr $= \varrho_L j_L^2 / \varrho_L g d = j_L^2 / g d$), which is the ratio of the horizontal component of momentum of liquid phase, $\varrho_L j_L^2$, to the gravity force in the vertical direction, $\varrho_L g d$. The critical velocity j_L was determined on the basis of experiments and theoretical analysis, and the result is expressed by

$$j_L / \sqrt{gd} = 0.538 \tag{4.1}$$

where ϱ_L is the density of liquid, j_L is the volumetric flux of liquid, g is the gravitational acceleration, and d is the tube diameter (equivalent diameter of channel cross-section) (Ishigai et al. 1969). For example, to avoid stratification, the minimum inlet velocity in a horizontal tube of 2 cm inner diameter (ID) is determined as $j_L = 0.538 \times 9.8 \times 0.02 = 0.238$ m/s by Eq. (4.1).

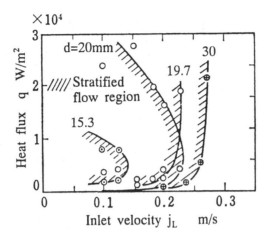

Figure 4.5 Dryout region (stratified region; Ishigai et al. 1969).

The critical mass flux, G_{cr} (kg/m^{2}s) is also given by Styrikovich's formula (Styrikovich et al. 1969), expressed by

$$G_{cr} = 0.38K \frac{\left(\varrho_L/\varrho_G\right)^{1/2}}{\left\{\sigma/\left(\varrho_L - \varrho_G\right)g\right\}^{1/4}} \frac{\varrho_L}{\left\{1 + x\left(\frac{\varrho_L}{\varrho_G} - 1\right)\right\}} \left(\frac{x}{1-x}\right)^{3/4} d^{1/2} \qquad (4.2)$$

where σ is the surface tension, and K is a flow parameter defined as

$$K = \alpha \Big/ \left(\frac{j_G}{j_G + j_L}\right) \qquad (4.3)$$

where α is the void fraction, and j_G is the volumetric flux of gas. The value of K proposed by Styrikovich is shown in Figure 4.6 and is discussed again in Section 4.3. The critical mass flux G_{cr} calculated from Eq. (4.2) increases with increasing quality, as shown by the solid lines in Figure 4.7. The values calculated from Eq. (4.1) are represented by broken lines in the figure, and these results do not show dependence on the quality. Both curves intersect at $x = 0.15$, and the predicted value of G_{cr} from Eq. (4.1) is on the safe side in the range of $x < 0.15$.

Burnout induced by stratified flow can be avoided by arranging tubes at an inclined angle θ to the horizon. For example, the stratified flow and wavy flow regions decrease remarkably with increase in the upward inclination angle of the tube and disappear at $\theta = 10°$, as shown in Figure 4.8, presented by Barnea for air–water two-phase flow (Barnea et al. 1980). In contrast, a downward inclination angle extends the stratified flow region as the gravity force stabilizes the flow. For instance, the critical value of the volumetric flux of liquid at $\theta = -5°$ is about 1.05 m/s, which is much higher than the value of 0.12 m/s at $\theta = 0°$. According to

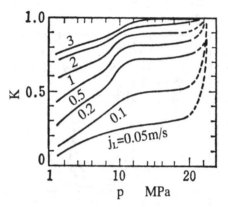

Figure 4.6 Values of K (Styrikovich et al. 1969).

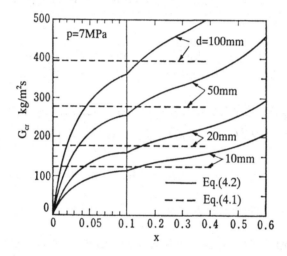

Figure 4.7 Critical mass flux of phase stratification.

Roko (1975) (D = 4.1 cm), the critical value decreases from 0.3 m/s in a horizontal tube ($\theta = 0°$) to 0.23 m/s at $\theta = 5.2°$, and the time-averaged void fraction in the upper part of the tube cross-section decreases remarkably with an increase in the inclination angle up to $\theta = 5°$, which is followed by a slight change in the range of $\theta > 5°$. The flow pattern in inclined tubes with $\theta > 25°$ is similar to that in vertical tubes.

4.1.3 Two-phase flow in vertical serpentine tubes

Vertical serpentine tubes are often used in conventional boilers and exhaust gas boilers. In serpentine tubes, as shown in Figure 4.9 (Akagawa et al. 1968), the flow patterns observed in the upward and downward flow tubes differ. Under certain

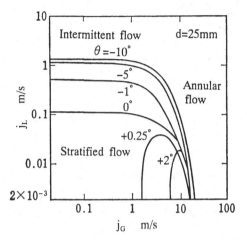

Figure 4.8 Stratified flow region (based on Barnea 1980).

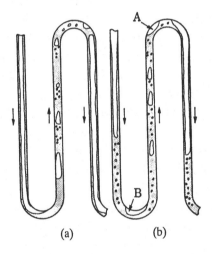

Figure 4.9 Flow configuration of two-phase flow in serpentine-tube system. (a) $j_L = 0.2$ m/s; (b) $j_L = 1.1$ m/s (Akagawa et al. 1968).

operating conditions, a long gas column is formed in the tube of downward flow, while the flow pattern in the upward flow tube is bubbly or slug flow. This results in rather a large flow resistance owing to gravity, which is followed by flow stagnation or flow oscillations. This stagnation of the vapor column at a bend may in some cases induce burnout, and the periodic alternation of the vapor column and the liquid column during flow oscillation may lead to thermal fatigue of the tube wall. Thus special attention should be focused on the design of vertical serpentine tube systems.

The time variation of the flow configuration during flow oscillation is demonstrated in Figure 4.9. At one instant, the flow in the upward flow tube is slug flow and that in the downward flow tube is "falling-film flow" (similar to annular flow), as shown in Figure 4.9a. Then the liquid of the falling film accumulates at the lower bend, and the length of the liquid column in the downward flow tube increases gradually, as shown in Figure 4.9b. The liquid column is abruptly accelerated toward the tube of upward flow owing to the decrease in the pressure head difference between the upward and downward flow tube. This process is repeated, and the flow oscillation is sustained.

The maximum pressure difference is postulated as (height) $\times$ $(\varrho_L - \varrho_G)g$ between the inlet of the downward flow tube and the outlet of the upward flow tube in the state of Figure 4.9a. This pressure difference is much higher than the frictional pressure drop; therefore, a sufficient driving head of the water circulation pump is needed to ensure the safe circulation of water. The flow regime map for vertical serpentine tubes is shown in Figure 4.10 (Akagawa et al. 1968), with the coordinate system of volumetric flux ratio (volume flow quality), $j_G/(j_G + j_L)$, and Froude number, $\mathrm{Fr} = (j_G + j_L)^2/gd$, which is the same coordinate system as is used in flow pattern maps for upward vertical flow proposed by Kozlov (1952) and by Griffith (1964). The flow patterns in Figure 4.10 are represented by the combination of flow patterns in the upward and downward flows. The flow patterns in the downward flow are classified into slug flow (S), falling-film flow (F), mixture of falling-film and slug flow (F.S), and mixture of falling-film and gas flow with droplets in the core (F.D). The classification of flow patterns is the same as defined in Section 4.1.1. Thus the flow patterns in the serpentine tube are classified into S–F flow, S–F.S flow, S–S flow, and S–F.D flow. The S–F flow occurs in the region of $\mathrm{Fr} < 0.57$. The S–F and S–S flows are stable, but the S–F.S flow is oscillatory and is accompanied by pressure fluctuations of large amplitude. The S–F.D flow is also oscillatory, with small amplitude pressure fluctuations. The boundary between S–F.S flow and S–F.D flow shifts, depending on the tube height L, as shown in Figure 4.10.

When a Taylor bubble (long vapor column) stagnated in the downward flow tube penetrates the upward flow tube, the minimum volumetric flux of liquid, j_{L1}, is given by the following empirical formula:

$$w_B = 1.2(j_L + j_G) + 0.35\sqrt{gd} \qquad (4.4)$$

This value coincides approximately with the value obtained from the following equation of the rising velocity w_B of a Taylor bubble in a vertical tube by substituting $w_B = 0, j_G = 0$:

$$j_{L1} \cong 0.29\sqrt{gd} \qquad (4.5)$$

An experimental value of j_{L1} in air–water two-phase flow (tube diameter $d = 3\,\mathrm{cm}$) is 0.16–0.2 m/s (mean value 0.175 m/s), independent of the gas slug length

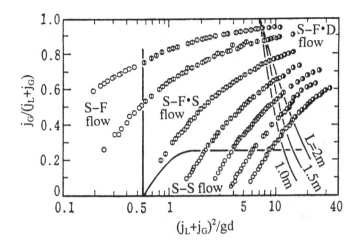

Figure 4.10 Flow pattern map in serpentine tube system (Akagawa et al. 1968).

and the radius ratio D/d in the upstream bend (D represents the radius of curvature of the bend), which agrees approximately with the value of 0.157 m/s from Eq. (4.4) (Akagawa et al. 1968).

The minimum volumetric flux of liquid, j_{L2}, which sweeps away a stagnant large bubble in the upper bend, is given by the following empirical formula:

$$j_{L2} = 0.61\sqrt{gd} \tag{4.6}$$

According to the experiment mentioned above, the minimum velocity becomes $j_{L2} = 0.33$ m/s and is independent of both the volume of the bubble and D/d.

The minimum volumetric flux of liquid, j_{L3}, which sweeps away a stagnant large bubble in the lower bend, is given by

$$j_{L3} = 1.37\sqrt{gd} - 0.024(D/d) \tag{4.7}$$

In the case of $d = 30$ mm and $D/d = 9.1$, the value becomes $j_{L3} = 0.5$ m/s in the air–water two-phase flow. When the velocity is higher than this minimum value j_{L3}, the large bubble is swept away into the upward flow tube and split into small bubbles. In general, these volumetric fluxes have the order of magnitude $j_{L3} < j_{L2} < j_{L1}$. This means that a large bubble stagnates most easily in the lower bend.

The flow rates of the liquid film and droplets change along the tube length in the vertical serpentine tube as shown in Figure 4.11 (Akagawa et al. 1982). The flow rate of the droplets, G_E, decreases abruptly at the upper and lower bends to about 10–20% of the inlet flow rate, and increases in the upward flow and downward flow tubes. Such characteristics are caused by the deposition and generation of droplets along the tube length as shown in Figure 4.12. The droplets deposit on the outer side of the bend wall owing to inertial force, and droplets

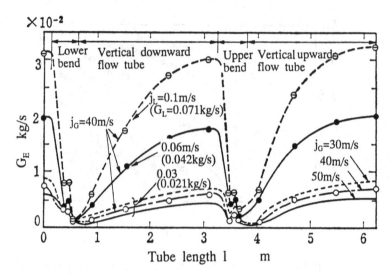

Figure 4.11 Distribution of droplet flow rate along tube length (Akagawa et al. 1982).

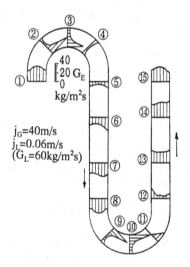

Figure 4.12 Distribution of droplet flow rate in tube cross-section (Akagawa et al. 1982).

are generated from the liquid film in the straight tube sections owing to the shear force of the gas flow on the interface.

The separation efficiency of droplets, η_s, which is defined by the second term of the following equation (4.8), is expressed by the empirical formula of that equation as a unique function of D/d in the range of $j_G = 30$–60 m/s, $j_L = 0.03$–0.15 m/s, and $D/d = 4$–10:

$$\eta_s = \frac{G_{in} - G_{out}}{G_{in}} = 1 - 0.47\left(\frac{D}{d}\right)^{0.71} \qquad (4.8)$$

where G_{in} and G_{out} are droplet flow rates at the inlet and outlet of the bend, respectively. According to this equation, the efficiency η_s is about 0.8 for $D/d = 6.7$, which means that most of the droplets deposit on the bend wall.

4.2 Boiling heat transfer in steam-generating tubes

The heat transfer coefficient in steam-generating tubes is a very important factor in the thermohydraulic design of the water wall in boiler furnaces and boiling water reactors in which the heat fluxes are very high. The flame in a boiler furnace emits thermal energy independently of the heating surface temperature; similarly, the heat generated in a nuclear reactor core is independent of the cooling channel wall temperature. This type of heat exchanger is denoted the "heat-flux-dominated type" (heat flux is independent of heating surface temperature). Therefore, when the value of the heat transfer coefficient is low and the heat flux is high, the tube wall temperature rises remarkably and burnout may occur. On the other hand, in the steam generator in a fast breeder reactor, for example, where thermal energy is transferred from high-temperature sodium (primary coolant) to the water (secondary coolant) through the tube wall, the magnitude of the heat flux is determined by the temperature difference. This type of heat exchanger is denoted the "temperature-difference-dominated type." In this case, the heat flux is affected by the heat transfer coefficients on the inner and outer surfaces of the tube, though the heat transfer coefficient on the water side is much lower than that on the sodium side. The heat flux and the area of heating surface are therefore mainly dominated by the heat transfer coefficient on the water side. Thus, prediction of the heat transfer coefficient with high accuracy is needed for the thermohydraulic design of steam-generating channels.

4.2.1 Axial distributions of wall temperature and heat transfer coefficient

The characteristics of flow and heat transfer phenomena in steam-generating tubes will be explained first by using an experimental result for an axial temperature distribution along the tube, which is heated by electric joule heating with uniform heat flux as shown in Figure 4.13 (Roko 1980). The thermodynamic equilibrium quality x_e and the thermodynamic equilibrium temperature of the water rise linearly from section A to section C, where the mean enthalpy in the cross-section is equal to the saturated liquid enthalpy ($x_e = 0$). The thermodynamic equilibrium temperature between C and F is approximately constant at the sat-

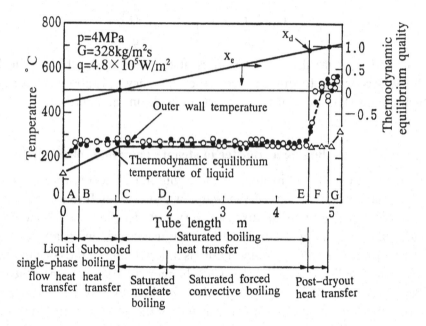

Figure 4.13 Temperature distribution along steam-generating tube (Roko 1980).

uration temperature, and the temperature rises linearly between F and G, where the steam is superheated.

The wall outer surface temperature rises linearly in the water single-phase heat transfer region (A–B), and that is approximately constant in the subcooled boiling region (B–C) and in the saturated boiling region (C–E). In the post-dryout heat transfer region (E–F) the wall temperature rises sharply, and in the steam single-phase heat transfer region (F–G) it rises parallel to the superheated steam temperature. This figure indicates that the temperature difference between the wall surface and the fluid is kept lower than 15°C in almost the whole region (A–E) of the tube length; however, in the post-dryout region (E–F), the temperature difference becomes very high, on the order of 300°C, and the tube may burn out. Therefore, it is very important to predict the location of the dryout point (F) and the wall temperature in the safety design of steam-generating systems.

The heat transfer coefficient h at the inner surface of the tube wall is defined by

$$h = q/(T_w - T_b)$$

where T_w is the surface temperature and, T_b is the mean fluid temperature at the tube cross-section, or the bulk temperature. The values of h can be obtained from the temperature difference $(T_w - T_b)$ along the tube length. According to the experiment, the values of h are about 4800 W/m²K in the liquid single-phase flow region (A–B), about 48,000 W/m^{2}K in the subcooled boiling region (B–C), and 48,000 W/m^{2}K in the saturated boiling or saturated evaporation region. These values are relatively high and the wall temperature is then quite safe; however, the value of h in the post-dryout region (liquid-deficient region) decreases remarkably to about 1600 W/m^{2}K and the wall temperature rises considerably.

The mechanisms of these heat transfer phenomena will be explained briefly, referring to the flow configuration as follows.

- Subcooled boiling: The core liquid is not saturated but bubbles are generated on the wall owing to the high wall temperature, and the bubbles collapse in the core of liquid.
- Saturated nucleate boiling: The fluid is in the saturated state, and bubbles generated on the wall are distributed in the whole tube cross-section.
- Saturated forced convective boiling: The flow pattern is an annular flow or an annular–mist flow and the liquid layer is thin; nucleate boiling then does not occur and the steam is generated by evaporation from the liquid layer surface.
- Post-dryout heat transfer: The flow pattern is a mist flow and most of the heat is transferred from the wall surface to the steam; the droplets are heated by the superheated steam, and evaporation occurs on the droplet surface.

The heat transfer characteristics in steam-generating tubes of the temperature-difference-dominated type will be explained. Figure 4.14 shows a typical example of experimental results such as the distribution of water and sodium temperatures, thermodynamic equilibrium quality, heat flux, and heat transfer coefficient in a double-tube countercurrent heat exchanger in which the primary fluid (heat sink side) is sodium and the secondary fluid (heat source side) is water (Roko 1980). As the water flows toward the right and the sodium toward the left in the figure, the equilibrium quality of the water rises along the tube length and the sodium temperature drops toward the left. The magnitude of the heat flux exchanged between both fluids can be determined by the axial gradient of the sodium temperature along the tube length. The figure indicates that the value of heat flux q increases from that in the water single-phase flow region (A–B), in the subcooled boiling region (B–C), and in the saturated boiling region (C–E); at the location of the dryout point (E) the heat flux decreases abruptly (E–F) and takes an approximately constant value in the post-dryout (F–G) and the superheated steam (G–H) regions, where the thermodynamic equilibrium quality x_e is 0.5–1.2. The value of the heat transfer coefficient h in the subcooled and saturated boiling regions is considerably higher than that in the water single-phase flow and takes the maximum value of 10^5 W/m^{2}K at location C; it then decreases abruptly at the post-dryout point E ($x_e = 0.55$) to 2×10^3 W/m^{2}K, and takes a rather low value of 2×10^3 to 5×10^3 W/m^{2}K in the post-dryout heat transfer region. It should therefore be noticed, for the design of the steam generator, that

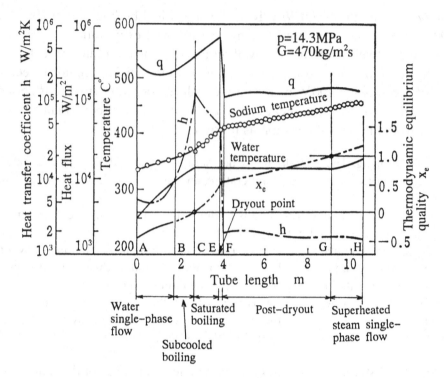

Figure 4.14 Heat transfer characteristics in steam-generating tube heated by sodium (Roko 1980).

the value of the heat transfer coefficient in the post-dryout region is only 2–4% of that in the saturated boiling region.

In both cases, the temperature-difference-dominated and the heat-flux-dominated conditions, the heat transfer mechanisms are the same in respective regimes such as subcooled boiling, saturated nucleate boiling, and saturated forced convective boiling, and post-dryout heat transfer. Therefore, in the following section, methods of predicting the region of existence of each heat transfer mode and the values of the heat transfer coefficient will be briefly described.

4.2.2 Heat transfer in the subcooled boiling region

An example of the incipient condition of subcooled boiling is shown in Figure 4.15 as a function of heat flux and mass flow rate G. The ordinate is the subcooling, $\Delta T_{\text{sub}} = (T_s - T_b)$, where T_s is the saturation temperature and T_b the bulk temperature. The value of ΔT_{sub} increases with increasing heat flux q and decreasing mass flux G. According to the figure, for instance, subcooled boiling occurs at $\Delta T_{\text{sub}} = 78\,\text{K}$, that is, $T_b = T_s - 78 = 207.8°\text{C}$, and at the thermody-

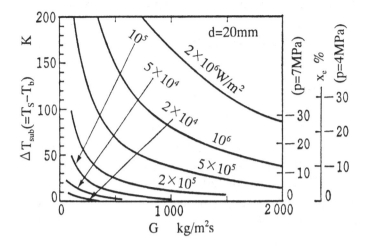

Figure 4.15 Incipient condition of subcooled boiling.

namic quality $x_e = -25\%$, for the condition of $p = 7\,\text{MPa}$, $G = 1000\,\text{kg}/(\text{m}^2\text{s})$, and $q = 10^6\,\text{W}/\text{m}^2$.

The value of ΔT_{sub} in Figure 4.15 was obtained by the following method. Just ahead of the incipient point of subcooled boiling the heat flux q and the heat transfer coefficient h are expressed by Eqs. (4.9) and (4.10) (Dittus–Boelter equations), respectively, owing to single-phase heat transfer:

$$q = h\left(T_w - T_b\right) \tag{4.9}$$

$$\frac{hd}{\lambda_L} = 0.023\,\text{Re}^{0.8}\,\text{Pr}_L^{0.4} \tag{4.10}$$

From these equations

$$T_w - T_b = q/h = q\bigg/\left(0.023\frac{\lambda_L}{d}\,\text{Re}^{0.8}\,\text{Pr}_L^{0.4}\right) \tag{4.11}$$

where d is the inner diameter of the tube, λ_L is the heat conductivity of the liquid, Pr_L is the Prandtl number of the liquid, and Re is the Reynolds number. Thus the value of $(T_w - T_b)$ is determined by the heat flux and the mass flux. On the other hand, the value of $(T_w - T_s)$ is expressed by the Jens–Lottes equation (Eq. 4.12), which is a function of the heat flux and is independent of the mass flux and the bulk temperature T_b:

$$T_w - T_s = 25e^{-\frac{p}{6.17}}\left(q/10^6\right)^{0.25} \tag{4.12}$$

where p is expressed in MPa, q in W/m², and T in °C.

At the incipience of subcooled boiling, the value of q in Eq. (4.11) must also be equal to that in Eq. (4.12). Also, ΔT_{sub} is expressed by Eq. (4.13); consequently, the value of ΔT_{sub} can be obtained as a function of q, G, and p (Eq. (4.14)), by substituting the values of Eqs. (4.11) and (4.12) into Eq. (4.13); the resulting value is shown in Figure 4.15.

$$\Delta T_{sub} = T_s - T_b = \left(T_w - T_b\right) - \left(T_w - T_s\right) \tag{4.13}$$

$$\Delta T_{sub} = f\left(q, G, p\right) \tag{4.14}$$

The heat flux at subcooled boiling depends on the temperature difference $(T_w - T_s)$, as shown by Eq. (4.12); therefore, when the heat flux q is given, the wall temperature T_w can be predicted as described above.

For heat exchangers of the temperature-difference-dominated type, the heat flux at subcooled boiling is determined using Eq. (4.12) as follows. The heat flux is given by

$$q = \left(T_f - T_w\right)\Big/\left(\frac{d}{d_0}\frac{1}{h_0} + \frac{d}{\lambda}\ln\frac{d_0}{d}\right) \tag{4.15}$$

where d_0 is the outside diameter of the tube, h_0 is the heat transfer coefficient on the outer surface, T_f is the temperature of the fluid on the outer side, and T_s is the saturation temperature in the inner fluid. Thus, when T_f and T_s are given, the heat flux q and wall temperature T_w can be determined by Eqs. (4.12) and (4.15), respectively.

4.2.3 Heat transfer in saturated nucleate boiling and saturated forced convection boiling regions

Heat transfer performance is very high in the saturated nucleate boiling region and the saturated forced convective region, because a liquid layer exists on the tube wall. But the heat transfer mechanisms in these two regions are different. In this section, for simplicity, a correlation of the heat transfer coefficient that applies to both regions is introduced. The heat transfer coefficient is expressed by Schrock and Grossman (1962) as

$$h = 170\frac{\lambda_L}{d}\text{Re}_{L0}{}^{0.8}\text{Pr}_L{}^{1/3}\left\{\frac{q}{Gr} + 1.5 \times 10^{-4}\left(\frac{1}{X_{tt}}\right)^{0.66}\right\} \tag{4.16}$$

$$X_{tt} = \left(\frac{1-x}{x}\right)^{0.9}\left(\frac{\varrho_L}{\varrho_G}\right)^{0.5}\left(\frac{\mu_L}{\mu_G}\right)^{0.1} \tag{4.17}$$

where G is the mass flux (kg/m²s), r is the latent heat of evaporation, and X_{tt} is a parameter denoted as the Martinelli parameter defined by Eq. (4.17). The term (constant $\times$ (λ_L/d) Re_{L0} Pr_L) in the equation represents the heat transfer coefficient in the liquid single-phase flow (Dittus–Boelter equation), and the term

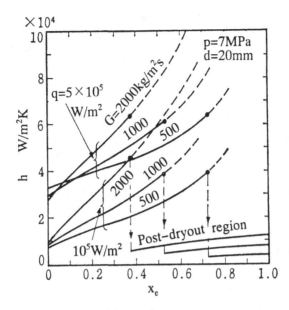

Figure 4.16 Heat transfer coefficient in saturated nucleate boiling and post-dryout regions.

{ } represents a two-phase friction multiplier. In the equation, the first term (q/Gr) shows the effect of nucleate boiling and the second term shows that of convective heat transfer. According to the equation, in the low-quality region (the saturated nucleate boiling region) the first term dominates, that is, the heat transfer coefficient increases with the heat flux, and in the high-quality region (the saturated forced convective boiling region) the effect of the second term increases, that is, the heat transfer coefficient increases due to the increase in the liquid layer velocity.

An example of values of the heat transfer coefficient calculated by Eq. (4.16) is shown in Figure 4.16. The figure indicates that the values at $x_e = 0$ are approximately equal to those in the nucleate pool boiling and are approximately 8000 and 30,000 W/m²K at $q = 10^5$ and 5×10^5 W/m², respectively. The heat transfer coefficient increases with increasing quality x_e, but at a dryout point shown by a mark, which is explained in Section 4.2.4, it drops abruptly to the value at post-dryout heat transfer given by Eq. (4.23), described below. In these saturated nucleate boiling and forced convective heat transfer regions the value of h is very high, being 30,000–60,000 W/m²K for a considerably high heat flux of $q = 5 \times 10^5$ W/m²; thus the temperature rise of the tube wall, $(T_w - T_s) = \Delta T_{sat} = q/h$, is quite small, on the order of 8–17 K.

Chen correlates the heat transfer coefficient with higher accuracy but via a rather complex method (1966); however, it is not introduced in this book.

4.2.4 Heat transfer in the post-dryout region

In once-through boilers, if all the water supplied into the evaporator tube is evaporated and superheated, then a dryout region inevitably exists in the tube. As the heat transfer coefficient in the region is rather low and the wall temperature may become too high, it is necessary to predict with high accuracy the value of the heat transfer coefficients and the location of the incipient point of dryout in the design of once-through boilers.

The wall temperature rise owing to dryout in the high-quality region is called "second-order burnout." But even if dryout occurs at a low heat flux condition, burnout of the tube does not necessarily occur because of a small rise in the wall temperature. In contrast to second-order burnout, a burnout by film boiling at a high heat flux condition accompanied by a high wall temperature rise is called "first-order burnout." Both burnouts should be clearly distinguished in the design of steam-generating systems.

The quality at the dryout point, x_d, depends on the pressure and the mass flux and not on the heat flux, as shown in Figure 4.17. According to Doroshchuk and Konjkov (1972), however, x_d is slightly affected by the heat flux in the low heat flux and high mass flux ($G > 2000\,kg/m^2s$) region. In ordinary steam-generating tubes in boilers in which the maximum heat flux is usually lower than $4 \times 10^5\,W/m^2$, the value of x_d is not practically affected by the heat flux. In Figure 4.17 Roko's (1980) data (low mass flux region), Levitan and Lantsman's (1975) data (high mass flux region), and Doroshchuk and Nigmatulin's (1970) data ($G = 500$–$2000\,kg/m^2s$) are shown by rearranging them. The figure clearly indicates that x_d decreases with increasing mass flux. This tendency is similar to that of the flow pattern transition from annular flow to mist flow. The reason can be explained by the flow characteristics, namely, that the shear stress on the interface between the liquid layer and the core steam stream in annular flow increases, and then the rate of droplet generation increases with increasing mass flux and the liquid layer disappears; consequently, the heat transfer performance is lowered.

Three correlations on x_d are shown in this section, though many correlations have been proposed. Roko's (1980) correlation is given by

$$\frac{1 - x_d}{x_d} = \left\{ 12.01\left(\frac{p}{p_{cr}}\right)^2 - 5.628\left(\frac{p}{p_{cr}}\right) + 0.8623 \right\}\left(\frac{G}{1000}\right)^{1.2} \tag{4.18}$$

This equation was obtained from the data of $p = 2$–$14\,MPa$, $G = 150$–$800\,kg/m^2s$, and $d = 12.5\,mm$ in a vertical tube. The equation is also applicable in the range of $p = 7$–$12\,MPa$ and $G < 3000\,kg/m^2s$.

Doroshchuk and Nigmatulin's (1970) correlation is given by

$$x_d = 8.92(10 \cdot p)^{0.15} G^{-0.45} \quad (p < 0.5\,MPa) \tag{4.19}$$

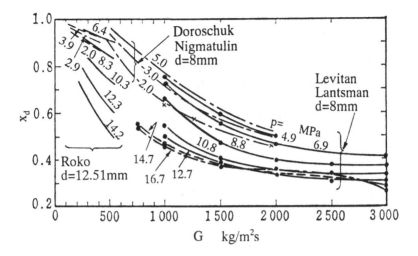

Figure 4.17 Dryout quality vs. mass flux.

$$x_d = \left[0.0031 \frac{\sigma}{v_L} \frac{\varrho_L}{\varrho_L - \varrho_G} \frac{1}{G}\right]^{1/2} \quad (p > 0.5\,\text{MPa}) \tag{4.20}$$

where σ is the surface tension and v is the kinematic viscosity. Equation (4.19) was obtained by correlating data of $p = 0.6$–$5\,\text{MPa}$, $G = 500$–$2000\,\text{kg/m}^2\text{s}$, $d = 8\,\text{mm}$, and tube length $L = 1.5\,\text{m}$ of a vertical tube, and Eq. (4.20) by correlating data of $p = 9$–$18.6\,\text{MPa}$, $G = 750$–$4000\,\text{kg/m}^2\text{s}$, $d = 8\,\text{mm}$, and $L = 1\,\text{m}$ and $6\,\text{m}$ tubes. Levitan and Lantsman's (1975) correlation is expressed by the following equation, which is obtained from the experimental values of $p = 4.9$–$16.7\,\text{MPa}$, $G = 750$–$3000\,\text{kg/m}^2\text{s}$, $L = 1.5\,\text{m}$ and $3\,\text{m}$, and $d = 8\,\text{mm}$ of vertical tubes as shown in Figure 4.17:

$$x_{d(8)} = \left[0.39 + 1.57\left(\frac{p}{9.8}\right) - 2.04\left(\frac{p}{9.8}\right)^2 + 0.68\left(\frac{p}{9.8}\right)^3\right]\left(\frac{G}{1000}\right)^{-0.5} \tag{4.21}$$

In the case of tubes whose diameter is not equal to $8\,\text{mm}$, the value of x_d is given by

$$x_d / x_{d(8)} = (8/d)^{0.15} \tag{4.22}$$

This equation shows that dryout quality decreases with increasing diameter.

Figure 4.18 shows the effects of mass flux and pressure on the value of x_d given by these three correlations. This figure shows that x_d by these correlations coincides and that x_d can be predicted with sufficient accuracy over a wide range of mass flux and pressure. As the dryout is a phenomenon related to flow configura-

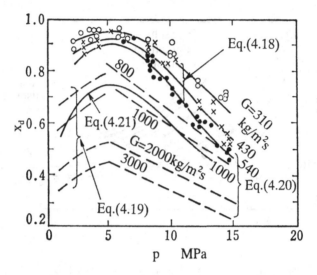

Figure 4.18 Dryout quality vs. pressure.

tion, the value of x_d is little affected by the heat flux as described above, though the heat flux affects the distance to the dryout point from the tube inlet.

Macbeth's (1963) and Biasis et al.'s (1967) correlations, however, contain a term for the effect of heat flux on x_d. But according to the results calculated from these correlations, the effect of heat flux is very weak in the range of ordinary heat flux in boiler furnaces. Therefore these correlations are not described in this book.

Bishop et al.'s (1965) correlation for the heat transfer coefficient in the post-dryout region is expressed by the following equation, which can be used in the wide range of $x_e = 0.1–1.0$, $G = 700–3400 \, \text{kg/m}^2\text{s}$, and $p = 4–21.5 \, \text{MPa}$:

$$\text{Nu} \equiv \frac{hd}{\lambda_{Gf}} = 0.0193 \, \text{Re}_f^{0.8} \, \text{Pr}_{Gf}^{1.23} \left[x_e + \frac{\varrho_G}{\varrho_L}(1 - x_e) \right]^{0.8} \left(\frac{\varrho_G}{\varrho_L} \right)^{0.068}$$

$$\text{Re}_f = Gd_i / \mu_{Gf} \tag{4.23}$$

In this equation, the term of (constant) $\times \text{Re}^m\text{Pr}^n$ is equivalent to the heat transfer coefficient in steam single-phase flow (Dittus–Boelter equation) when the total mass of the two-phase mixture flows as the state of steam, and the term of [ ] represents the effect of mass flux of steam at each location along the tube length. The subscript f means the property at a temperature $T_f = (T_w + T_s)/2$. The influence of droplets in the core steam flow on the heat transfer is apparently not represented in this equation.

An example of the distribution of heat transfer coefficient along the tube length, calculated by Eq. (4.23), is shown in Figure 4.19. For a case of $q = 5 \times 10^5 \, \text{W/m}^2$ and $G = 1000 \, \text{kg/m}^2\text{s}$, for instance, the value of heat transfer coefficient

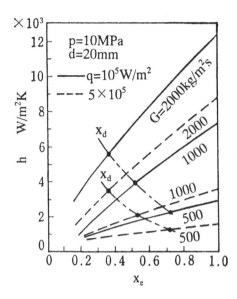

Figure 4.19 Post-dryout heat transfer coefficient.

at dryout point ($x_d = 0.52$) is $h = 2000\,\text{W/m}^2\text{K}$, and this increases along the tube length and reaches $h = 3500\,\text{W/m}^2\text{K}$ at the location of $x_e = 1$. In this case, the value of the wall temperature rise at the dryout point is $\Delta T_{\text{sat}} = T_w - T_s = 5 \times 10^5/2000 = 250°\text{C}$. As can be seen from the figure, the heat transfer coefficient in the post-dryout region is on the order of $10^3\,\text{W/m}^2\text{K}$. This value is very low, 1/10 to 1/100 of that in the saturated boiling region. This characteristic can also be seen in Figure 4.16.

Figure 4.19 indicates that the heat transfer coefficient is significantly affected by the heat flux, though Eq. (4.23) contains no term allowing for the effect of heat flux. This can be explained as follows. The effect of heat flux is included in the evaluation of fluid properties by the temperature T_f. The relationship between the heat transfer coefficient h at arbitrary temperature T and h_s, at the saturation temperature T_s is expressed by

$$\frac{h}{h_s} = \left(\frac{\lambda_G}{\lambda_{Gs}}\right)\left(\frac{\text{Pr}_G}{\text{Pr}_{Gs}}\right)^{1.23}\left(\frac{\mu_{Gs}}{\mu_G}\right)^{0.8} \tag{4.24}$$

where the subscript s denotes the properties at saturation temperature. The normalized heat transfer coefficient h/h_s decreases remarkably with increase of the temperature T_f, as shown in Figure 4.20. This means that the heat transfer coefficient decreases with increasing heat flux, which induces a rise of the temperature T_f.

For the lower mass flux region ($G < 500\,\text{kg/m}^2\text{s}$), it is preferable to use Roko's

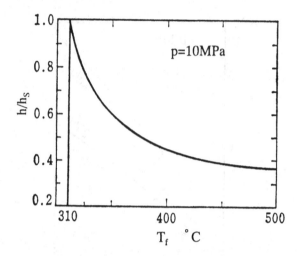

Figure 4.20 Influence of temperature on post-dryout heat transfer.

(1980) approximation method, which has higher accuracy. The actual quality x_a (thermodynamic nonequilibrium quality) is less than the thermodynamic equilibrium quality x_e, because superheated steam and droplets exist simultaneously in the tube cross-section in the post-dryout region. Figure 4.21 shows the relationship between x_a and x_e. The location where x_a becomes 1 is larger than the location of $x_e = 1$, which is denoted as x_0. The value of x_0 is a function of the mass flux and the heat flux and is given by Roko's (1980) empirical formula:

$$\left.\begin{array}{l} x_0 = 1 + 1.09 F^{-0.536} \\[2mm] F = 2.08 \times 10^4 \left(\dfrac{p}{p_{cr}}\right)^{1.6} \left(\dfrac{G}{1000}\right)^{2.5} x_d^2 d \left(q \cdot 10^{-5}\right)^{-1} \end{array}\right\} \qquad (4.25)$$

The distribution of x_a versus x_e is approximately expressed by an empirical formula:

$$x_a = \left\{(1 - x_d)x_e + (x_0 - 1)x_d\right\} / (x_0 - x_d) \qquad (4.26)$$

The heat transfer coefficient is also given for the regions where $x_e \gtrless 1$:

$$\left.\begin{array}{l} x_e \leq 1: \quad \dfrac{h}{h_0} = 0.9 \Big/ \left(1 + \dfrac{h_0}{q}\dfrac{r}{c_p}\dfrac{x_e - x_a}{x_a}\right) \\[4mm] x_e > 1: \quad \dfrac{h}{h_0} = 0.9 \Big/ \left(1 + \dfrac{h_0}{q}\dfrac{r}{c_p}\dfrac{x_e(1 - x_a)}{x_a}\right) \\[4mm] h_0 = 0.0073 \left(\dfrac{\lambda_{Gf}}{d}\right)\left(\dfrac{Gd}{\mu_{Gf}}\right)^{0.886} Pr_{Gf}^{0.61} x_a^{0.886} \end{array}\right\} \qquad (4.27)$$

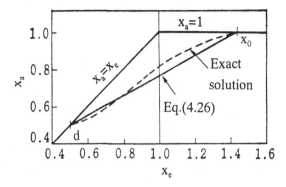

Figure 4.21 Actual quality distribution x_a.

Figure 4.22 shows the experimental results for the distribution of the heat transfer coefficient and the values calculated from Eqs. (4.25)–(4.27). The figure indicates that the heat transfer coefficient h decreases along the tube length from the value at the dryout point to a minimum and then increases a little; the calculated results agree well with the experimental. The tendency of the distribution of h differs from the distribution of h at high mass flux that was shown in Figure 4.19.

Furthermore, for a low mass flux condition, Remizov's empirical formula, which is applicable for ranges of $p = 7$–$14\,\text{MPa}$, and $G = 350$–$700\,\text{kg/m}^2\text{s}$, is given below, and this agrees with Roko's experimental data:

$$h = 1.16\left[\frac{1.25 + 0.025G}{(x_e + 0.001) - x_d} - (4650 - 8G)(x_e - x_d) + 1240\right] \quad (4.28)$$

A transition region, of length 4–7 cm, exists between the saturated boiling region and the fully post-dryout region. In the transition region, the axial temperature gradient of the tube wall becomes quite large and temperature oscillations occur in some conditions. Figure 4.23 shows, for instance, the temperature rise ΔT_{sat} (the difference between the inner wall temperature and the saturation temperature) and the standard deviation of the amplitude of the wall temperature oscillation δT_w, which were measured by Yamauchi (1980) in a vertical tube of ID $= 4\,\text{mm}$ heated by electric joule heating. According to the experiment, the maximum values of the transition region length, L_{tr}, ΔT_{sat}, and δT_w are 170 mm, about 200°C, and about 30°C, respectively.

The transition region length L_{tr} is about 40–170 mm in a heat exchanger tube heated by sodium, and it decreases with increasing pressure and mass flux as shown in Figure 4.24 (Roko 1980). Also, the axial gradient of the tube wall temperature in the transition region, $\Delta T_w/L_{tr}$, is about 0.5°C/mm at 8 MPa and

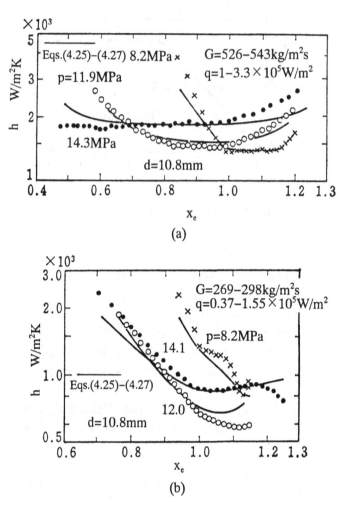

Figure 4.22 Heat transfer coefficient at low mass flux in post-dryout region (Roko 1980).

increases with the pressure to about 1°C/mm at 12 MPa, taking the constant value of 1°C/mm over 12 MPa. Here, ΔT_w is the temperature difference between the inner wall temperature at the inlet and that at the outlet of the transition region.

It has been experimentally shown that there are two modes of oscillation in exchangers of the temperature-difference-dominated type; one is an oscillation of relatively large amplitude with a long period of 10–30 s (A mode), the other is of small amplitude with a short period of 1–5 s (B mode). The A mode is caused by flow rate oscillation of under 2% of the mean flow rate, and the B mode by irregular displacement of the dryout point in the axial direction.

The amplitudes of wall temperature oscillation are shown against the mass flux

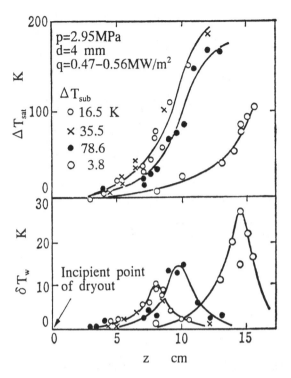

Figure 4.23 Wall superheat and amplitude of temperature fluctuation in transition region of dryout (Yamauchi 1980).

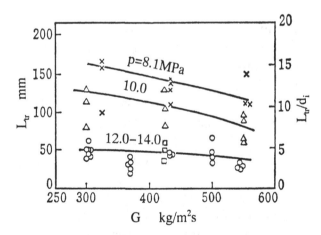

Figure 4.24 Length of transition region of dryout (Roko 1980).

by the dimensionless expression in Figure 4.25 (Sudakov et al. 1979; Roko 1980). The figure indicates that the amplitude decreases with increasing mass flux and with increasing pressure and that the amplitude at A mode is larger than that at B mode.

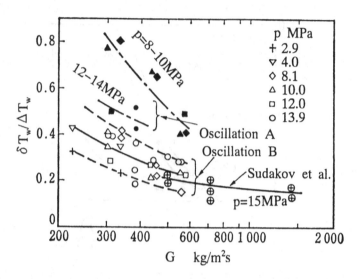

Figure 4.25 Amplitude of wall temperature fluctuation in transition region (Roko 1980).

4.2.5 Heat transfer at high heat flux

Prediction of the value of the allowable maximum heat flux, which is denoted critical heat flux (CHF), is very important for the safety design of steam-generating systems, because some parts of the heating surfaces in boilers and especially in nuclear reactors are in a condition of very high heat flux. In this section, boiling heat transfer characteristics at very high heat flux are described for pool boiling and convective boiling in tubes or channels.

A typical example of the "boiling curve" in pool boiling is shown in Figure 4.26, which shows the relationship between the heat flux and the surface overheat temperature $\Delta T_{sat} = (T_w - T_s)$. The boiling curve was first obtained from an experiment in which a thin platinum wire heating surface was electrically heated at atmospheric pressure by Nukiyama (1934). When the heat flux is increased gradually from point O at $q = 10^3 \, W/m^2$, the temperature of the heating surface or ΔT_{sat} and the heat transfer mode change as follows: With increasing heat flux ΔT_{sat} increases in the natural convective heat transfer region (O to A), and reaches 7 K at a point (A) where nucleate boiling occurs with a heat flux of about $10^4 \, W/m^2$; with further increase of heat flux ΔT_{sat} reaches about $50 \, K$ at $q = 10^6 \, W/m^2$ at point C, where part of the heating surface is covered by a vapor film; in the region C–D film boiling and nucleate boiling exist partially. Then a small increase of the heat flux induces an abrupt rise of the wall temperature and ΔT_{sat} reaches a very high value, that is, $900–1000 \, K$ at point D, where the whole surface is covered by the vapor film (this is denoted as "film boiling"). Such a high wall temperature may cause burnout in ordinary steel tubes.

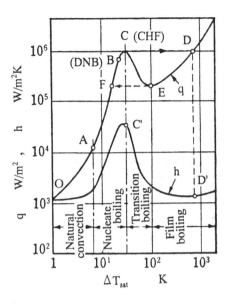

Figure 4.26 Boiling curve in pool boiling.

The phenomenon of abrupt temperature rise is called burnout. The heat flux at point C is referred to as CHF or burnout heat flux. Also, point B is an incipient point from nucleate boiling to transition boiling and is called the "departure from nucleate boiling" (DNB) point. When the heat flux is decreased from the point D, film boiling continues to a point E, an abrupt temperature drop from E to F occurs and, at point F, nucleation boiling is restored.

The heat transfer coefficient obtained from $h = q/\Delta T_{sat}$ is shown in Figures 4.26 and 4.27. These figures show that the value of h increases from 2000 W/m²K at point A' (the end of liquid natural convection) to 35,000 W/m²K at point C' (the burnout point) and then drops remarkably to the small value of 1200 W/m²K at point D'.

The critical heat flux q_{cr} at atmospheric pressure is about 10^6 W/m² as shown above, but is affected by pressure and subcooling. The values of q_{cr} are given by Eq. (4.29) (Kutatelaze 1959) for saturated liquid ($\Delta T_{sub} = 0$) and by Eq. (4.30) (Zuber et al. 1961) for subcooled liquid:

$$\left(q_{cr}/\varrho_G r\right)\Big/\left[\frac{\sigma g(\varrho_L - \varrho_G)}{\varrho_G^2}\right]^{1/4} = 0.16 \tag{4.29}$$

$$\frac{q_{cr(sub)}}{q_{cr}} = 1 + 5.3\left(\frac{\varrho_L}{\varrho_G}\right)^{3/4}\left[\left(\frac{\varrho_L a_L}{\sigma}\right)^2\Big/\sqrt{\frac{\sigma}{g(\varrho_L - \varrho_G)}}\right]^{1/4}\frac{c_{pL}}{r}\Delta T_{sub} \tag{4.30}$$

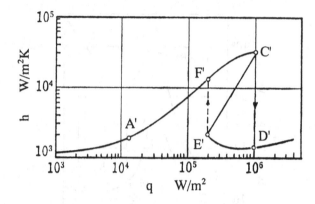

Figure 4.27 Heat transfer coefficient in pool boiling.

where q_{cr} in Eq. (4.30) is evaluated by Eq. (4.29).

The values of critical heat flux at pressures of 0.1–22 MPa calculated from Eqs. (4.29) and (4.30) are shown in Figure 4.28. This figure indicates that q_{cr} takes a maximum value at a pressure of 0.6–0.7 MPa, which is about one-third of the critical pressure, and increases with increasing subcooling.

Next, the critical heat flux in convective boiling in tubes or channels will be described. The critical heat flux depends on the mass flux, tube diameter, tube length, subcooling, and fluid properties. Many correlations of critical heat flux have been presented; however, in the following the prediction of q_{cr} based on Katto's (1981) method will be described. The value of q_{cr} is expressed by

$$q_{cr} = q_{cr0}\left(1 + K\frac{\Delta i_{sub}}{r}\right) \tag{4.31}$$

where K is a constant as shown below, Δi_{sub} is the subcooling at the tube inlet, and r is the latent heat of vaporization. q_{cr0} is the critical heat flux at $\Delta T_{sub} = 0$ and is given by a function of the dimensionless parameters ϱ_G/ϱ_L, Weber number $M = \varrho_L/(G^2\, l)$ and l/d as

$$\frac{q_{cr0}}{Gr} = f\left(\frac{\varrho_G}{\varrho_L}, \frac{\sigma\varrho_L}{G^2 l}, \frac{l}{d}\right) = f(p, G, l, d) \tag{4.32}$$

where l is the tube length and d is the tube inner diameter.

The relationships expressed by Eq. (4.32) are given for four regions, denoted L, H, N, and NP. Figure 4.29 shows these regions on a coordinate system composed of l/d and Weber number M. The region L corresponds to the vicinity of the dryout point, the region N to the vicinity of the DNB point, H to a transition region between L and N, and HP to a high-pressure region that is equivalent to N. The boundaries are expressed by

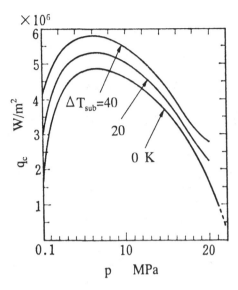

Figure 4.28 Critical heat flux in pool boiling.

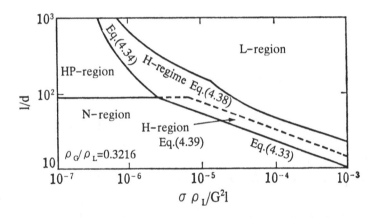

Figure 4.29 Regions of critical heat flux (Katto 1981).

$$\text{H} \sim \text{N:} \, l/d = 77M^{-0.37} \tag{4.33}$$

$$\text{H} \sim \text{HP:} \, l/d = \left[0.384\left(\frac{\varrho_G}{\varrho_L}\right)^{0.467} - M^{0.16}\right] \Bigg/ \left[0.286M^{0.293} - 0.00119\left(\frac{\varrho_G}{\varrho_L}\right)^{0.467}\right] \tag{4.34}$$

The critical heat fluxes q_{cr0} are expressed for each region by the following equations:

$$\text{L:} \, q_{cr0}/Gr = 0.25/(l/d), \quad (l/d > 150) \tag{4.35}$$

$$q_{cr0}/Gr = cM^{0.043}/(l/d)$$
$$c = 0.25 \quad (l/d < 50)$$
$$c = 0.34 \quad (l/d > 150)$$
(4.36)

$$K = 1.043/4cM^{0.043}$$
(4.37)

$$\text{H}: \frac{q_{cr0}}{Gr} = 0.10\left(\frac{\varrho_G}{\varrho_L}\right)^{0.133} M^{1/3}\left(1 + 0.0031\frac{l}{d}\right)^{-1}$$
(4.38)

$$\frac{q_{cr0}}{Gr} = 0.098\left(\frac{\varrho_G}{\varrho_L}\right)^{0.133} M^{0.433}\left(\frac{l}{d}\right)^{0.27}\left(1 + 0.0031\frac{l}{d}\right)^{-1}$$
(4.39)

$$K = \frac{5}{6}\left(0.0124 + \frac{d}{l}\right)\left(\frac{\varrho_G}{\varrho_L}\right)^{-0.133} M^{-1/3}$$
(4.40)

$$K = 0.416\left(0.0221 + \frac{d}{l}\right)\left(\frac{d}{l}\right)^{0.27}\left(\frac{\varrho_G}{\varrho_L}\right)^{-0.133} M^{-0.433}$$
(4.41)

$$\text{N}: \frac{q_{cr0}}{Gr} = 0.98\left(\frac{\varrho_G}{\varrho_L}\right)^{0.133} M^{0.433}\left(\frac{l}{d}\right)^{0.27}\left(1 + 0.0031\frac{l}{d}\right)^{-1}$$
(4.42)

$$\text{HP}: \frac{q_{cr0}}{Gr} = 0.0384\left(\frac{\varrho_G}{\varrho_L}\right)^{0.60} M^{0.173}\left(1 + 0.28M^{0.233}\frac{l}{d}\right)^{-1}$$
(4.43)

$$K = 0.0138\left\{215M^{0.54} + \frac{d}{l}\right\}\left(\frac{\varrho_G}{\varrho_L}\right)^{-0.65} M^{-0.456}$$
(4.44)

Figure 4.30 shows an example of the results for q_{cr} calculated by these equations and also the quality at the tube outlet $x_{e(out)}$. This figure indicates clearly that the critical heat flux increases with increasing mass flux and subcooling, and, in a case of low mass flux of $G = 500\,\text{kg/m}^2\text{s}$ and $\Delta T_{sub} = 0$ (in the L region), the critical heat flux q_{cr0} is $0.85 \times 10^6\,\text{W/m}^2$ and the outlet quality $x_{e(out)}$ is 0.86. That is rather a high quality; therefore burnout is expected to be induced by dryout, not by film boiling caused by high heat flux. On the other hand, in the case of a high mass flux of $3000\,\text{kg/m}^2\text{s}$ and $\Delta T_{sub} = 0$ (in the H or N region), q_{cr0} is increased to $1.65 \times 10^6\,\text{W/m}^2$ and $x_{e(out)}$ decreases to 0.3; therefore this burnout is due to DNB.

Channels in the core of a boiling water reactor are composed of fuel rod bundles and a square channel box as shown below in Figure 4.67. The critical heat flux in these channels can be predicted by the above correlation using the equivalent hydraulic diameter of the channel instead of the tube diameter. For the severe requirement of accurate prediction of the critical heat flux in boiling water reactor (BWR) channels, a large number of experiments have been conducted in many laboratories as described in Section 4.8.4. Here, we first introduce the Hench–Levy correlation, which was used in the early period of BWR design

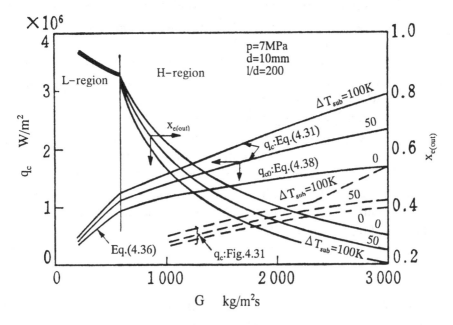

Figure 4.30 Critical heat flux in forced convective boiling.

(Healzer et al. 1966). The critical heat flux q_{cr} is expressed as a function of the mass flux, channel outlet quality, and pressure as:

$$q_{cr} = f(G, x_e, p) \tag{4.45}$$

Values of q_{cr} are expressed for three regions of the thermodynamic equilibrium quality x_e by the following equations:
Region 1

$$x_e < x_1: \quad q_{cr} = \left\{3.47 - 8.86 \times 10^{-2}(p - 4.14)^{1.25}\right\} \times 10^3 q_{cr0} \tag{4.46}$$

Region 2

$$x_1 < x_e < x_2: \quad q_{cr} = \left\{1.9 - 3.3x_e - 0.7\tan h^2\left(\frac{2.21G}{10^3}\right)\right\} q_{cr0} \tag{4.47}$$

Region 3

$$x_2 < x_e: \quad q_{cr} = \left\{0.6 - 0.7x_e - 0.09\tan h^2\left(\frac{1.47G}{10^3}\right)\right\} q_{cr0} \tag{4.48}$$

$$\left.\begin{aligned} x_1 &= 0.273 - 0.212\tan h^2\left(2.21G/10^3\right) \\ x_2 &= 0.50 - 0.269\tan h^2\left(2.21G/10^3\right) \end{aligned}\right\} \tag{4.49}$$

where x_e is the thermodynamic equilibrium quality at the channel outlet and x_1 and x_2 are values defined by Eq. (4.49). This correlation is applicable for $G = 0.27 \times 10^3$ to $2.17 \times 10^3 \, \text{kg/m}^2\text{s}$, $p = 4.1$–$10 \, \text{MPa}$, equivalent hydraulic diameter $d_e = 8.2$–$12.3 \, \text{mm}$, and the minimum clearance between rods is larger than $1.5 \, \text{mm}$. Figure 4.31 shows an example of the values of q_{cr} calculated from these equations. Region 1 where q_{cr} is constant corresponds to the state of low quality and high heat flux, region 3 where q_{cr} decreases slightly with x_e corresponding to the state near the dryout point with high quality and low heat flux, and region 2 where q_{cr} decreases with x_e is the transition state between 1 and 3. The tendency for q_{cr} to decrease with increase of mass flux which is apparently seen in Figure 4.31 may be considered unreasonable, but this is not so in practice, as is explained below.

The outlet quality x_e is determined by the heat balance and expressed as a function of the mass flux G and heat flux q_{cr} by

$$x_e = \left(\frac{4q_{cr}l}{Gd} - \Delta i_{sub} \right) \Big/ r \tag{4.50}$$

Consequently, at a channel of definite length, the increase of mass flux causes x_e to decrease, then q_{cr} increases in regions 2 and 3. This tendency can be seen in Figure 4.31, which shows the result calculated from the Hench–Levy correlation. Furthermore, the tendency agrees well with the result calculated from Katto's correlation (Eqs. (4.31)–(4.38)), but the former values are one-half to one-third times smaller than the latter. The reason for this is that the Hench–Levy correlation (Eqs. (4.46)–(4.48)) is made for the safety design to express the lower limit of the experimental data for q_{cr}, as shown below in Figure 4.82, whereas Katto's correlation (Eqs. (4.31) and (4.38)) corresponds to the average value of a number of experimental data.

Next, a correlation on the critical heat flux based on "boiling length" (L_B) proposed by Hewitt et al. (1970) will be explained. The method has been used widely and is applicable for not only a uniform heat flux distribution but also an arbitrary distribution. The boiling length is also used in the GEXL (General Electric critical quality (x_{cr}) – boiling length (L_B)) correlation described in Section 4.8.4 concerning the design of boiling water reactors. The boiling length is defined as the distance from the incipient point of bulk boiling to the burnout point, or the heated length over which the steam quality is greater than zero. The framework of the correlation is as follows. The critical quality x_{cr} is presumed to be expressed, using the boiling length L_B, as

$$x_{cr} \times K_1(G) \times K_2(p) = f\big(L_B \times K_3(G) \times K_4(d) \big) \tag{4.51}$$

The relationship is given by a curve determined experimentally or by a five-order polynomial, where $K_1(G)$, $K_2(p)$, $K_3(G)$, and $K_4(d)$ are functions of only G, p, G, and d, respectively, and are expressed by a curve determined experimentally or

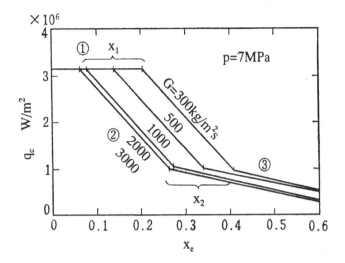

Figure 4.31 Critical heat flux by Hench–Levy's correlation.

by polynomials (K_1, K_2, K_3: five-order, K_2: seven-order). The correlation is therefore expressed by using 35 constants. These are not shown in this book; only the process of predicting the critical heat flux is explained in the following example.

For the case of $p = 7\,$MPa, equivalent hydraulic diameter $d = 10\,$mm, channel length $L_T = 3\,$m, $G = 2000\,$kg/m^{2}s, inlet temperature $T_{in} = 235.8°$C (subcooling $\Delta i_{sub} = 248.6\,$kJ/kg), and $r = 1506\,$kJ/kg, the critical quality x_{cr} and the critical heat flux q_{cr} will be obtained in the following manner. For these conditions the following values are obtained from the original curves or polynomials:

$$K_1(G) = 1.15, \qquad K_2(p) = 1.0, \qquad K_3(G) = 1.07, \qquad K_4(d) = 3.12$$

Next, using the value of L_T as a first approximation for L_B, the value of $L_B \times K_3(G) \times K_4(d) = 3 \times 1.07 \times 3.12 = 10\,$m is obtained. Then the value of $x_{cr} \times K_1(G) \times K_2(p) = 0.58$ is obtained from Eq. (4.51). The critical quality x_{cr} can therefore be obtained as

$$x_{cr} = 0.58 / \big(K_1(G) \times K_2(p)\big) = 0.504$$

and the preheating length L_s is given from the heat balance as

$$L_s = \Delta i_{sub} \cdot L_B / x_{cr} r \tag{4.52}$$

The boiling length L_B is then determined as

$$L_B = L_T - L_s = 3.0 - 248.6 \times 3 / (0.504 \times 1506) = 3.0 - 0.98 = 2.02\,\text{m}$$

Using the value of 2.02 m as the second approximation for L_B, values of $x_{cr} = 0.504$ and $L_s = 0.619\,$m are obtained. By repeating the calculation, the converged values of $L_B = 2.26\,\text{m}$ and $x_{cr} = 0.504$ can be obtained. Thus the critical heat flux q_{cr} is given from the heat balance as

$$q_{cr} = x_{cr}Grd/4L_B \tag{4.53}$$

Finally, the critical heat flux is determined as

$$q_{cr} = 0.504 \times 2000 \times 0.01 \times 1506 \times 2.26 = 1679\,\text{kW/m}^2 = 1.679 \times 10^6\,\text{W/m}^2$$

4.3 Basic equations and flow characteristics of gas–liquid two-phase flows

4.3.1 Conservation equations of mass and momentum

Basic equations of gas–liquid two-phase flow, which are available for the thermal and hydraulic planning and design of steam-generating systems, are briefly reviewed in this section. Because the flow configuration of two-phase flow shows significant complexity, as in the flow pattern described in Section 4.1, theoretical models have been intensively developed for this flow pattern in the last two or three decades (Akagawa 1974; Ueda 1981; Hetstroni 1982; Handbook of Fluid Mechanics 1987; Handbook of Gas–Liquid Two-Phase Flow Technology 1989). In this book, however, the detailed description of such models is bypassed and, instead, simplistic yet basic equations that can be universally applicable are described.

The relevant models discussed in this section are the homogeneous model and separated flow model of two-phase flow. The homogeneous flow model is derived from the separated flow model as a special case. In the separated flow model (referred to as the two-fluid model in recent advanced modeling in this class), two phases are assumed to be separated by a hypothetical smooth interface as shown in Figure 4.32. The ratio of the cross-sectional area occupied by the gas phase to the total flow area is referred to as the void fraction. A similar definition is usually used generally, even in the homogeneous model in which an average volumetric concentration of the gas phase is also defined as the void fraction.

The mass conservation equations are expressed by

$$\left.\begin{aligned}
\frac{\partial}{\partial t}(\varrho_G \alpha) + \frac{\partial}{\partial z}(\varrho_G \alpha u_G) &= m_G/A \\
\frac{\partial}{\partial t}\{\varrho_L(1-\alpha)\} + \frac{\partial}{\partial z}\{\varrho_L(1-\alpha)u_L\} &= m_L/A
\end{aligned}\right\} \tag{4.54}$$

where u_G and u_L represent the average velocities of the gas and liquid phases, respectively, A is the cross-sectional area, and m_G and m_L are the mass flow rates through the gas–liquid interface, that is, owing to the phase change, per unit

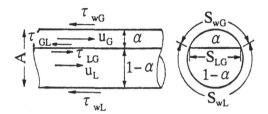

Figure 4.32 Separated flow model.

length. These two terms appear as a source term in the mass conservation equation, and $m_G = -m_L$.

The momentum equations for the steady state of each phase are expressed by the following equations:

$$\frac{\partial}{\partial t}(\varrho_G \alpha u_G) + \frac{1}{A}\frac{\partial}{\partial z}(\varrho_G \alpha u_G^2 A) + \alpha\frac{\partial p}{\partial z} + \varrho_G g \alpha \sin\theta$$

$$+ \frac{S_{wG}}{A}\tau_{wG} + \frac{S_{GL}}{A}\tau_{GL} + \frac{M_{GL}}{A} = 0 \tag{4.55}$$

$$\frac{\partial}{\partial t}\left\{\varrho_L(1-\alpha)u_L\right\} + \frac{1}{A}\frac{\partial}{\partial z}\left\{\varrho_L(1-\alpha)u_L^2 A\right\} + (1-\alpha)\frac{\partial p}{\partial z}$$

$$+ \varrho_L g(1-\alpha)\sin\theta + \frac{S_{wL}}{A}\tau_{wL} + \frac{S_{GL}}{A}\tau_{LG} + \frac{M_{LG}}{A} = 0 \tag{4.56}$$

where S is the peripheral length of the interface between the gas and the liquid, τ is the shear stress, M is the momentum transfer due to the phase change, that is, evaporation or condensation, and θ is the inclination angle of the tube relative to the vertical axis. Subscripts wg, wl, and gl represent the interface between wall and gas, wall and liquid, and gas and liquid, respectively.

Since, in steam-generating tubes, M_{GL} is equal to $-M_{LG}$, the momentum equation for the two-phase mixture can be expressed by the following equation, which is the sum of Eqs. (4.55) and (4.56):

$$\frac{\partial}{\partial t}\left\{\varrho_G \alpha u_G + \varrho_L(1-\alpha)u_L\right\} + \frac{1}{A}\frac{\partial}{\partial z}\left\{\varrho_G \alpha A u_G^2 + \varrho_L(1-\alpha)A u_L^2\right\}$$

$$+ \frac{\partial p}{\partial z} + \left\{\varrho_G \alpha + \varrho_L(1-\alpha)\right\}g\sin\theta + \frac{1}{A}\left(S_{wG}\tau_{wG} + S_{wL}\tau_{wL}\right) = 0 \tag{4.57}$$

The terms represent, respectively, acceleration in time, acceleration in space, pressure difference, gravity force, and shear force. The shear force is the sum of the shear forces on the interface between wall and gas, and that between wall and liquid, which is usually expressed as the frictional pressure drop denoted by $(-\partial p/\partial z)_f$. That is,

$$\left(-\frac{\partial p}{\partial z}\right)_f = \frac{1}{A}\left(S_{wG}\,\tau_{wG} + S_{wL}\,\tau_{wL}\right) \tag{4.58}$$

For homogeneous flow, the momentum equation is reduced, substituting $\alpha = 0$, $\varrho_L = \varrho$ and $u_L = u$ into the equation, to

$$\frac{\partial(\varrho u)}{\partial t} + \frac{\partial}{\partial z}\left(\varrho u^2\right) + \frac{\partial p}{\partial z} + \varrho g \sin\theta + \left(-\frac{\partial p}{\partial z}\right)_f = 0 \tag{4.59}$$

This equation coincides with the well-known momentum equation of single-phase flow. Consequently, Eq. (4.59) can be applied to homogeneous two-phase flow, in which ϱ is defined as the average density of the two-phase mixture and u as the average velocity of the mixture across the tube cross-section.

4.3.2 Void fraction

Exact prediction of the void fraction is quite important for the thermal and hydraulic design of steam-generating systems, because the gravity term and the acceleration term in the momentum equations are functions of the void fraction, and also because the reactivity in nuclear reactors is affected by the void fraction (void reactivity).

The void fraction is a function of the volume flow rates (volumetric flux) j_G, j_L, the densities ϱ_G and ϱ_L, the viscosities μ_G and μ_L, the surface tension σ, the geometrical dimensions of the channel, and the mass fluxes of steam and water. This complexity results in a number of correlations linking these parameters. In a steam–water two-phase flow, however, the properties of steam and water are functions of pressure; consequently the void fraction reduces to a function of the mass fluxes, the quality, and the channel dimension.

In this section, three correlations that are convenient for practical application will be reviewed. Smith's (1969) empirical formula for the void fraction is given by Eq. (4.60), which was obtained from experimental data over wide ranges of pressure and flow rates:

$$\alpha = \left[1 + 0.4\frac{\varrho_G}{\varrho_L}\left(\frac{1-x}{x}\right) + 0.6\frac{\varrho_G}{\varrho_L}\left(\frac{1-x}{x}\right)\left\{\frac{\dfrac{\varrho_L}{\varrho_G} + 0.4\left(\dfrac{1-x}{x}\right)}{1 + 0.4\left(\dfrac{1-x}{x}\right)}\right\}^{1/2}\right]^{-1} \tag{4.60}$$

According to this formula, the void fraction is determined only by the quality and the density ratio, ϱ_L/ϱ_G, which is a function of the pressure independent of the mass flow rate and the channel dimensions. The values given by Eq. (4.60) are represented by solid lines in Figure 4.33.

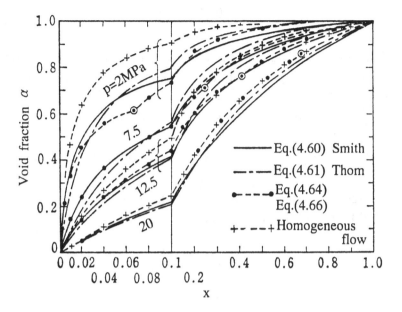

Figure 4.33 Void fraction.

Thom's (1964) formula is given by

$$\left.\begin{array}{l} \alpha = \left[1 + \dfrac{\varrho_G}{\varrho_L}\dfrac{1-x}{x}K\right]^{-1} \\[3mm] K = 0.93\left(\dfrac{\varrho_L}{\varrho_G}\right)^{0.11} + 0.07\left(\dfrac{\varrho_L}{\varrho_G}\right)^{0.56} \end{array}\right\} \tag{4.61}$$

Thom's correlation is shown by dot–dash lines in Figure 4.33. These curves by Eqs. (4.60) and (4.61) agree well in the figure, and are well verified by comparison with the experimental data.

Furthermore, it should be noted that the void fraction is linearly proportional to the volume flow fraction β, as is expressed by Eqs. (4.63) and (4.64) in a very simple form (Armand's expression) (Armand 1946, 1947, 1950):

$$\beta \equiv j_G/\left(j_G + j_L\right) \tag{4.62}$$

$$\alpha = K\beta = K\dfrac{j_G}{j_G + j_L} \quad \left(\beta < 0.86\right) \tag{4.63}$$

$$= K\left[\dfrac{\varrho_L}{\varrho_G}x \middle/ \left\{1 + x\left(\dfrac{\varrho_L}{\varrho_G} - 1\right)\right\}\right] \tag{4.64}$$

in which the constant K is a unique function of the pressure, as proposed by Bankoff's (1960) formula:

$$K = 0.71 + 0.0143p \quad (p, \text{MPa}) \tag{4.65}$$

Thus the void fraction can be predicted by the simple expression of Eq. (4.64), and the values shown in Figure 4.33 also agree approximately with those of Eqs. (4.60) and (4.61). Here, Eq. (4.63) holds in the range of $\beta < 0.86$, and the void fraction in the range of $\beta > 0.86$ is calculated by Eq. (4.66), which is derived on the basis of a linear relationship to meet two distinct points, $\alpha = \alpha_{0.86}$ at $\beta = 0.86$ and $\alpha = 1$ at $\beta = 1$:

$$\alpha = \alpha_{0.86} + \frac{1 - \alpha_{0.86}}{1 - 0.86}(\beta - 0.86) \quad (\beta > 0.86) \tag{4.66}$$

The void fraction can also be expressed in terms of the slip ratio, $S = u_G/u_L$; this is referred to as the "slip flow model":

$$\alpha = \left[1 + \left(\frac{1}{\beta} - 1\right)S\right]^{-1}$$

$$\alpha = x \Big/ \left[\frac{\rho_L}{\rho_G}(1 - x)S + x\right] \tag{4.67}$$

Many correlations of the slip ratio can be found in the literature on two-phase flow, and the void fraction can be predicted by Eq. (4.67) together with these correlations. In general, the slip ratio ranges from $S = 1–5$, being 1 at $\alpha = 0$ ($x = 0$) and increasing with the increase in the quality x.

The void fraction in the homogeneous two-phase flow model, that is, $S = 1$, is shown in Figure 4.33 in comparison with those of the other correlations mentioned above. The figure indicates the applicability of such a simple homogeneous model in the high-pressure region.

4.3.3 Pressure drops

Frictional pressure drop in a two-phase flow is affected by a number of parameters, as can be easily understood from the complicated configuration of two-phase flows. It is therefore quite difficult to evaluate the frictional pressure drop by a purely theoretical approach. Thus a number of empirical correlations have been proposed for practical use. In this section, the representative prediction method of pressure drop, that is, Thom's (1964) method, is reviewed; this has been verified with sufficient accuracy for practical use.

The frictional pressure drop in a steam–water two-phase flow, $(-dp/dz)_f$, is correlated in dimensionless form with pressure and quality, as shown in Figure 4.34:

$$\phi_{L0}^2 \equiv \frac{\left(-\dfrac{dp}{dz}\right)_f}{\left(-\dfrac{dp}{dz}\right)_{L0}} = f(p, x) \tag{4.68}$$

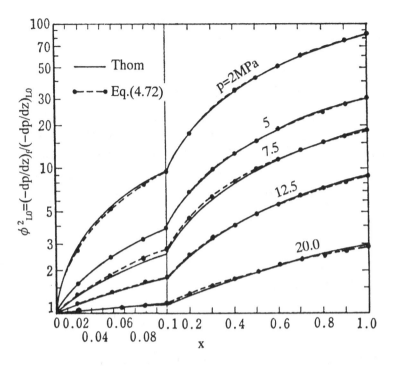

Figure 4.34 Frictional pressure drop in two-phase flow.

where ϕ_{L0}^2 is referred to as the two-phase frictional multiplier, and $(-dp/dz)_{L0}$ is the single-phase pressure drop of the liquid flow when the mass flow rate is equal to the total mass flow rate of the two-phase mixture. The single-phase pressure drop is calculated from Eq. (4.69) by using the volumetric flux of the liquid j_{L0} and the friction factor for turbulent flow λ_{L0} that is expressed by the following term of [ ]:

$$\left(-\frac{dp}{dz}\right)_{L0} = \frac{\lambda_{L0}}{d}\frac{\varrho_L}{2}j_{L0}^2 = \left[\frac{0.316}{\mathrm{Re}_0^{0.25}}\right]\frac{1}{d}\frac{\varrho_L}{2}j_{L0}^2 = \frac{0.316}{\left(j_{L0}d\varrho_L/\mu_L\right)^{0.25}}\frac{1}{d}\frac{\varrho_L}{2}j_{L0}^2 \quad (4.69)$$

Since the total mass flux is constant along a steam-generating tube, that is, $\varrho_L j_{L0}$ = constant, the single-phase pressure drop $(-dp/dz)_{L0}$ is also constant along the tube length. Therefore, the value of ϕ_{L0}^2 in Figure 4.34 gives the two-phase flow friction multiplier at a given location of the quality x. For example, the two-phase friction multiplier $\phi_{L0}^2 = 6$ for $p = 7.5\,\mathrm{MPa}$ and $x = 0.3$. This means that the frictional pressure gradient at the given location is six times larger than that in single-phase flow of the liquid.

Values of ϕ_{L0}^2 can be expressed approximately by simple forms under an assumption of the homogeneous flow model as follows. The frictional pressure drop is postulated to be expressed by a form of the second term in Eq. (4.70a):

$$\left(-\frac{dp}{dz}\right)_f = \frac{\lambda_{TP}}{d}\frac{\varrho_{TP}u_{TP}^2}{2} = \frac{\lambda_{TP}}{d}\frac{1}{\varrho_{TP}}\frac{\left(\varrho_{TP}u_{TP}\right)^2}{2} = \frac{\lambda_{TP}}{d}\frac{G^2}{2}v_{TP} \tag{4.70a}$$

$$= \frac{\lambda_{TP}}{d}\frac{G^2}{2}v_L\left\{1 + \left(\frac{\varrho_L}{\varrho_G} - 1\right)x\right\} \tag{4.70b}$$

where λ_{TP}, ϱ_{TP}, and u_{TP} are the friction factor, average density, and mean velocity of the homogeneous two-phase mixture, respectively. Rearrangement of the equation gives the fourth term of Eq. (4.70a) and also Eq. (4.70b), where $G = u_{TP}\varrho_{TP}$ (kg/m²s) is the total mass flux, $v_L = (1/\varrho_L)$ is the specific volume, and $v_{TP} = (1/\varrho_{TP})$ is the specific volume of the two-phase mixture, that is,

$$v_{TP} = \frac{1}{\varrho_{TP}} = v_L\left\{1 + \left(\frac{\varrho_L}{\varrho_G} - 1\right)x\right\}$$

Thus ϕ_{L0}^2 is expressed by Eq. (4.71) from the combination of Eq. (4.69), Eq. (4.70b), and the relationship $G = \varrho_L u_{L0} = \varrho_L j_{L0}$:

$$\phi_{L0}^2 = \frac{\lambda_{TP}}{\lambda_{L0}}\left\{1 + x\left(\frac{\varrho_L}{\varrho_G} - 1\right)\right\} \tag{4.71}$$

Assuming that the friction factor of two-phase flow, λ_{TP}, is nearly equal to that of the liquid, λ_{L0}, Eq. (4.71) is reduced to

$$\phi_{L0}^2 = \frac{\left(-\dfrac{dp}{dz}\right)_f}{\left(-\dfrac{dp}{dz}\right)_{L0}} = 1 + x\left(\frac{\varrho_L}{\varrho_G} - 1\right) \tag{4.72}$$

Thus the two-phase friction multiplier ϕ_{L0}^2 can be expressed in such very simple form. The values by Eq. (4.72) agree well with those by Thom's correlation, as shown in Figure 4.34.

According to Idsinga et al.'s (1977) assessment of the accuracy of each correlation, Thom's (1964) correlation, the correlation of Eq. (4.72), Baroczy's (1966) correlation, and the correlation of Eq. (4.74) are classified in the best rank of accuracy among 18 correlations. This assessment was conducted through a comparison with 3450 experimental data points in the ranges of $p = 1.7$–10.3 MPa, $G = 270$–4340 kg/m²s, $x = 0$–1.0, and equivalent hydraulic diameter $d_e = 2.3$–33 mm. The friction multiplier ϕ_{L0}^2 is expressed by Eq. (4.74), including the effect of the viscosity under the assumptions that λ_{TP} is expressed in a way similar to that of single-phase flow, that is, $\lambda_{TP} = 0.3164/(Gd_e/\mu_{TP})^{0.25}$, and that the viscosity of the two-phase mixture μ_{TP} is expressed by Eq. (4.73):

$$\mu_{TP} = \mu_L\left\{1 + x\left(\frac{\mu_L}{\mu_G} - 1\right)\right\} \tag{4.73}$$

$$\phi_{L0}{}^2 = \left\{1 + x\left(\frac{\mu_L}{\mu_G} - 1\right)\right\}^{0.25} \left\{1 + x\left(\frac{\varrho_L}{\varrho_G} - 1\right)\right\} \tag{4.74}$$

Thus the multiplier $\phi_{L0}{}^2$ is expressed by a function of only the quality, independently of the mass flux. The influence of the mass flux is taken into account in Baroczy's (1966) correlation, but the accuracy using this method is on the same order as Thom's correlation. Therefore, Baroczy's correlation is not descubed here.

Local pressure drops by a contraction, an enlargement of channel cross-section, or by bends, $(-dp/dz)_b$, can be evaluated, similarly to Eq. (4.72), by

$$\phi_{L0}{}^2 \equiv \frac{\left(-\dfrac{dp}{dz}\right)_b}{\left(-\dfrac{dp}{dz}\right)_{b0}} = 1 + Cx\left(\frac{\varrho_L}{\varrho_G} - 1\right) \tag{4.75}$$

where $(-dp/dz)_{b0}$ is the single-phase pressure drop when the total mass of the two-phase mixture flows as liquid phase, and C is an empirical constant: $C = 1$ for enlargement or contraction of the channel or the cylindrical nozzle, $C = 1.5$ for gate valves, $C = 1.6$ for tees, $C = 2.2$–4 for bends ($p/p_{cr} > 0.1, x = 0$–0.5), and $C = 1.5$ for bends of small curvature ($p/p_{cr} < 0.015, x = 0$–0.1), where p_{cr} is the critical pressure of the fluid (Rohsenow and Hartnett 1973).

The accelerational pressure drop in a steam-generating tube, which is expressed in the second term of Eq. (4.57), is obtained by integrating between two locations represented by subscripts 1 and 2, and is expressed, independently of the tube length, by

$$\Delta p_a = \frac{G^2}{\varrho_L}\left[\left(\frac{x_2{}^2}{\alpha_2} - \frac{x_1{}^2}{\alpha_1}\right)\frac{\varrho_L}{\varrho_G} + \left\{\frac{(1 - x_2)^2}{1 - \alpha_2} - \frac{(1 - x_1)^2}{1 - \alpha_1}\right\}\right] \tag{4.76}$$

In a steam generating tube of $x_1 = 0$ at the inlet and $x_2 = x$ at the exit, the accelerational pressure drop is expressed by

$$\Delta p_a = \frac{G^2}{\varrho_L}\left[\frac{\varrho_L}{\varrho_G}\frac{x^2}{\alpha} + \frac{(1 - x)^2}{1 - \alpha} - 1\right] \tag{4.77}$$

The accelerational pressure drop can also be obtained as a function of x at the tube exit by substituting Eq. (4.60), (4.61), (4.64), or (4.66) into Eq. (4.77).

4.4 Thermohydraulic aspects of the evolution of electricity-generating plants

The increase in electricity demand for the development of people's livelihood, industry, and transportation has been very rapid in the past 50 years. The growth rate of total output in electricity-generating power plants in Japan, for example, is shown in Figure 4.35. The growth rate of the capacity was about 12% per year from 1960 to 1973, when the first oil shock occurred. Since then, the growth rate is still large but is linear with time. Most of the electricity is generated by steam power plants, that is, fossil-fuel-fired and nuclear power plants.

The increase in demand for electricity generation has resulted in power plants with larger unit capacity and higher efficiency. The unit capacity has reached 1000 MW per unit in fossil-fuel-fired plants and 1300 MW in nuclear plants, as shown in Figure 4.36. Also, the steam pressure in boilers has risen so as to improve the thermal cycle efficiency, as explained in Section 2.3. The transition to the highest steam pressure in boilers is also shown in Figure 4.36.

The evolution of boiler type, classified by the water circulation in a broad sense (see Chapter 1), as shown schematically in Figure 4.36, is the outcome of the rise in unit capacity and boiler pressure. The historical transition from the natural-circulation to the forced-circulation and finally to the once-through type is clearly seen in the figure. The steam pressure in boilers rose year by year from 6 MPa in 1955 to 19 MPa in 1961. In such a high pressure the density difference between the steam and the water diminishes and forced flow is adopted to supply enough water to the evaporator tubes. The first forced-circulation boiler (16.9 MPa, 522 t/h, 220 MW) in Japan, for an electric power plant, was built in 1959.

In this period, the evolution of boiler type was very fast in Japan, and only three years later, in 1962, the first once-through boiler of subcritical pressure (17.4 MPa, 569° C, 500 t/h, 220 MW) for a large electric power plant started operating. With further high pressure, that is, in the supercritical pressure range, the boiler inevitably becomes the once-through flow type, because the phase change from water to steam takes place gradually. In 1967, the first supercritical pressure once-through power plant (24.6 MPa) was put into operation. The forced-circulation boiler made further progress and a controlled-circulation boiler (16.9 MPa, 566° C, 250 MW) was built, in which the circulation loop was designed to utilize natural circulation force to help the forced circulation to enhance the water circulation, as presented in detail in Sections 4.7.1 and 4.7.2. The circulation ratio in this boiler is 2.72, which is considerably smaller than the value of 4 in the former 16.9 MPa class of forced-circulation boilers.

The large power plants of the once-through boiler had usually been used for base load. However, due to the increase in nuclear power plants that have been operated under a constant base load, fossil-fuel-fired plants have been assigned to meet the load variation. Therefore, for even the supercritical pressure once-through boiler, partial load operation down to 10–25% was required, and the supercritical sliding pressure plant was developed to improve the efficiency at such a low partial load. In 1980 the first plant (25.5 MPa, 541, 566° C, 1950 t/h, 600 MW) was put into operation.

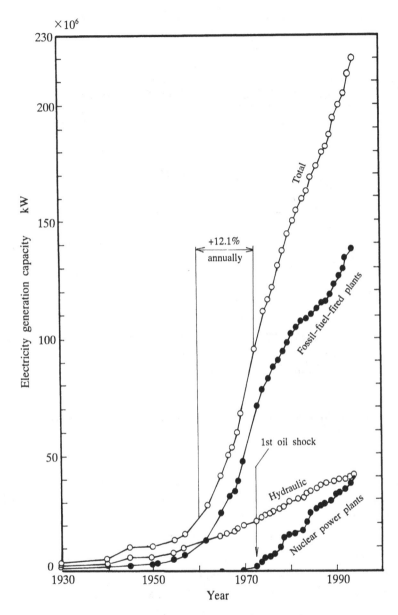

Figure 4.35 Transition of electricity-generation capacity in Japan.

The worldwide transition of the boiler type mentioned above is more or less similar to the Japanese example presented in Figure 4.37. In the drawing, boilers classified by pressure are shown over a period from 1955 to 1975 and boilers classified by water circulation are shown over a period from 1976 to 1995. The ordinate indicates the ratio of the output of each group to the total output in

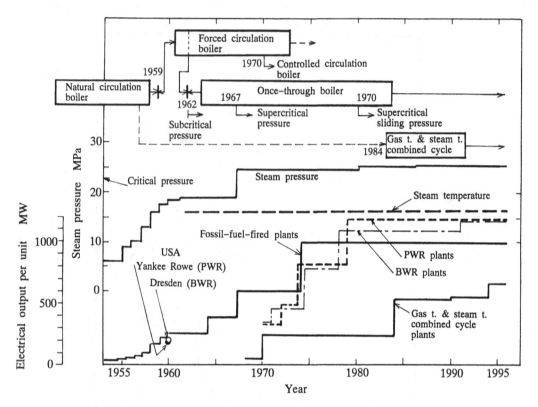

Figure 4.36 Evolution of electricity generation plants in Japan.

plants built in the year. Also, the total output of plants built each year is shown. It can be clearly seen that all the plants built before 1957 were of 12.7 MPa class (natural circulation boiler). After 1958 the ratio of the 16.9 MPa class boiler increased, and finally in 1966 all boilers built were of this class. Since the late 1960s, the ratio of the supercritical pressure class (once-through boiler) increased rapidly, and after 1976 most of the newly built power plants were once-through boilers, either of the fixed pressure type or of the sliding pressure type. The figure also indicates that forced-circulation boilers, including the controlled-circulation type, were built from 1977 to 1989 and were not built thereafter for the large electric power plants.

In Chapter 1, the logical sequence of development of boiler technology is described according to the intrinsic law of boiler development, which states that, after the era of the natural-circulation water-tube boiler, the transition to the once-through boiler via the forced-circulation boiler as the intermediate type is inevitable. The transition and its evolution can be clearly demonstrated in the history of large power plants shown in Figures 4.36 and 4.37.

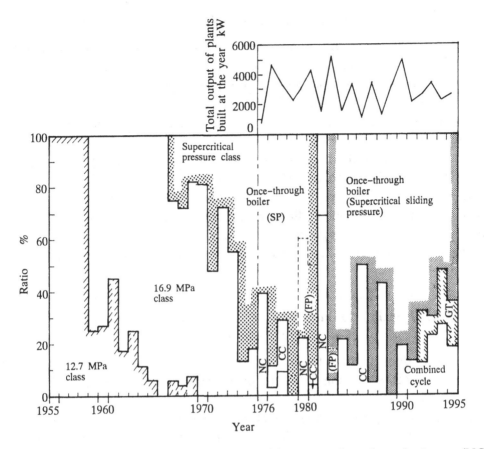

Figure 4.37 Transition of boiler types in electricity generation plants in Japan. (NC) Natural-circulation boiler; (CC) controlled-circulation boiler; (FP) supercritical pressure boiler (fixed pressure); (SP) supercritical sliding pressure boiler; (GT) gas turbine plants.

The demand for still higher efficiency of power plants led to the development of power plants that combine gas turbines with steam turbines as described in Section 2.4.8. The plant efficiency of the recent combined cycle plants is 45–47% and will reach 50–55% by elevating the gas turbine inlet temperature, whereas the highest plant efficiency in the present supercritical pressure once-through boiler power plants is 40–42%. Thus, combined cycle power plants are increasingly adopted as large electric power plants, as shown in Figure 4.37. Gas turbine plants are also increasingly being adopted. In a recent combined cycle power plant built in 1994, for example (steam turbine output, 58,750 kW; gas turbine output, 96,210 kW per unit), the steam pressure is 7.46 MPa, the temperature is 530°C, and the boiler capacity is 210 t/h. As the boiler pressures are low in combined cycle power plants, in the range of 5.7–7.46 MPa, it is sensible to select boilers of the natural-circulation type.

4.5 Thermal and hydraulic design of natural-circulation boilers

In the natural-circulation boiler, water is supplied to the evaporator tubes by natural circulation in the circulation loop without the use of a water-circulating pump. This type is widely used for medium- and small-capacity boilers because of its simple construction and easy operation. For large-capacity boilers in electric power stations, however, the use of natural-circulation boilers is limited, as mentioned in Section 4.4.

In this section, the thermal and hydraulic planning and design of the natural-circulation loop is described. The discussion is devoted to determining the arrangement of evaporator tubes, tube diameter, and number of tubes, and also the downcomer arrangement, diameter and number of downcomers corresponding to the design of the evaporator system, at given conditions of parameters such as the evaporation rate, heat load in the furnace, and heat fluxes at heating surfaces.

4.5.1 Principles of natural water circulation and characteristics of circulation velocity

The average density of the two-phase mixture in evaporator tubes is lower than that in a downcomer owing to the steam generation. This density difference produces a driving force that circulates water from an upper drum through the downcomer to the evaporator tubes, as shown in Figure 4.38. The velocity of the circulating water at the inlet of the evaporator tube, referred to as the "circulation velocity," can be predicted as follows.

Under the assumption of a steady flow condition, integration of the momentum equation (4.59) along the downcomer is expressed by

$$p_2 - p_1 = -\left(\zeta_1 + \zeta_2 + \lambda_F \frac{L_F}{d_F} \right) \frac{\varrho_L u_F^2}{2} + \varrho_L g H \qquad (4.78)$$

where p_1 and p_2 are pressures in the upper drum and the lower header, respectively; ζ_1, ζ_2, ζ_3, ζ_4 represent local pressure loss factors at the inlet and outlet of the downcomer and the inlet and outlet of the evaporator tube, respectively; λ_F is a friction factor of the downcomer; and H is the height difference between the upper drum level and the lower header level. On the other hand, the integration of Eq. (4.57) along the evaporator tube gives

$$p_2 - p_1 = \zeta_4 \left\{ \frac{x_e^2}{a_e \varrho_G} + \frac{(1 - x_e)^2}{(1 - a_e)\varrho_L} \right\} \frac{(\varrho_L u_0)^2}{2}$$

$$+ \zeta_3 \frac{\varrho_L u_0^2}{2} + \int_0^{L_s} \left(-\frac{dp}{dl} \right)_f dl + g\int_0^H \{a\varrho_G + (1 - a)\varrho_L\} dz \qquad (4.79)$$

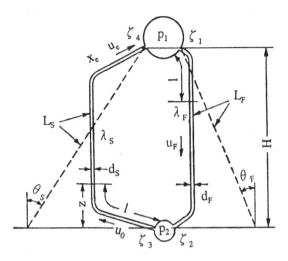

Figure 4.38 Natural water circulation loop.

where l is a space coordinate along the evaporator tube; z is the vertical height from the lower header; and x_e and α_e are the quality and void fraction at the evaporator tube outlet, respectively. Elimination of $(p_2 - p_1)$ from Eqs. (4.78) and (4.79) results in the momentum equation through the circulation loop:

$$
g\left[\varrho_L H - \int_0^H \left\{(1 - \alpha)\varrho_L + \alpha\varrho_G\right\}dz\right]
$$

$$
= \zeta_3 \frac{1}{2}\varrho_L u_0^2 + \left(\zeta_1 + \zeta_2 + \lambda_F \frac{L_F}{d_F}\right)\frac{\varrho_L u_F^2}{2} + \int_0^{L_s}\left(-\frac{dp}{dl}\right)_f dl
$$

$$
+ \zeta_4 \frac{(\varrho_L u_0)^2}{2}\left\{\frac{x_e^2}{\alpha_e\varrho_G} + \frac{(1 - x_e)^2}{(1 - \alpha_e)\varrho_L}\right\} \tag{4.80}
$$

The right-hand side of this equation represents the driving force of water circulation due to the density difference between the downcomer and the evaporator tube, and the left-hand side represents the flow resistance in the circulation loop composed of the frictional pressure losses in a single-phase water flow, from the upper drum through the downcomer to the evaporator tube inlet (1st term), the frictional pressure loss in two-phase flow in the evaporator tube (2nd term), and the pressure drop by acceleration of two-phase flow in the evaporator tube (3rd term).

The circulation velocity u_0 is, in principle, determined from Eq. (4.80) by the following procedure:

1. First, assume a value for circulation velocity. Then the distribution of quality x along the evaporator tube can be obtained from the heat balance at the given heat flux distribution.
2. The distribution of void fraction along the tube and the frictional pressure gradient $(-dp/dl)$ are calculated according to the method described in Sections 4.3.2 and 4.3.3, respectively.
3. Each term of Eq. (4.80) can be calculated, and then the equality of the equation is checked in comparison with the first quality obtained. If it is not correct, the true value is determined by iteration of the calculation, modifying the value of u_0. The exact solution of Eq. (4.80) can thus be obtained.

A further method of obtaining the approximate solution and the characteristic of the circulation velocity will now be described. The homogeneous flow model gives a formula for the circulation velocity in a closed form, with assumptions of constant physical properties, uniform heat flux, straight evaporator tubes with uniform cross-section and nondependency of friction factors on the Reynolds number. The circulation loop of the model is composed of a representative evaporator tube, a representative downcomer, a lower header, and an upper drum, as shown in Figure 4.38. On the basis of the assumption of homogeneous flow, where the slip ratio $S = 1$, the void fraction and two-phase frictional pressure gradient $(-dp/dl)$ are given by Eqs. (4.67) and (4.72), respectively, as follows:

$$\alpha = \left[1 + \frac{\varrho_G}{\varrho_L}\left(\frac{1}{x} - 1 \right) \right]^{-1} \tag{4.81}$$

$$\left(-\frac{dp}{dl} \right)_f = \phi_{L0}^2 \cdot \left(-\frac{dp}{dl} \right)_{L0} = \left(-\frac{dp}{dl} \right)_{L0}\left\{ 1 + \left(\frac{\varrho_L}{\varrho_G} - 1 \right)x \right\} \tag{4.82}$$

Substitution of these equations into Eq. (4.80) gives the circulation velocity u_0 as a function of the exit quality x_e, and the dimensionless expression is given as (Nakanishi 1986):

$$U = \left[2\left\{ 1 - \frac{\ln(1 + X)}{X} \right\} \middle/ \left\{ \frac{X}{2} + M \right\} \right]^{1/2} \tag{4.83}$$

where U is the dimensionless circulation velocity defined as

$$U \equiv u_0 \middle/ \sqrt{gd_s \cos\theta_s / \lambda_s} \tag{4.84}$$

where θ_s is an equivalent angle of the evaporator tube, defined as $\cos\theta_s = H/L_s$, as shown in Figure 4.38, and λ_s is the friction factor of single-phase flow in an evaporator tube, X is the phase change number, which is a characteristic quality at the evaporator tube outlet, defined as

$$X \equiv \left\{ (\varrho_L / \varrho_G) - 1 \right\}x_e \tag{4.85}$$

and M is a dimensionless quantity characterizing the flow resistance of the down-comer, defined as

$$M \equiv 1 + \left\{ \left(\frac{u_F}{u_0}\right)^2 \left(\zeta_1 + \zeta_2 + \lambda_F \frac{L_F}{d_F}\right) \middle/ \left(\frac{\lambda_s}{d_s} L_s\right) \right\} \tag{4.86}$$

where L_s is the total length of the evaporator tube.

The outlet quality x_e is related to the evaporation rate per evaporator tube, G_{ds}, by

$$G_{ds} = (\pi/4)d_s^2 \varrho_L u_0 x_e \tag{4.87}$$

A characteristic parameter of evaporator tubes, K_s, is introduced here as

$$K_s \equiv \frac{G_{ds}/\varrho_G}{d_s^{2.5}\sqrt{g\cos\theta_s/\lambda_s}} \tag{4.88}$$

Substitution of Eqs. (4.84) and (4.87) into Eq. (4.88) gives

$$K_s = \frac{\pi}{4}\frac{\varrho_L}{\varrho_G}\frac{x_e d_s^2}{d_s^2}\left(\frac{u_0}{\sqrt{gd_s\cos\theta_s/\lambda_s}}\right) = \frac{\pi}{4}\frac{\varrho_L}{\varrho_G}Ux_e \tag{4.89}$$

For simplicity of the formula a new parameter D is introduced as

$$D \equiv \left(1 - \frac{\varrho_G}{\varrho_L}\right)K_s \tag{4.90}$$

The parameter D is also interpreted as a dimensionless heat flux. These dimensionless parameters D, X, and U can then be related by

$$D = \frac{\pi}{4}\left(\frac{\varrho_L}{\varrho_G} - 1\right)x_e U = \frac{\pi}{4}XU \tag{4.91}$$

Combination of Eqs. (4.83) and (4.91) gives the final expression of U as a function of D:

$$U = \left[2\left\{1 - \frac{1 - \ln\left(1 + \frac{4}{\pi}\frac{D}{U}\right)}{\frac{4}{\pi}\frac{D}{U}}\right\} \middle/ \left(\frac{1}{2}\frac{4}{\pi}\frac{D}{U} + M\right)\right]^{1/2} \tag{4.92}$$

The relationship between U and D with M as parameter is shown in Figure 4.39, with equi-quality lines, that is, $x_e = $ constant, drawn in the figure. The figure indicates that the dimensionless circulation velocity U increases sharply at first with the dimensionless heat flux or the dimensionless evaporation rate D, rises to a maximum, and then decreases gradually, and that the dimensionless down-comer flow resistance M suppresses the value of U. The void fraction at the evap-

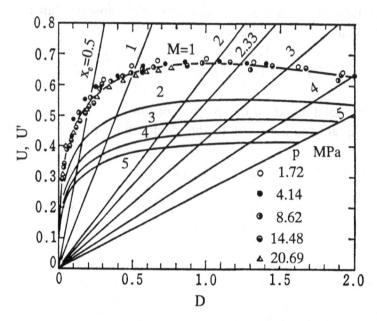

Figure 4.39 Normalized relationship between circulation velocity and evaporation rate (Nakanishi 1986).

orator tube outlet α_e is one of the important design parameters explained in Section 4.5.2. The value of X for $\alpha = \alpha_e$, X_e, is given by

$$X_e = \left(1 - \frac{\varrho_G}{\varrho_L}\right) \Big/ \left\{\left(\frac{1}{\alpha_e} - 1\right) + \frac{\varrho_G}{\varrho_L}\right\} \qquad (4.93a)$$

At lower pressures, the density ratio ϱ_G/ϱ_L is small enough to be negligible, and the above formula then becomes

$$X_e \cong \frac{\alpha_e}{1 - \alpha_e} \qquad (4.93b)$$

A design criterion for the natural-circulation boiler is that the void fraction at the evaporator tube outlet α_e be smaller than 0.7, which, from Eq. (4.93b), is equivalent to $X_e < 2.33$. Therefore, it is indicated in the figure that when $D = 0.8$, for example, the downcomer should be designed so as to ensure the condition $M < 4$.

The characteristics of natural circulation based on the homogeneous model described above are almost valid for the slip model, as is shown below. When Thom's correlation (4.61) is used as the slip model, numerical results from Eq. (4.80) can be approximated with good accuracy for $M = 1$ by Eq. (4.94). Figure

4.39 shows that the plotted data of the numerical calculation at various pressures agree well with Eq. (4.94).

$$U' = U\left/\left\{1 - 0.28\left(1 - \frac{p}{p_{cr}}\right)^2\right\}\right.$$ (4.94)

where U' is the dimensionless circulation velocity by the slip flow model, U is the circulation velocity for the homogeneous flow model according to Eq. (4.92), and p_{cr} is the critical pressure. This formula is also applicable for M other than unity and is very useful in the preliminary planning of water-circulation loops.

4.5.2 Design criteria and typical values for the circulation loop

There are two aspects of the design of natural water circulation loops. One is to ensure a sufficient mass flux of circulating water to avoid burnout of evaporator tubes. The other is to avoid tube wall temperature fluctuation and tube vibration due to oscillation of circulation velocity. The design criteria are therefore reduced, in principle, to those of critical heat flux, critical flow rate for burnout, and flow instability. In practical design, however, the circulation velocity and the void fraction at the evaporator tube outlet are used as the design criteria as follows:

1. Circulation velocity u_0: In vertical tubes and/or inclined tubes close to the vertical, $u_0 > 0.7$ m/s. In horizontal tubes, u_0 should be larger than the critical velocity of phase stratification shown in Section 4.1.2, and, in general, $u_0 > 1.2$ m/s.
2. Void fraction at evaporator tube outlet α_e: $\alpha_e < 0.7$.
3. Quality at evaporator tube outlet x_e: The dimensionless characteristic quantity X_e is expressed as a function of x_e by Eq. (4.85). The stability criterion for the density wave oscillation is $X_e < 11$, according to various researchers (Yamauchi et al. 1981; Furutera and Furukawa 1985; Yamauchi and Nakanishi 1985).

The characteristic parameters K_S (or D), K_F, and M surveyed in actual boilers are shown below. This gives useful information for the initial planning stage of water-circulation loops. The characteristic parameter for downcomers, K_F is defined similarly to K_S of Eq. (4.88) as

$$K_F = \frac{G_{dF}}{\varrho_G d_F^{2.5}\sqrt{g\cos\theta_F/\lambda_F}}$$ (4.95)

where G_{dF} is the evaporation rate per downcomer, θ_F is the equivalent angle of the downcomer defined as $\cos\theta_F = H/L_F$, as shown in Figure 4.38, and L_F is the total length of the downcomer.

The values K_S, K_F, and M of boilers built after 1984 are plotted against the pressure in Figure 4.40 (Akagawa et al. 1979; Ishigai and Akagawa 1984; Nakanishi

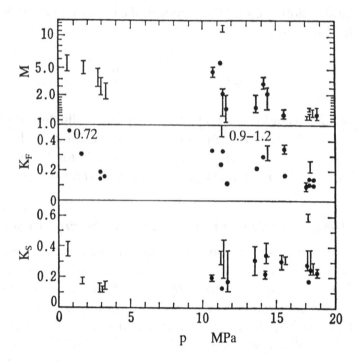

Figure 4.40 Surveyed data of K_s, K_F, and M of boilers in the market.

1986). As the heat fluxes differ at each evaporator tube bank, the K_S in a boiler has a certain scatter. The average value is represented by a dot within the scatter range. It is indicated that the maximum value is about 0.4 and the average value is about 0.3, independent of the boiler pressure. The value of M decreases with the pressure and is generally smaller than 5. This value becomes smaller than 2, however, at pressures higher than 15 MPa. Values of K_F are also plotted for three groups, divided according to year of construction, in Figure 4.41 (Ishigai and Akagawa 1984). The figure indicates that the values of K_F are in the range of 0.1 to 0.4 except in low-pressure boilers built before 1970.

These values can be used as good guides for the initial design stage of natural water circulation loops. Figure 4.42 is the design chart for downcomers based on K_F, showing the relationship between downcomer diameter d_F and G_{dF}/ϱ_G $\sqrt{(\cos\theta_F)}$ with K_F as the parameter obtained from Eq. (4.95) (Ishigai et al. 1969b; Akagawa et al. 1979). Using this chart, the initial planning of the downcomer can be carried out easily, as in the following example. For a boiler with two down-comers at pressure 4.5 MPa and evaporation rate 50 t/h, the value of the abscissa is $G_{dF}/\varrho_G \sqrt{\cos\theta_F} = (5000/2)/22.7 = 110\,\text{m}^3/\text{h}$ (where $\varrho_G = 22.7\,\text{kg/m}^3$ and $\cos\theta_F$ is assumed to be unity). When the value of $K_F = 0.2$ is selected as the design value, the downcomer diameter d_F is thus determined to be 140 mm from the figure. When K_F is selected as 0.3, d_F becomes 115 mm.

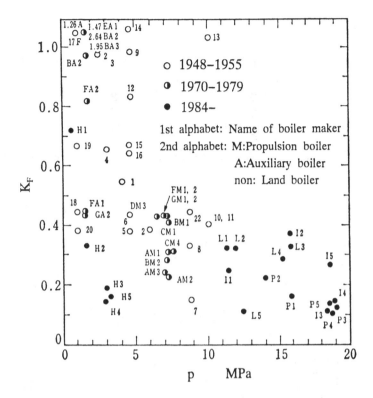

Figure 4.41 Surveyed data of K_F of boilers in the market.

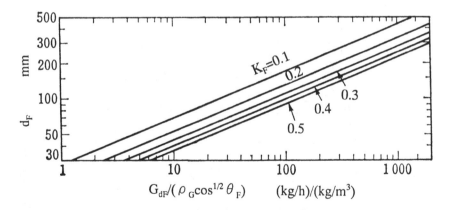

Figure 4.42 Simplified design chart for downcomers (Ishigai et al. 1969b).

For the design of water-circulation loops, the number of downcomers and their diameters can be determined by solving Eq. (4.80) by computers; however, the simple method described above, obtained by combining the analysis of a physical model of water circulation with the surveyed values of the

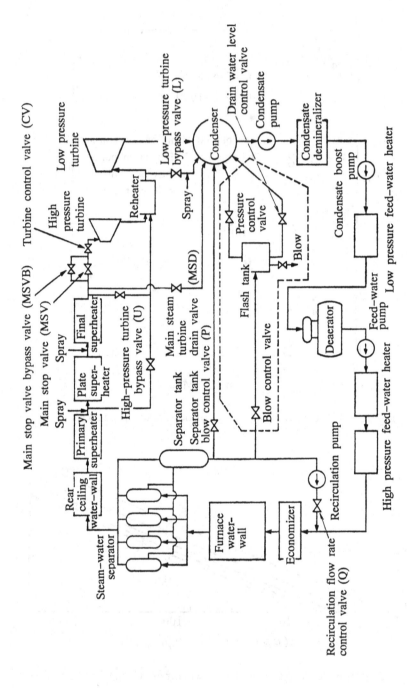

Figure 4.43 Flow system of supercritical sliding-pressure once-through boiler (Ishikawajima Harima Heavy Industries).

characteristic parameters, is also useful for designers concerned with the initial planning.

4.6 Thermal and hydraulic design of once-through boilers

In the once-through boiler, subcooled water, supplied by a feed-water pump at the inlet of the tubes, is heated, evaporated, and superheated in the long evaporation tubes to produce superheated steam at a prescribed pressure and temperature at the outlet of the tubes. Since the state of the superheated steam at the tube outlet is determined by the total amount of heat input, independent of the heat flux distribution along the tube at a defined mass flow rate, the once-through boiler is a fairly simple device to obtain superheated steam of a prescribed state at any partial load.

In contrast, in natural-circulation and forced-circulation boilers, the evaporation rate is determined by the heat load in the evaporator tubes and the superheated steam temperature by the heat load in the superheater. The superheated steam temperature therefore varies with the evaporation rate. In once-through boilers, on the other hand, the pressure is very sensitive to the load because the heat capacity of the working fluid in the small diameter tube is very small. Thus high-grade control is essential in this type of boiler.

Fundamental subjects to be considered in the thermal and hydraulic design of water-tube systems in once-through boilers are: (1) avoidance of flow instabilities; (2) avoidance of burnout at the high-quality region, that is, keeping the tube wall temperature below the permitted maximum; (3) avoidance of burnout at any partial load; and (4) hardware and software of safety control and easy start-up and shut-down systems.

4.6.1 Constitution of the flow system

A typical flow system and the side view of a supercritical sliding-pressure once-through boiler (pressure 24.6 MPa, temperature 538/566°C) are shown in Figures 4.43 (Matumoto and Ohki 1978; Matumoto et al. 1979) and 4.44 (Yanuki et al. 1985), respectively. At loads over 25% of rated load, the water fed by a feed-water pump flows through the high-pressure feed-water heater, economizer, furnace water wall, steam–water separator, rear water-wall tubes at the ceiling, and superheaters. The superheated steam produced is supplied to the turbine.

At rated and relatively high loads the boiler is operated as a purely once-through type. At partial loads, however, the boiler is operated by sliding the pressure as a function of load in the manner shown in Figure 4.45 (Matumoto and Ohki 1978). During such operation, the pressure is determined by the turbine system and the boiler is operated by fully opening the turbine inlet valve. At

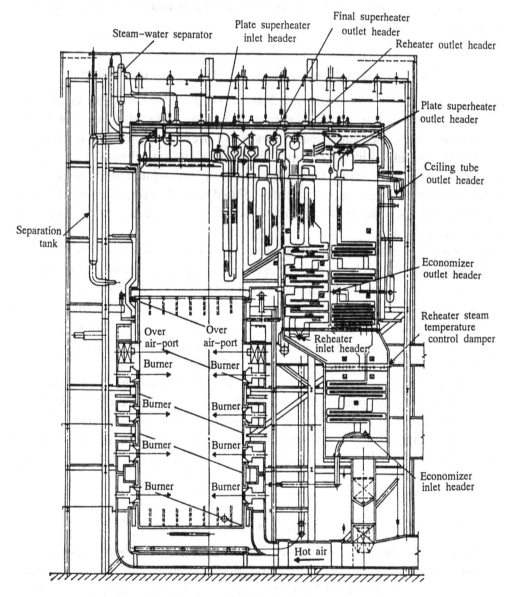

Figure 4.44 Once-through boiler (Ishikawajima-Harima Heavy Industries).

constant-pressure operation, on the contrary, the turbine inlet pressure is controlled by throttling the turbine inlet valve, which results simultaneously in the control of the flow rate. This induces a reduction in cycle efficiency owing to the irreversibility of the throttling. On the other hand, in sliding-pressure operation, the efficiency drop at partial loads is small because of the nonexistence of throttle loss.

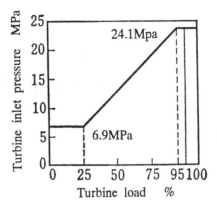

Figure 4.45 Sliding pressure.

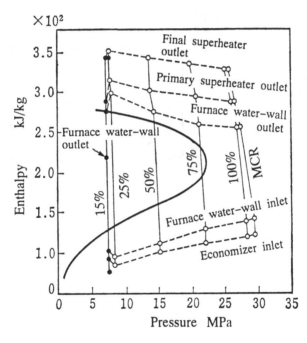

Figure 4.46 Enthalpy change along steam-generating tube.

Figure 4.46 demonstrates the change of thermodynamic state of the working fluid along the evaporation tube (Yanuki et al. 1985). At loads of 100–77%, the system pressures are in the supercritical region. In contrast, at 77–25% loads, the system pressures are in the subcritical region and boiling occurs in the water-wall tubes, while the state of the fluid at the outlet of the water wall is superheated steam. This means that once-through operation is possible without the

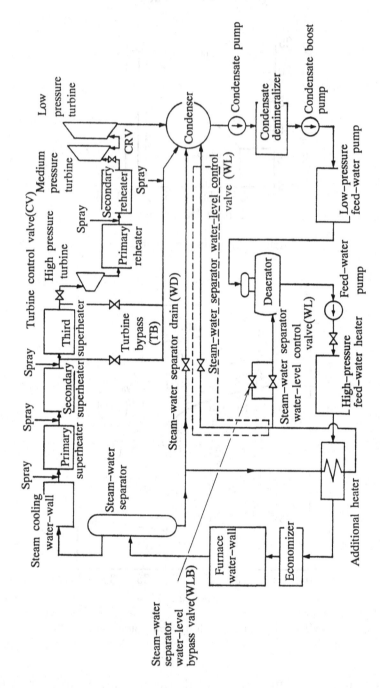

Figure 4.47 Flow system of supercritical sliding-pressure once-through boiler (Mitsubishi Heavy Industries).

steam–water separator. At loads under 25%, a certain critical flow rate (the minimum allowable flow rate for safe operation) should be supplied to avoid burnout in the water-wall tube. Thus the actual plant is designed to supply a sufficient amount of water into the water wall by circulating the water separated at the steam–water separator through the separator tank and recirculation line with a recirculation pump. Thus the once-through boiler is operated in a manner similar to the forced-circulation boiler. For instance, at 15% load the steam at the outlet of the tube is wet steam, and the water separated in the separators is recirculated. In this mode of operation, water separation is conducted by the steam water separator while water level control is conducted in the separator tank by a recirculation control valve (Q) and a separation tank blow control valve (P).

Since the late 1970s, daily start and stop (DSS) operation has been required to correspond to the daily load variation, even for once-through boiler power stations. A necessary condition for DSS is, for instance, that the start-up time from a state of having stopped for 11 ± 3 hours to the full load be within 100 minutes, which is actually a very severe condition. Therefore, a superimposed recirculation line, superheater bypass line, and high-pressure and low-pressure turbine bypass lines are installed, and a computer control system is introduced to regulate the feed-water, fuel, and water spray flow rates at each stage of the superheater. The starting operation is also conducted by using the recirculation pipe line, superheater bypass line, and turbine bypass lines shown in Figure 4.33 (Ishimoto and Ooki 1977; Nakano 1982).

There are a few types of flow system in once-through boilers other than that described above. For instance, some plants have a flash tank line represented by the dotted line in Figure 4.43, and others have the flow line shown in Figure 4.47 (Shiojima and Haneda 1978). In the latter boiler, the flow line at loads over 25% of maximum continuous rate (MCR) is similar to that in Figure 4.43, whereas at partial loads under 25% a recirculation loop, consisting of the steam–water separator, additional heater, deaerator, feed-water pump, high-pressure feed-water heater, additional heater, economizer, and furnace water wall, is used to ensure that the mass flow rate is above the critical flow rate. In this case, the recirculation loop can be constituted by utilizing the feed-water pump without a water-circulation pump.

4.6.2 Design of the furnace water wall

The fundamental concept of the thermohydraulic design of the furnace water wall will be described in this section, under the assumption that the geometrical dimensions of the furnace have been determined by means discussed in Section 3.2, in which the determination of various parameters, for example, the furnace volume, shape of furnace cross-section, and area of heating surface, was discussed with regard to the type of fuel, type of burner, flame size, and heat flux on the heating surface.

The fundamental conditions required for the design of furnace water wall are:

1. sufficient mass flow rate to avoid burnout, that is, to keep the tube wall temperature below the allowable limit at any load;
2. uniform distribution of flow rate in parallel tubes;
3. stable flow without flow oscillation;
4. safe operation at any partial load;
5. maintenance of the pressure drop in tubes at an appropriate but not too high level.

Among these requirements, the problems of critical mass flux, (1) and (4), are quite important for suitable design.

A typical arrangement of the furnace water wall in large- and medium-capacity boilers (300–1000 MWe) constructed since the late 1970s is a combination of inclined spiral tubes in the lower part of the furnace and vertical tubes in the upper part, as shown in Figures 4.44 and 4.48. Those lower and upper tubes are connected by Y-shaped pieces or headers. Provided that the number of lower spiral tubes is less than that of the upper tubes, sufficient mass flux can be given in the lower tubes relative to the critical mass flow rate.

The arrangement of spiral tubes is determined as follows. For a furnace of square cross-section with width B and height H of the spiral tube section, the total cross-sectional area $F = (\pi/4)d_i^2 n$ of evaporator tubes and the evaporation rate G_T are given by

$$G_T = F \cdot G_{cr} = (\pi/4)d_i^2 n G_{cr} \tag{4.96}$$

where G_{cr} is the critical mass flux predicted for the given heat flux in the furnace, as described in Sections 4.2 and 4.6.3, d_i is the inner diameter of a tube, and n is the number of tubes. Equation (4.96) is then rearranged as

$$(\pi/4)d_i^2 n = G_T/G_{cr} \tag{4.97}$$

Thus the number of tubes can be determined for a given tube diameter.

The projected area of a tube in the water wall is equal to the pitch S as shown in the schematic diagram of Figure 4.49. When, for simplicity, S is assumed to be proportional to d_i, $S = c_1 d_i$ (c_1 being a constant), the width of the tube panel, B_s, is determined as

$$B_s = nS = \frac{G_T}{G_{cr}} \frac{1}{(\pi/4)d_i^2} c_1 d_i = c_1 \frac{4}{\pi} \frac{G_T}{G_{cr}} \frac{1}{d_i} \tag{4.98}$$

If, on the other hand, the width B of the furnace is given as described above, then the inclination angle θ of the tubes can be obtained as

$$\sin \theta = B_s / B \tag{4.99}$$

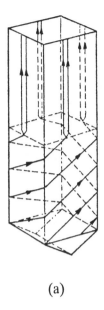

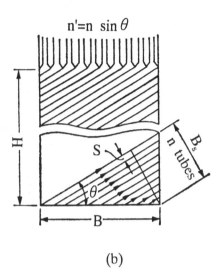

$n' = n \sin \theta$

(a)　　　　　　　　　　(b)

Figure 4.48 Water-wall tube arrangement. (a) Arrangement and (b) connection of spiral and vertical tubes.

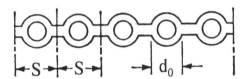

Figure 4.49 Water-wall tubes.

For instance, when the inclination angle θ is 30°, $\sin \theta = 1/2$, and the number of spiral tubes becomes half that of vertical tubes. The tube arrangement in this case is that one spiral tube is connected to two vertical tubes by a Y-shaped piece as shown in Figure 4.48a. The inclination angle used in conventional boilers is 15–30°, and headers are often used for the connection. In water-wall spiral tubes, the total amount of heat absorbed at each tube becomes fairly uniform, even in the case that the heat flux distribution is not uniform in the lateral direction in the furnace. Thus the flow rates become uniform between each tube, which makes it unnecessary to adjust orifices or distribution headers for uniform distribution of flow.

The historical aspect of furnace water-wall configuration in once-through boilers will be explained briefly in the following. Since the late 1920s, once-through boilers had been developed as the Benson boiler in Germany, the Sulzer

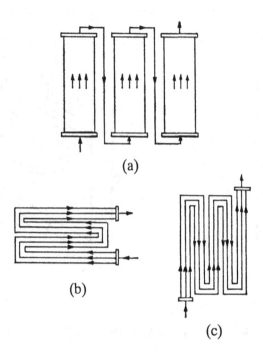

Figure 4.50 Types of water-wall construction. (a) Initial Benson boiler type; (b) horizontal meander and (c) vertical meander tube type.

mono-tube boiler in Switzerland, and the Ramzin boiler in the USSR. This happened mainly because of the simplicity of the steam-generation principle, whereas difficult problems exist in water treatment and plant control. In the Benson boiler, the furnace water wall was constructed from sectionalized vertical evaporation tubes and downcomers, as shown in Figure 4.50a. Therefore, a suitable design was needed to give appropriate flow distribution in parallel tubes, because of the confluence and redistribution of the steam–water mixture in the headers. In the Sulzer mono-tube boiler, horizontal or vertical meander tubes were used, as shown in Figure 4.50b and c, in which throttles were installed at the inlet of the tubes to adjust the flow rate in each tube. In the Ramzin boiler, an inclined spiral tube configuration was used, which is similar to the spiral tube sections shown in Figure 4.48. At present the inclined spiral tube configuration is used in most large- and medium-capacity boilers as a result of studies on the rational configuration of the furnace water wall.

4.6.3 Critical heat flux and critical mass flow rate

Critical heat flux and critical mass flow rate at supercritical and near-critical pressures will be discussed in this section; those at subcritical pressures have been

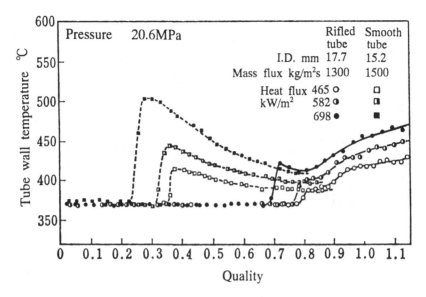

Figure 4.51 Wall temperature distribution along tube length (20.6 MPa; Kawamura et al. 1980).

treated in Section 4.2.5. Figures 4.51 and 4.52 show tube wall temperature distribution along the tube length in a smooth tube and a rifled tube at 18.6 and 20.6 MPa, respectively (Kawamura et al. 1980; Iwabuchi et al. 1982). The geometrical configuration of a rifled tube, for example, is: outside diameter 28.6 mm, inside diameter at its groove 17 mm, rib height 0.83 mm, rib number at tube cross-section 4, and angle against tube axis 30°. It should be noted in these figures that a sharp temperature rise, that is, burnout, occurs at a low quality in the smooth tube, whereas in the rifled tube this does not occur and only a small temperature rise is observed at a high quality. At supercritical pressures, burnout occurs as in the case of subcritical and low pressures, although the boiling phenomenon does not exist, as can be seen in Figure 4.52. The critical mass flux for the burnout at a given heat flux can be determined from these figures.

In the case of half-periphery heating (180° heating), which is similar to the actual situation in furnace water-wall tubes, the tube wall temperatures are 0–30°C lower than in the case of full-periphery heating (360° heating). The heat transfer coefficients in a rifled tube at near-critical pressures are plotted against fluid enthalpy in Figure 4.53 (Iwabuchi et al. 1982). This curve shows a peculiarity at near-critical pressures: at pressures lower than 21.1 MPa, the heat transfer coefficients are very high except in the high enthalpy region ($i > 2300$ kJ/kg), while at 21.3 MPa, very close to the critical pressure (the reduced pressure ratio $p/p_{cr} = 0.96$, p_{cr} being the critical pressure), the heat transfer coefficient decreases

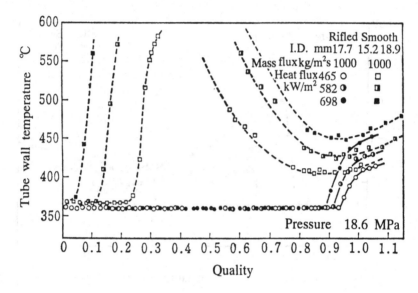

Figure 4.52 Wall temperature distribution along tube length (18.6 MPa; Kawamura et al. 1980).

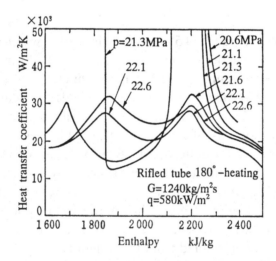

Figure 4.53 Heat transfer coefficient in rifled tube of 180° heating (Iwabuchi et al. 1982).

abruptly to about $100 \, \text{W/m}^2\text{K}$ at $i = 1850 \, \text{kJ/kg}$, rises at $i = 2100 \, \text{kJ/kg}$ to more than $5 \times 10^4 \, \text{W/m}^2\text{s}$, and again decreases at $i = 2250 \, \text{kJ/kg}$. Consequently, the design of the water wall should take into account that the abrupt decrease of heat transfer coefficient, that is, rise of tube wall temperature, would occur during

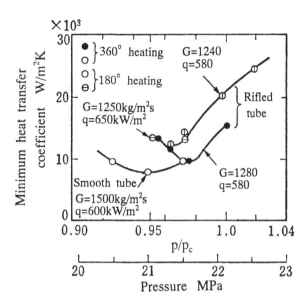

Figure 4.54 Minimum heat transfer coefficient (Iwabuchi et al. 1982).

operation near critical pressures. Figure 4.54 shows the relationship between the minimum heat transfer coefficient and the reduced pressure ratio p/p_{cr} in the smooth and rifled tubes, which indicates that the heat transfer coefficient has its minimum value over the reduced pressure ratio range 0.95–0.98.

Design values of the mass flux in water-wall tubes are determined according to the data of heat transfer performance presented, for example, in Figures 4.51–4.54. Examples of mass fluxes in each tube section in a supercritical sliding-pressure boiler are demonstrated in Figure 4.55 (Kawamura et al. 1980). The operating pressure at loads above 78% is supercritical, and at loads below 78% the boiler is operated with sliding pressure at subcritical pressures, as shown by curve 1. The critical mass flux G_{cr} can be predicted on the basis of the operating pressure and the heat flux for a given boiler load in the same manner as described above, and the predicted values are shown by curves 2 and 3 for inclined-spiral smooth tubes located in the lower part of the furnace and for vertical rifled tubes located in the upper part of the furnace, respectively. The design values of mass flux G are determined, taking the safety margin into account, as curves 4 and 5, respectively, which are considerably larger than G_{cr} (curves 3 and 5). The values of curves 3 and 5 are lower than those of curves 2 and 4 because of higher heat transfer performance in the rifled tube, as shown in Figures 4.51 and 4.52. At loads below 25%, the mass fluxes are kept constant by operating the superimposed recirculation line described in Section 4.6.1.

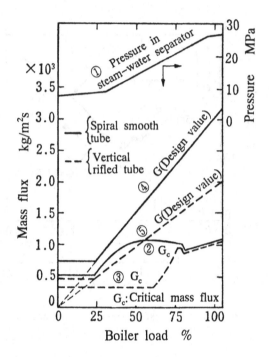

Figure 4.55 Design values of mass flux and critical mass flux, example in Mitsubishi Heavy Industries (Kawamura et al. 1980).

4.6.4 Mass flux and pressure drop in each flow system

The mass flux in furnace water-wall tubes at full load is $3300\,kg/m^2s$ in inclined-spiral smooth tubes and $2000\,kg/m^2s$ in vertical rifled tubes, as shown in Figure 4.55. Under these operating conditions, mass fluxes and pressure drops in other parts of the boiler, such as the economizer and superheater, are listed in Table 4.1. These values have been determined by considering both the heat transfer problem (allowable metal temperature at the wall) and the fluid flow problem (pressure drop in tubes), as described below.

The heat flux at the outer surface of the tube is given by

$$q = K(T_G - T_f) \tag{4.100}$$

$$K = \left[\frac{1}{h_G} + \frac{\delta_1}{\lambda_1} + \frac{\delta}{\lambda_m} + \frac{\delta_2}{\lambda_2} + \frac{1}{h_f} \right]^{-1} \tag{4.101}$$

where K is the overall heat transfer coefficient, T_G the gas temperature, T_f the fluid temperature, R_1 the inner radius of the tube, R_2 the outer radius, h_G, h_f the heat transfer coefficient at the outer and inner surfaces of the tube, respectively,

Table 4.1 *Design values of mass flux and pressure drop*
(courtesy of Mitsubishi Heavy Industries)

	Design example of supercritical sliding-pressure boiler at 100% load			Planned value of extrasupercritical pressure boiler at MCR (pressure drop (MPa))
	Velocity (m/s)	Mass flux (kg/m^{2}s)	Pressure drop (MPa)	
Main steam pipe				0.9
Superheater tube	10–20 (Outlet)	1000–2500	0.5–1.0	1.2
Flue evaporator	5–10 (Outlet)	1000–1500	0.1–0.2	0.2
Wall tube in downstream flue	10–15 (Outlet)	2000–3000	0.3–0.4	0.6
Furnace water wall				
Smooth tube	−6.6 (Inlet)	−3300		
Rifled tube	3–5 (Inlet)	2000–2500	1.0–1.5	1.6
Economizer	1.5–2.0 (Inlet)	1000–1500	0.3–0.4	0.3
Main feed-water tube				0.1
Connecting tubes				0.6
Total			1.3–3.5[a]	5.5

[a] From economizer inlet to superheater outlet.

δ_1 and δ_2 the thickness of scales on the inner and outer surfaces, respectively, δ the thickness of the tube, λ_1 and λ_2 the thermal conductivity of the scales, and λ_m the thermal conductivity of the tube metal.

The heat transfer coefficient h_G depends on the gas velocity, the pitch of the tube arrangement, the flow direction relative to the tube, and so on, which are determined in the gas-side design described in Chapter 3. The heat transfer coefficient h_f depends on the fluid velocity in the tube and is proportional to the 0.8 power of the velocity. Thus a water flow with high velocity results in a drop in the tube wall temperature, while it induces an increase in the pressure drop that is proportional to the 1.8 power of the velocity. Therefore, the appropriate value of water velocity should be determined taking into account the balance of both effects. The values listed in Table 4.1 are determined on this basis.

The pressure drops in once-through boilers are relatively high compared with the system pressure, and reach 11–20% of the system pressure, as shown in Table 4.1 and Figure 4.56. The figure shows the pressure drops and the boiler exit pressures in recent supercritical sliding-pressure boilers using coal firing, liquified natural gas firing, and mixed firing of coal and oil. At a power station where the

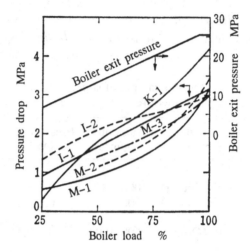

Figure 4.56 Pressure drop in supercritical sliding-pressure boilers.

turbine inlet pressure is 24.6 MPa, a relatively high delivery pressure of 31–33 MPa is needed in the feed-water pump owing to the large pressure drop. At a plant, for instance, with an output of 700 MWe and an evaporation rate of 2300 t/h, two feed pumps of 14 kW each are installed, and the power ratio relative to the output is 4–4.5%.

4.6.5 Small-capacity once-through boilers

A typical example of a flow system in a small-capacity once-through boiler (0.7 MPa, 840 kg/h) is shown in Figure 4.57. The water supplied by a feed-water pump first flows into an outer coil tube (28 mm ID) of downward flow and then flows through an inner coil tube (45 mm ID) of upward flow that constitutes the furnace, and finally flows into a steam–water separator as wet steam with a quality of about 70%. The steam separated is used as the process steam, and the water is recirculated into a hot well tank so as to mix with the feed water to retain the feed-water temperature of 63.4°C.

Examples of design values of mass fluxes, heat fluxes on the heating surface, furnace heat generation rates, pressure drops, and tube diameters in once-through boilers rated from 100 to 2000 kg/h are shown in Figure 4.58. According to the curves in the figure, for a boiler of rated output 1000 kg/h, for example, the mass fluxes are 700 and 210 kg/m²s in the outer and inner coils, respectively. These fluxes are considerably lower than the values of 2000–3000 kg/m²s in large-capacity boilers shown in Table 4.1. This is mainly because the quality at the exit of the evaporator tube is kept at about 70%, and dryout does not occur in the downstream region (see Sections 4.2.3 and 4.2.4). Moreover, the mass flux is

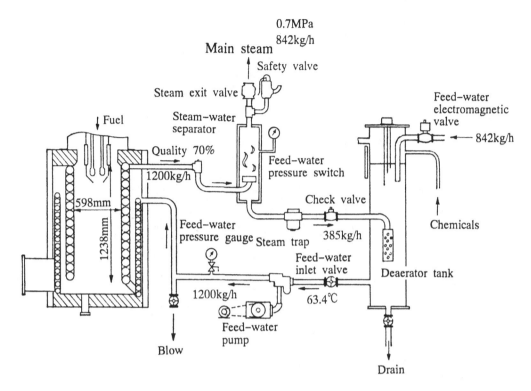

Figure 4.57 Flow system in small-capacity once-through boiler (courtesy of Takuma Co.).

designed to be proportional to the boiler capacity (evaporation rate), because the heat flux q_s (W/m^2) at the inner surface of the furnace increases with the increase in the boiler capacity, as discussed in Chapter 3.

The pressure drops in evaporator tubes are about 0.2 MPa in boilers under 500 kg/h rated output and about 1 MPa in boilers of 1500 kg/h. This means that the pressure drop is considerably larger than the system pressure of about 0.7 MPa in small-capacity once-through boilers.

4.7 Thermal and hydraulic design of forced-circulation boilers

In a forced-circulation boiler water is circulated by means of a circulating pump in loops composed of evaporator tubes and downcomers. Since enough water can be supplied into the evaporator tubes at any load, any type of constitution of the circulation loop can be freely selected. Thus not only the conventional vertical straight-tube system but also systems of vertical or horizontal meander tubes can be adopted. Moreover, small-bore tubes with diameters of 50.8–38.1 mm can be used as evaporator tubes, while in natural-circulation boilers evaporator tubes of 70–80 mm ID have been used to reduce the pressure drop to create sufficient mass flux.

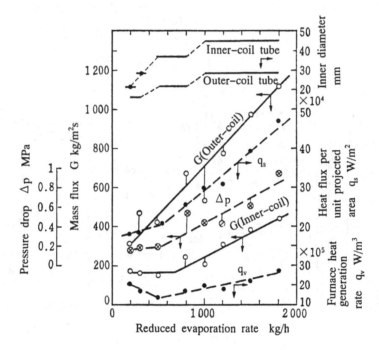

Figure 4.58 Design parameters in small-capacity once-through boilers (courtesy of Takuma Co.).

4.7.1 Constitution of the flow system

A typical example of a sliding-pressure forced-circulation boiler with medium capacity (350 MWe, 17.7 MPa) and a schematic diagram of the flow system are shown in Figures 4.59 and 4.60, respectively. The furnace water wall consists of vertical straight tubes to use natural circulation in addition to forced circulation to enhance the water circulation. This type of boiler is referred to as a "controlled-circulation boiler."

The water-circulation loop is composed of the steam–water drum (1; 1680 mm ID), downcomer (2; 355.6 mm OD, 6 tubes), circulation pump manifold (3), circulation pump suction tubes (4; 558.3 mm OD, 3 tubes), circulation pumps (5; 3 sets), circulation pump delivery tubes (6; 318.5 mm OD, 6 tubes), distribution drum (7; 864 mm), furnace water walls (8–10), headers (11–13; 2674 mm ID), and riser tubes (14–16; 139.8 mm, 14, 18, and 22 tubes, respectively). The furnace water wall is composed of front wall tubes (8; 45 mm OD, 814 tubes), rear wall tubes (9; 45 mm OD, 814 tubes), and side wall tubes (10; 45 mm OD, 376 evaporator tubes, and 50.8 mm, 224 lower nonheated tubes). Moreover, other circulation loops

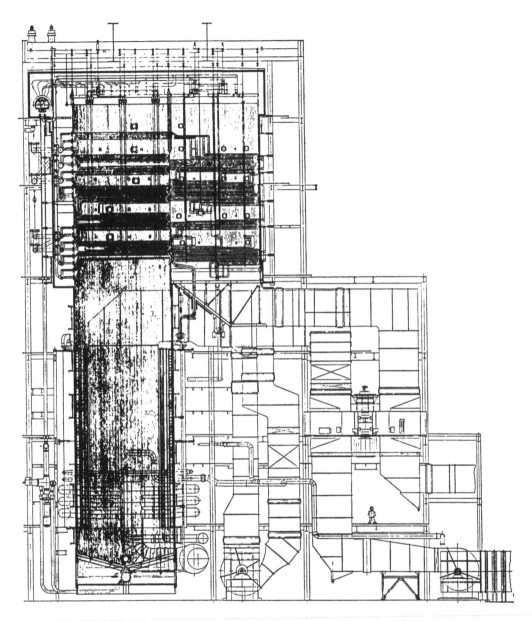

Figure 4.59 Forced-circulation boiler (courtesy of Mitsubishi Heavy Industries).

parallel to the water-wall loops are installed; these comprise the distribution drum (7; 864 mm OD), evaporator inlet tubes (17; 139.8 mm OD, 16 tubes), flue evaporator tubes (18; 45 mm OD, 324 tubes), evaporator hanger tubes (19; 45 mm OD, 324 tubes), evaporator outlet header (20; 216.3 mm OD), evaporator riser tubes (21; 139.8 mm OD, 18 tubes), and steam–water drum (1). Three circulation pumps,

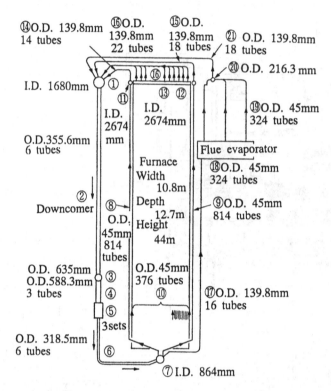

Figure 4.60 Circulation loop in forced-circulation boiler (courtesy of Mitsubishi Heavy Industries).

one of which is the reserve pump, with a capacity of $2810\,m^3/h$ ($1537\,t/h$, $326\,kW$) per unit, are installed to cope with an evaporation rate of $1130\,t/h$.

Since the heat loads at each circulation loop differ from each other, the mass fluxes at each loop should be adjusted corresponding to the load for safe operation. These parallel tubes at each loop are therefore divided into a few groups, and orifices are installed at the inlet of each tube to ensure the required flow rate. To enhance natural circulation, six downcomers of large diameter ($355.6\,mm$) are installed. The total cross-sectional area of the downcomers reaches about $1.3\,m^2$, which is larger than the total cross-sectional area (about $1\,m^2$) of the evaporator tubes.

4.7.2 Circulation ratio as a design criterion

Fundamental subjects to be considered in the design of flow systems in forced-circulation boilers are: (1) mass flux sufficient to avoid burnout at the furnace water-wall tubes; (2) proper flow distribution in parallel tubes corresponding to heat load distribution; and (3) appropriate mass flux so as not to increase the pump power, which reduces the plant efficiency.

Subject (1) is related to the critical mass flux. What has been described for once-through boilers in Section 4.6.3 can also be applied to forced-circulation boilers, as will be discussed in Section 4.7.3 below. In the present section, a design criterion for forced-circulation boilers is described in terms of the problem of "circulation ratio."

The circulation ratio R is defined as

$$R = \left(\text{mass flow rate of circulation water}\right)\big/\left(\text{evaporation rate}\right)$$

$$= G_R/G_S \tag{4.102}$$

$$\cong G_R/x_e\, G_R \doteq 1/x_e \tag{4.103}$$

where G_R (kg/s) is the mass flow rate of circulation water, G_S (kg/s) is the evaporation rate, and x_e is the quality at the outlet of the evaporator tube. The evaporation rate G_S is expressed as $x_e \times G_R$ under the assumption that the water at the tube inlet is in the saturated state. Consequently, the circulation ratio is converted to a reciprocal of the quality at the tube outlet, $1/x_e$, as shown in Eq. (4.103).

The recommended values of the circulation ratio and related factors are listed in Table 4.2. For boilers of 2-MPa class, for instance, the circulation ratio is 5–9, which corresponds to a volumetric flow quality β of 0.82–0.9 and a void fraction α_e of 0.61–0.78.

The void fraction at the outlet of the evaporator tubes, α_e, is used as a design criterion in natural-circulation boilers as described in Section 4.2, where, for instance, the criterion is $\alpha_e < 0.7$. On the other hand, in forced-circulation boilers the corresponding void fraction is 0.48–0.88, as shown in Table 4.2. This does not mean that the design criterion is of the same order for both types of boiler, because the problem of the absolute mass flux still exists. In natural-circulation boilers the mass flux is in a certain limited range owing to the natural-circulation principle, while in forced-circulation boilers the required value of mass flux can be given by the water-circulating pump and is much higher than that in natural-circulation boilers.

The circulation ratio R generally used for 17-MPa class conventional forced-circulation boilers is about 4 and the ratio for controlled-circulation boilers of medium capacity (350–600 MWe) is 2–3 ($x_e = 0.33$–0.5). This difference is mainly because rifled tubes having a high heat transfer performance are installed and also because natural circulation assists the forced circulation, as shown in Figure 4.61.

As an example of design values related to the water circulation in a sliding-pressure-controlled circulation boiler, the circulation ratios, water mass flux, water velocities at tube inlets, and so on are shown against boiler loads in Figure

Table 4.2 *Recommended values of circulation ratio and void fraction in forced-circulation boilers*

Boiler pressure	1 MPa class	2 MPa class	3 MPa class	17 MPa class	17 MPa class
Circulation ratio R	4–8	5–9	6–11	ca. 4	2–3[a]
Quality at outlet of evaporator tube					
$\quad x_e = 1/R$	0.12–0.25	0.1–0.2	0.09–0.17	0.25	0.33–0.55
$\quad \varrho_G/\varrho_L$	0.005	0.012	0.018	0.21	0.33–0.5
$\quad \beta$	0.96–0.98	0.82–0.90	0.77–0.87	0.61	0.70–0.82
α_e, Eq. (4.60)	0.75–0.88	0.61–0.78	0.53–0.70	0.48	0.59–0.74

[a] Controlled-circulation boilers with rifled tubes.

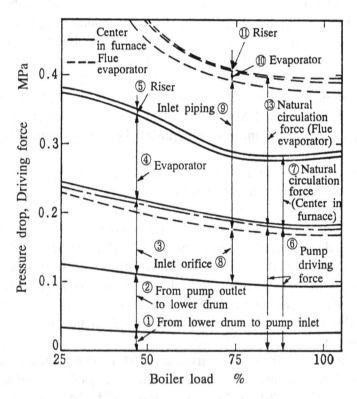

Figure 4.61 Pressure drop and driving force of circulation pump.

4.62. According to the plotted data, the circulation ratio is 1.94 at MCR in the furnace water wall, and 1.74 in the flue evaporator tubes with lower heat fluxes. The mass flux is about 1000 kg/m²s and 740 kg/m²s in the center part of the furnace water wall with high heat flux and in the corner part with lower heat flux,

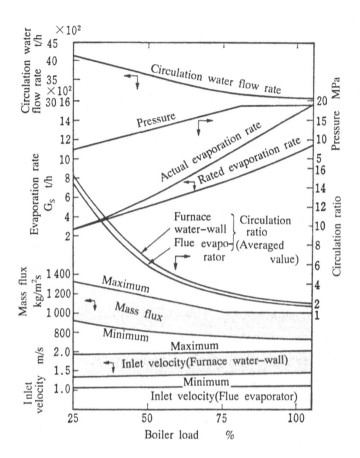

Figure 4.62 Characteristic values of water circulation in forced-circulation boilers (courtesy of Mitsubishi Heavy Industries).

respectively. At partial loads the circulation ratio is increased with load, and at 25% load the circulation ratio reaches 15.

The mass flux is determined by the balance between the driving forces of the circulating pump, the driving head of natural circulation, and the pressure drops in the loop. Figure 4.62 shows the driving forces and pressure drops against the boiler loads in the loops. At MCR, for instance, the pressure drops in the downcomer (1 and 2), the inlet orifice (3), and the evaporator and riser (4 and 5) are all about 0.1 MPa, and the total pressure drop is about 0.3 MPa. Thus a relatively large pressure drop is imposed in the orifice. On the other hand, the driving head by the pump is 0.2 MPa and that by natural circulation is 0.1 MPa, which reaches one-third of the total head. The total pumping head of 0.3 MPa is balanced by the total pressure drop of 0.3 MPa.

Selection of the tube diameter of the furnace water wall is conducted by the following thermal and hydraulic considerations. The quantity of heat absorbed at an evaporator tube Q and the evaporation rate G_S' are given under the assumption that the water at the tube inlet is in the saturated state:

$$Q = qSL, \qquad G_S' = Q/r$$

where q is the heat flux, determined from combustion in the furnace as described in Section 3.2, r is the latent heat of evaporation, L is the length of the tube (height of furnace), and S is the pitch of the water-wall tube as shown in Figure 4.49 and is assumed to be expressed by $S \propto d_0$ in this section. Since the heating surface area of the water wall A is determined as described in Section 3.2, the following relation holds:

$$A = nSL \propto nd_0 = \text{constant}$$

This indicates that the tube diameter is inversely proportional to the number of tubes. The circulating water flow rate G_R is then expressed, with a given critical mass flow rate G_{cr}, by

$$G_R = n(\pi/4)d_i^2 G_{cr} \propto nd_0^2 G_{cr} = d_0(nd_0)G_{cr} \propto d_0 \qquad (4.104)$$

This means that the circulating water flow rate is proportional to the tube diameter. Thus the total evaporation rate G_S is expressed by

$$G_S = nqSL/r \propto nd_0L/r \propto nd_0 = \text{constant}$$

The circulation ratio R can then be expressed as

$$R = G_R/G_S = \left(nd_0^2\right)G_{cr}r/(nd_0L) \propto G_{cr}d_0 \propto d_0 \qquad (4.105)$$

This means that the circulation ratio is proportional to the tube diameter. Consequently, when tubes of large diameter are used, a large amount of water should be circulated. The volume of the tubes V is expressed as

$$V = n(\pi/4)d_i^2 L \propto nd_0^2 \propto d_0(nd_0) \propto d_0 \qquad (4.106)$$

Then the volume of water stored in the boiler is proportional to the tube diameter. In forced-circulation boilers, tubes are used with a diameter smaller (38–50 mm) than in natural-circulation boilers, and therefore the stored water volume and heat capacity become small. This results in a short start-up time. On the other hand, the pressure drop increases with reducing tube diameter, as shown below.

The pressure drop Δp_f is given by the following equation as described in Section 4.3.3:

$$\Delta p_f = \lambda \frac{L}{d_i} \frac{G^2}{2} v_m$$

where G is the mass flux, $G = \varrho_L u_{L0}$, u_{L0} is the inlet velocity, v_m is the average specific volume of the two-phase mixture, $v_m = v_L + x_m(v_G - v_L) \doteq x_m v_G$, x_m is the average quality, and λ_m is the friction factor in two-phase flow. The above equation is converted into Eq. (4.107) by substituting a relationship, $x_m = 0.5x_e$, where x_e is the quality at the tube outlet:

$$\Delta p_f \cong \lambda \frac{L}{d_i} \frac{G^2}{4} v_G x_e \propto \frac{G^2}{d_0} x_e \qquad (4.107)$$

Rearranging Eqs. (4.103), (4.105), and (4.107), one can obtain the following equations:

$$\Delta p_f \propto \frac{G^2}{d_0} \frac{1}{R} \propto \frac{G^2}{d_0} \frac{1}{G_{cr} d_0} \propto \frac{G^2}{d_0^2} \qquad (4.108)$$

or

$$\Delta p_f \propto \frac{G^2 d_0^2}{d_0^3} \frac{1}{R} \propto \frac{R}{d_0^3} \qquad (4.109)$$

This relationship indicates that the pressure drop varies inversely as two powers of the tube diameter at a constant mass flow rate and as three powers of the tube diameter at a constant circulation ratio. The required power P of the water circulation pump is proportional to $nG(\pi/4) d_i^2 \Delta p_f$; the power can then be expressed by the following relationships:

$$P \propto nGd_0^2 \Delta p_f \propto (nd_0)Gd_0 \frac{G^2}{d_0^2} \propto \frac{G^3}{d_0} \qquad (4.110)$$

$$P \propto nGd_0^2 \Delta p_f \propto (nd_0) \frac{(Gd_0)^2}{d_0^3} \propto \frac{R^2}{d_0^3} \qquad (4.111)$$

Thus the required power of the pump varies inversely as the tube diameter at a constant mass flow rate and as three powers of the tube diameter at a constant circulation ratio. A reduction in the tube diameter results in an increase in the pressure drop (negative effect for the required power of pump) and in a decrease in the weight of the boiler and the water capacity (positive effect). Therefore, the selection of tube diameter has to be made considering thermal and hydraulic effects as well as the structural problems of the water wall. As a result of practical experience, tubes of 38–50 mm OD have generally been used in recent boilers.

4.7.3 Determination of mass flux in tubes

The design criterion for the water-circulation problem in forced-circulation boilers is based on the critical mass flux, similar to that described in Section 4.6.3 for once-through boilers. The circulation ratio mentioned above is also a crite-

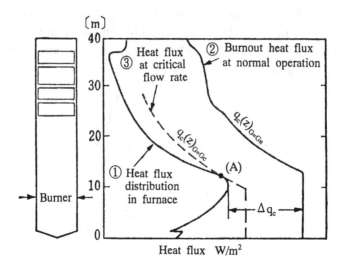

Figure 4.63 Determination of mass flux based on critical heat flux (Nakano et al. 1972).

rion based on the critical mass flux. In this section, two methods of determining the value of mass flux will be described.

One method will be explained with reference to Figure 4.63 (Nakano et al. 1972). Curve 1 represents the vertical distribution of heat flux in the furnace, and curve 2 the distribution of burnout heat flux at the given mass velocity, which is obtained from the following discussion. The burnout heat flux q_{cr} is expressed as a function of the mass flux G, quality at the tube exit x_e, and pressure p, as shown by Eq. (4.45), namely:

$$q_{cr} = f(x_e, G, p)$$ (4.112)

This equation is also applicable to any local point, that is, the burnout heat flux at the local point where the quality x is given by $q_{cr} = f(x, G, p)$. Since the distribution of quality can be calculated from the heat flux distribution, and is expressed by $x = f(z)$ as a function of distance from the tube inlet z, the burnout heat flux at the local point along the tube, q_{cr}, can be expressed as

$$q_{cr} = f(x(l); G, p) = f_G(l, p)$$ (4.113)

Thus, curve 2 of q_{cr} was obtained for the design mass flux G_n from Eq. (4.113). Next, the calculation of distribution of q_{cr} for various values of the mass velocity gives distribution curve 3, which is tangential to curve 1. The value of critical mass flux G_{cr} is then determined as the mass flux of curve 3. At the critical mass flux G_{cr}, burnout would occur at the contact point of both curves (A). The distance between curves 1 and 2, Δq_{cr}, corresponds to the safety margin. Thus the design

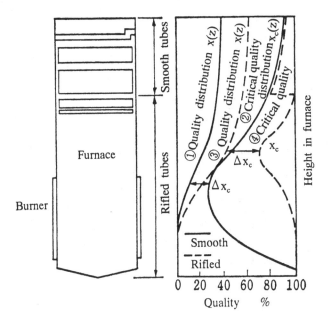

Figure 4.64 Determination of mass flux based on critical quality (Kawamura et al. 1980).

value of the mass flux G_n is determined from the appropriate value of Δq_{cr} given by practical experience.

The other method of determining the mass velocity is based on the critical quality x_{cr}, as shown in Figure 4.64 (Kawamura et al. 1980). As is shown by curve 1, the distribution of quality $x(z)$ can be obtained from the heat flux distribution for a given vertical mass flux in the furnace, $q(z)$, in a smooth tube. The critical quality x_{cr} is found by solving Eq. (4.112) as

$$x_{cr} = g(q, G, p) = g(q(l), G, p) = g_1(l, G, p) \tag{4.114}$$

Thus the distribution of $x_{cr}(z)$ can be determined as curve 2 from the distribution of $q(z)$. The distance between curves 1 and 2, Δx_c, represents a safety margin. If the value of mass flux G decreases, curve 1 shifts toward the right and finally contacts curve 2 at mass flux G_c, which is the critical value. Similarly, for the rifled tube, curves 3 and 4 represent distributions of $x(z)$ and $x_{cr}(z)$, respectively, and Δx_{cr} is the safety margin. A comparison of the two cases shows that quality $x(z)$ for the rifled tube (curve 3) is larger than $x(z)$ for the smooth tube (curve 1), owing to the high heat transfer performance of the rifled tube as shown in Figures 4.51 and 4.52. The recommended values of mass flux shown in Figure 4.61 are thus determined by the method mentioned above, and the recommended values of circulation ratio are also determined from the result.

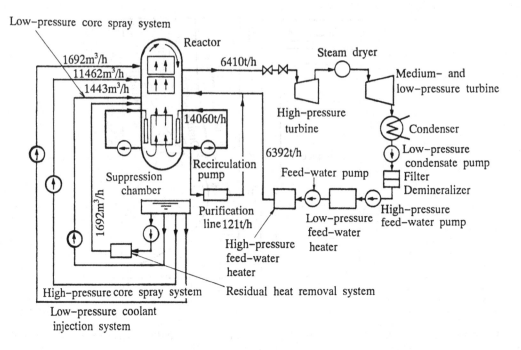

Figure 4.65 Flow system of boiling water reactor.

4.8 Thermal and hydraulic design of boiling water reactors

A typical flow system of the BWR is shown in Figures 4.65 and 4.66. Saturated steam generated in the core of a nuclear reactor flows through a high-pressure turbine, a mist separator, and a medium- and low-pressure turbine to a condenser. The condensate in the condenser is returned to the reactor core through low-pressure and high-pressure feed-water heaters. The flow system, in principle, is similar to that in a fossil-fuel electric power station, and only the heat sources differ; that is, the heat is generated in one system by nuclear fission and in the other by combustion of fossil fuels.

In nuclear reactors, a flow system for emergency core cooling against a loss of coolant accident (LOCA) is additionally installed. In Figure 4.65 these flow lines and capacities (for a 1100-MWe power station) are shown. At LOCA the water, stored in a suppression chamber located in the lower part of the pressure containment vessel, is supplied to the core through a low-pressure injection system (3 lines, capacity 1692 m³/h), a low-pressure core spray system (1 line, capacity 1443 m³/h), and a high-pressure core spray system (1 line, 1462 m³/h). Furthermore, a residual heat removal system (2 lines, capacity 1692 m³/h) is installed (Kono et al. 1984).

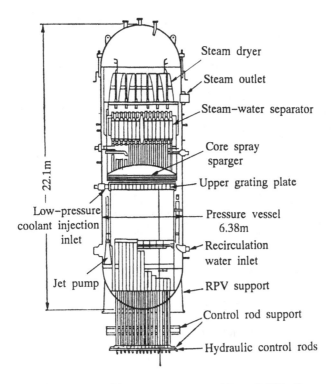

Figure 4.66 Boiling water reactor (Type BWR-5).

4.8.1 System overview and fuel assembly structure

The flow system in the core is constructed as a forced-circulation type. The circulating water (flow rate G_R), supplied by recirculation pumps installed outside the pressure vessel, is supplied to jet pumps installed at the annulus between the pressure vessel wall and the core shroud. The feed water (flow rate G_w) from the high-pressure feed-water heater and the saturated water from the steam–water separator in the pressure vessel are sucked and mixed with the water through the jet pumps. The total amount of circulating water, G_L, which is two to three times larger than G_R, is supplied to the fuel assembly inlet. The steam–water mixture generated in the core flows into the steam–water separator through the upper plenum; the humidity is further reduced by a steam dryer; and the saturated steam is fed into the turbine. Two recirculation pumps and 20 jet pumps are installed in a 1100-MW-class station.

Typical specifications of BWRs for 500-, 800-, 1100-MWe, and advanced boiling water reactors (ABWR) are shown in Table 4.3 (Thermal and Nuclear Power 1981; Takashima et al. 1984). The flow rates in a reactor core of the 1100-MWe class, for example, are as follows. For an evaporation rate G_S of 6430 t/h, the flow rate induced by the recirculation pump G_R and the total flow rate in the core G_L

Table 4.3. *Specification of BWRs*

	500-MW	800-MW	1100-MW	1350-MW (ABWR)
Reactor thermal power	1593 MWt	2436	3293	3926
Reactor pressure	7.03 MPa	7.03	7.03	71.7
Steam flow rate	2920 t/h	4750	6430	7480
Feed-water temperature	215°C	215	215	215
Number of fuel assemblies	368	560	764	872
Fuel rod arrangement	8×8	8×8	8×8	8×8 (Step II)
Number of control rods	89	137	185	205
Core dimension	$3.3 \text{m}\phi \times 3.7 \text{m}$	$4.1\phi \times 3.7$	$4.8\phi \times 3.7$	$5.2\phi \times 3.7$
Core power density	50 kW/l	50	50	50.6
Flow rate of core coolant	22,900 t/h	35,600	48,300	52,200
Number of jet pumps	16	20	20	10 (internal pump)
Circulation ratio	7.8	7.5	7.5	7.0
Power per unit fuel assembly	4.33 MW	4.35	4.31	4.5
Heat flux at fuel rod surface	487 kW/m^2	489	485	504
Reactor pressure vessel				
ID	4.7 m	5.6	6.4	7.1
Height	21 m	22	23	21
Reactor recirculation lines				
Line number	2	2	2	none
Flow rate/line	5300 m^3/h	7400	10,700	7700 (internal pump)
Turbine type				
50 Hz	TC4F-35	TC6F-35	TC6F-41	TC6F-52
60 Hz	TC4F-38	TC6F-38	TC6F-43	
Electric power output (MWe)	520–550	820–860	1100–1140	1356

are 15,800 t/h and 48,300 t/h, respectively. Consequently, a water flow rate of $(G_L - G_R) = 32,500$ t/h is additionally induced by the jet pumps. When the water circulation in a BWR is evaluated in terms of the circulation ratio, which gives a criterion for forced-circulation boilers, the circulation ratio becomes $R = G_L/G_S = 48,300/6430 = 7.5$, corresponding to the quality at the outlet of the core, x_e, of about 0.13. Table 4.3 indicates that the circulation ratio for BWRs of various capacities is 7.5–7.8.

In the ABWR plant, 10 internal pumps are installed in the shroud around the core, instead of having recirculation pumps outside the pressure vessel. This

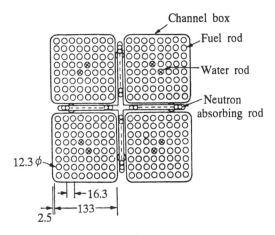

Figure 4.67 Fuel assembly (cross-section).

improves the safety of the reactor against LOCA, owing to the absence of an outer recirculation loop.

The cross-section of a fuel assembly is shown in Figure 4.67. In the fuel assembly currently used, fuel rods of 12.3 mm diameter are arranged in an 8 × 8 lattice in a square channel box of 133 mm width, and the boiling water flows through the subchannel outside the fuel rods. Figure 4.68 (Yamashita et al. 1984) illustrates the configuration of a fuel assembly. Cross-shaped control-rod plates are installed between fuel assemblies, and these are installed with approximately 0.3 m pitch in a core. The control-rod plate is composed of a number of stainless steel tubes filled with B_4C powder, which absorbs neutrons. The thermal output of a fuel assembly reaches about 4.3 MW, and this determines the number of assemblies for a required power of the plant. For example, the number of assemblies for a 1100-MWe (3293 MWt) plant is 3293/4.3 = 764. Numbers of fuel assemblies are also listed in Table 4.3.

Power control in BWRs is conducted by two different methods. One is by the penetration length of the control rods. The other is by the water flow rate; that is, an increase in the recirculating water flow rate reduces the void fraction in the core, which results in an increase in reactivity owing to the increase of the volume fraction of water as the moderator, and thus in an increase in the reactor power. Conversely, a decrease in the water flow rate induces a power decrease. Figure 4.69 (Research Committee on Nuclear Reactor Safety 1985b) shows the relationship between the reactor power and the recirculating water flow rate. At a certain penetration length of the control rods, the reactor power changes with the recirculating water flow rate as illustrated, for example, by curves 1–3. The penetration length of the control rods changes the reactor power as in curves 4–6,

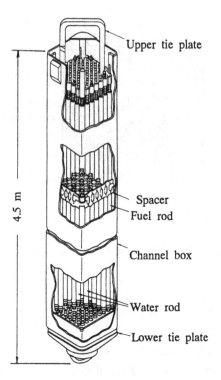

Figure 4.68 Fuel assembly (Yamashita et al. 1984).

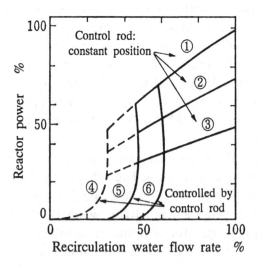

Figure 4.69 Relationship between power and recirculation water flow rate (Research Committee on Nuclear Reactor Safety 1985b).

under an approximately constant water flow rate. In general, control by recirculating water flow rate is conducted for a short-period power variation, and control for a large power change, such as plant shutdown and an adjustment for a long-term change of reactivity in the nuclear fuel, are conducted by the control rods.

4.8.2 Thermal conditions in the reactor core

The thermal conditions in reactor cores are listed in Table 4.3. The power density (heat generated per unit volume) is about 50 kW/l (50×10^3 kW/m³), which is defined as Q (thermal power) / V (core volume). For a 1100-MWe plant, for example, Q is 3293 MW and the core volume V, including fuel rods, control rods, and water, is $(\pi/4) \times 4.8^2 \times 3.7 = 66.9$ m³ (diameter and height of the core are 4.8 m and 3.7 m, respectively); then $Q/V = 3293 \times 10^3$ kW/66.9 m³ $= 50 \times 10^3$ kW/m³. This shows that the heat generation rate in nuclear reactors is more than 100 times larger than that in boiler furnaces in 500- to 1000-MWe-class fossil-fuel power plants. The average heat flux at the surface of a fuel rod is about 490 kW/m², as shown in Table 4.3. This value is evaluated as Q (thermal power) / S (total area of fuel rod surface); then $Q/S = 3293 \times 10^3/6787 = 485$ KW/m², where S = (number of fuel assembly) $\times$ (number of fuel rods in a fuel assembly) $\times \pi$ $\times$ (fuel rod outside diameter) $\times$ (fuel rod length) $= 764 \times 62 \times \pi \times 0.0123 \times 3.71 = 6787$ m². As the distribution of thermal power is not uniform along a fuel rod, the maximum value of heat flux reaches about 2.4 times the average value, that is, about 1120 kW/m². This value is considerably higher than the heat flux of 115–230 kW/m² at boiler furnaces in high-capacity plants. This means that the thermal and hydraulic design of the core is of great importance for reactor safety.

The heat generation rate per unit volume q (W/m³) is proportional to the mass of fission material, the cross-section for nuclear fission, and the neutron flux, and is expressed as

$$q = \left(kr\varrho_m N_A/A_f\right)E\sigma_f\phi \tag{4.115}$$

where k is the enrichment, r is the fraction of nuclear fission material in the fuel material, ϱ_m is the density of the fuel material, N_A is Avogadro's constant, A_f is the atomic number of the nuclear fission material, E is the heat generated per nuclear fission (J), σ_f is the microscopic cross-section, and ϕ is the neutron flux (n/cm²s). For UO_2, which is widely used as a nuclear material, $E = 3.17 \times 10^{-11}$J, $\sigma_f = 335$b (b $= 10^{-24}$ cm²), and $\varrho_m = 10{,}500$ kg/m³. For $k = 0.025$, for example, the heat generation rate q is expressed as

$$q = 6.3 \times 10^{-12}\phi \ \left(MW/m^3\right) \tag{4.116}$$

This equation indicates that the distribution of heat generation in the core is a function of the distribution of neutron flux, which can be obtained from the dif-

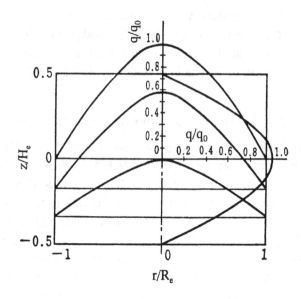

Figure 4.70 Distribution of heat generation rate in reactor core.

fusion equation and is expressed by a cosine distribution in the axial direction and a zero-order Bessel function in the radial direction for a homogeneous cylindrical core of radius R and height H. Thus the distribution of the heat generation rate q is expressed as

$$q = q_0(\phi/\phi_0) = q_0 \cos(\pi z/H_e) \cdot J_0(2.405r/R_e) \qquad (4.117)$$

where q_0 is the heat generation rate at the center of the core, H_e and R_e are the effective length, $H_e = H + 2\lambda_e$, $R_e = R + \lambda_e$, for the height and radius, respectively, λ_e is the extrapolation distance of neutrons, and $\lambda_e = 0.71\lambda_t$, where λ_t is the mean free path and equals 0.43 cm in light water and about 2.5 cm in heavy water. The distribution of the heat generation rate calculated from Eq. (4.117) is shown in Figure 4.70. In cases where a reflector is installed around the core, the heat generation rate in the outer part of the core increases, which leads to a more uniform distribution.

The profile of distribution mentioned above is an ideal case with a homogeneous core construction. In practice, however, cores are not homogeneous owing to variations in construction material as well as the void fraction distribution along the channel, and thus the distribution differs from the ideal case. Furthermore, the fuel concentration changes in space and time owing to the existence of burn-up distribution. Figure 4.71 illustrates such distributions of heat generation rate in actual reactors (Miki and Ohki 1984). In fuel rods used in the past, the heat generation rate in the lower part of the core decreases during the last period of the fuel cycle. This has recently been improved, however, so as to realize a fairly uniform distribution of heat generation rate, as shown in the figure

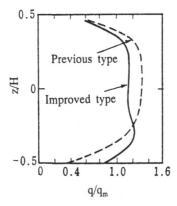

Figure 4.71 Distribution of heat generation rate in actual reactors (Miki and Ohki 1984).

(see also Table 4.5 below), by using fuel rods composed of highly enriched fuel in the upper part and low enriched fuel in the lower part.

The ratio of the maximum heat generation rate to the axially averaged value is referred to as the "power peaking factor." For the distribution expressed by Eq. (4.117), as an example, the power peaking factor is given by

$$q_0 \bigg/ \int_{-H/2}^{H/2} q_0 \cos\left(\frac{\pi z}{H}\right) dz = \frac{\pi}{2} = 1.57$$

To ensure the safety criteria against the most severe condition, a safety factor that is referred to as the "hot spot factor" or "hot channel factor" is introduced in the thermal and hydraulic design of the reactor core. These factors are determined taking into account the difference between the maximum heat generation rate and the average value, the manufacturing variation in the fuel and cladding material, and the variation of flow channel dimensions. For example, the hot channel factors for the heat generation rate in the radial and axial directions, F_R and F_Z, are given from Eq. (4.117) as

$$F_R = J_0(0)\int_{-H/2}^{H/2} \cos(\pi z/H)dz \bigg/ \left[\int_0^R J_0(2.405r/R)2\pi r\, dr \times \int_{-H/2}^{H/2} \cos(\pi z/H)dz\right]$$

$$= 2.32$$

$$F_Z = J_0(0)\bigg/ \left[\frac{J_0(0)}{H}\int_{-H/2}^{H/2} \cos(\pi z/H)dz\right] = 1.57$$

Consequently, the hot channel factor at the center of the core is $F_R \times F_Z = 2.32 \times 1.57 = 3.64$. Since the distribution of q is flatter in an actual reactor than that described above, values of, for example, $F_R = 1.4$ and $F_Z = 1.5$ are used. If a safety factor $F_L = 1.24$ for variation in the fuel properties is added, then the total hot channel factor F becomes $F = F_R \times F_Z \times F_L = 1.4 \times 1.5 \times 1.24 = 2.6$.

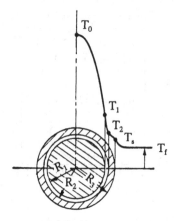

Figure 4.72 Radial distribution model of fuel rod temperature.

One of the thermal and hydraulic design criteria is the limitation of the maximum temperature of the fuel rod. The melting point of UO_2 is about 2800°C and it decreases as the reaction proceeds. Therefore, the maximum allowable temperature at the center of the fuel rod is 2500°C in an emergency and 1850°C during normal operation. On the basis of this criterion the design value of the heat generation rate per unit length of fuel rod, referred to as linear fuel rating, is chosen to be about 440 kW/m (see Table 4.5). The fuel rod is composed of a zircalloy-2 cladding tube of 12.3 mm OD, 0.86 mm thickness, and 3.71 m length, and UO_2 pellets of 10.3 mm OD and 11 mm length (the advanced type of 8 × 8 fuel rod in Table 4.5). Helium gas is pumped into the initial clearance of 0.24 mm between the cladding tube and fuel pellets. The radial distribution of temperature in a fuel rod is shown in Figure 4.72. The equation of heat conduction in a fuel pellet is

$$\frac{1}{r}\frac{d}{dr}\left(\lambda_F r \frac{dT}{dr}\right) = -q_F \tag{4.118}$$

where q_F (W/m³) is the heat generation rate per unit volume of fuel, and λ_F is the thermal conductivity of UO_2, which is a function of temperature, as shown in Figure 4.73. Under the assumptions that q_F is uniform in the fuel and that λ_F is constant, $\lambda_F = \lambda_{Fm}$, as well as the boundary condition of $T = T_1$ at $r = R_1$, the temperature distribution is expressed as

$$T = T_1 + \frac{q_F}{4\lambda_{Fm}}\left(R_1^2 - r^2\right) \tag{4.119}$$

The temperature difference between the center and the outer surface of the fuel pellet is thus

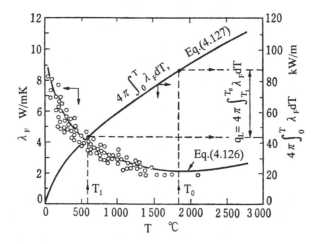

Figure 4.73 Thermal conductivity of UO$_2$ (Isshiki et al. 1967).

$$T_0 - T_1 = \pi q_F R_1^2 / 4\lambda_{Fm}\pi \qquad (4.120)$$

where T_0 is the temperature at the center. The temperature distributions in the clearance and cladding tube are also obtained from Eq. (4.118), with a condition that the heat generation rate be zero. The linear fuel rating q_L is the heat generated in the fuel rod of unit length $\pi R_1^2 q_F$, and is consequently expressed as

$$q_L = \pi R_1^2 q_F = 2\pi \left(T_1 - T_2 \right) \lambda_G \Big/ \ln \frac{R_2}{R_1} \qquad (4.121)$$

$$q_L = \pi R_1^2 q_F = 2\pi \left(T_2 - T_3 \right) \lambda_P \Big/ \ln \frac{R_3}{R_2} \qquad (4.122)$$

where λ_G is the thermal conductivity of the gas in the clearance and λ_P is that of the cladding tube. The heat transfer from the cladding tube surface to the coolant is expressed by

$$\pi R_1^2 q_F = 2\pi R_3 h \left(T_3 - T_f \right) \qquad (4.123)$$

where h is a heat transfer coefficient and T_f is the temperature of the coolant. The temperature difference between the center of the fuel rod and the coolant $(T_0 - T_f)$ is obtained from Eqs. (4.120)–(4.123) as

$$T_0 - T_f = \left(\pi R_1^2 q_F \right) \left\{ \frac{1}{4\pi\lambda_{Fm}} + \frac{\ln\left(R_2/R_1\right)}{2\pi\lambda_G} + \frac{\ln\left(R_3/R_2\right)}{2\pi\lambda_P} + \frac{1}{2\pi R_3 h} \right\} \qquad (4.124)$$

This equation indicates that the temperature difference $(T_0 - T_f)$ is proportional to the linear fuel rating q_L given by $\pi R_1^2 q_F$.

Next, on the basis of the equations mentioned above, the temperature at the center of the fuel rod T_0, which is a design criterion, will be examined. Since the dimension of clearance changes with the burn-up process of the fuel, it is difficult to predict the thermal resistance of the clearance exactly. Thus the following value is recommended for use. The thermal resistance $(T_2 - T_1)/q_L$ is approximately expressed, on the basis of a relationship that $\ln(R_2/R_1)$ is nearly equal to $(R_2 - R_1)/R_1$, as

$$\frac{T_2 - T_1}{q_L} = \ln\frac{R_2}{R_1}\bigg/2\pi\lambda_G \cong \frac{R_2 - R_1}{R_1}\frac{1}{2\pi\lambda_G} = \frac{1}{2\pi R_1}\left(\frac{\delta}{\lambda_G}\right)$$

where (δ/λ_G) is the thermal resistance, and the reciprocal (λ_G/δ) is the thermal conductance referred to as gap conductance. The value 5.7×10^3 (W/m^{2}K) is usually used for this gap conductance (δ = thickness of clearance, λ_G = thermal conductivity of clearance). The linear fuel rating is then expressed by

$$q_L = 2\pi R_1(\lambda_G/\delta)(T_1 - T_2) = 2\pi R_1 h_G(T_1 - T_2) \tag{4.121'}$$

Figure 4.74 illustrates the temperatures T_0–T_3 in a fuel rod against the linear fuel rating at $R_1 = 10.3/2 = 5.15$ mm, $R_2 = 5.29$ mm, $R_3 = 12.3/2 = 6.15$ mm, $h = 1.2 \times 10^4$ (W/m^{2}K), $\lambda_{Fm} = 3$ (W/m K), $h_G = 5.7 \times 10^3$ (W/m^{2}K), $\lambda_P = 17$ (W/m K), and $T_f = 286°$C. The figure indicates that the allowable maximum value of the linear fuel rating is dominated by the limitation of T_0. For instance, the maximum value 44 kW/m for q_L in current reactors corresponds approximately to $T_0 = 1850°$C.

The thermal conductivity λ_F is a function of the temperature, as shown in Figure 4.73, but was assumed to be constant in the above discussion. The temperature difference $(T_0 - T_1)$ for the practical case can be obtained as follows. Integration of Eq. (4.118) from $r = 0$ to R_1 gives

$$4\pi \int_{T_0}^{T_1} \lambda_F \, dT = \pi R_1^2 q_F \equiv q_L \tag{4.125}$$

It is indicated that the integration of λ_F with respect to temperature gives the linear fuel rating. In UO$_2$, this thermal conductivity λ_F is expressed as a function of the temperature as shown in Figure 4.73 (Isshiki et al. 1967):

$$\lambda_F = \frac{3.82 \times 10^3}{(T + 129)} + 0.0479\left(\frac{T}{100}\right)^3 \tag{4.126}$$

Equation (4.125) is rewritten as

$$4\pi\left(\int_0^{T_0} \lambda_F \, dT - \int_0^{T_1} \lambda_F \, dT\right) = q_L \tag{4.125'}$$

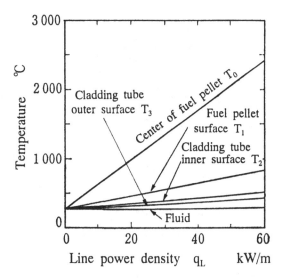

Figure 4.74 Temperatures in fuel rod.

The relationship between T_0, T_1, and q_L can then be obtained from the integration $\lambda_F dT$. The integration $4\pi\lambda_F dT$ is expressed by the following equation and is also shown in Figure 4.73:

$$4\pi \int_0^T \lambda_F \, dT = 48 \times 10^3 \ln\left(\frac{T + 129}{402}\right) + 0.015\left(\frac{T}{100}\right)^4 \qquad (4.127)$$

4.8.3 Thermal design principle and evolution of the boiling water reactor

In BWRs, 2–4% enriched ^{235}U is used as a nuclear fuel and light water as the moderator. The thermal energy is generated by the chain reaction of fission of ^{235}U. The reactivity, heat generation rate per unit volume (power density in the reactor core), economy, and safety of the plant are significantly affected by the dimensions, configuration, and enrichment of fuel and the arrangement of moderator and fuel rods.

The main principle of thermal design of the core is thus described as follows. In the case of fuel assembly in a square lattice as shown in Figure 4.75, for example, the selection of pitch s for a given fuel rod diameter d_0 is determined from the nuclear calculation. According to a simple ideal model, the nuclear reactivity is determined from the fuel enrichment and the volume ratio of moderator to fuel, V_M/V_F. The reactivity is expressed by the multiplication factor k, defined as the ratio of the number of fissions in a given period to that in the next period.

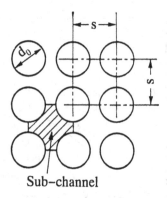

Sub–channel

Figure 4.75 Fuel rod lattice.

The relationship between the multiplication factor and the moderator to fuel ratio is shown with enrichment as a parameter in Figure 4.76, where k_∞ is the ideal multiplication factor in a core of infinite volume. The k_∞ value has a maximum at a particular V_M/V_F, for the following reason. In cases of low V_M/V_F, that is, high volume ratio of UO_2, the fraction of resonance absorption of fission neutrons by ^{238}U increases and the reactivity then decreases. With increasing V_M/V_F, on the other hand, the probability of resonance absorption decreases but the probability of collision of the thermal neutrons converted from the fission neutrons to ^{235}U decreases, which leads to a decrease in the reactivity. Therefore, k_∞ has a maximum at an appropriate value of V_M/V_F.

Here, the V_M/V_F shown in Figure 4.75, for example, is expressed by

$$V_M/V_F = \left(s^2 - \frac{\pi}{4} d_0^2 \right) \bigg/ \frac{\pi}{4} d_F^2 = \frac{4}{\pi} \left(\frac{s}{d_0} \right)^2 \left(\frac{d_0}{d_F} \right)^2 - \left(\frac{d_0}{d_F} \right)^2$$

$$= \left(\frac{d_0}{d_F} \right)^2 \left[\frac{4}{\pi} \left(\frac{s}{d_0} \right) - 1 \right] \tag{4.128}$$

where $d_F = 2R_1$ is the diameter of a pellet and $d_0 = 2R_3$ is the diameter of a fuel rod; thus the value of V_M/V_F is determined by the ratio of pitch to fuel rod diameter, s/d_0. In current reactors, d_0 is 12.3 mm, s is 16.3 mm, and s/d_0 is 1.32; then V_M/V_F is 1.9 from Eq. (4.128). In practice, however, the value of V_M/V_F in Type BWR-5 reactors is 2.5, because there is clearance between fuel rod assemblies and two water rods (flow channel of coolant) are installed in the fuel assembly. Thus the design point of k_∞ is to the right of the maximum point as shown in Figure 4.76. Further detail of the nuclear design is beyond the scope of this book.

In the optimum design of the core the balance between nuclear and thermo-hydraulic conditions should be considered. Here the design values for the core

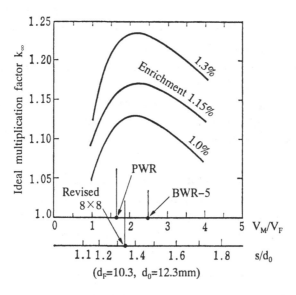

Figure 4.76 Ideal multiplication factor (curves from Handbook of Mechanical Engineering, JSME).

will be explained for the configuration shown in Figure 4.75. The following relationships exist between the core volume power density q_V (W/m³), heat generation per unit volume of fuel q_F (W/m³), linear fuel rating q_L (W/m), and heat flux at the fuel rod surface q_S (W/m²):

$$q_V = \frac{\pi}{4}\left(\frac{d_F}{s}\right)^2 q_F \qquad (4.129)$$

$$q_L = \frac{\pi}{4}d_F{}^2 q_F = \pi d_0 q_S \qquad (4.130)$$

The thermal criteria for fuel rod design are the limitation of the temperature at the center of the fuel rod T_0 and the temperature of the cladding tube, as mentioned in Section 4.8.2. For the former criterion, since T_0 is closely related to the linear fuel rating q_L as shown in Figure 4.74, the limitation of q_L is decided according to the following method. Equation (4.130) indicates that q_L is proportional to $d_F{}^2$ at a given value of q_F, and this corresponds to a limitation of the fuel pellet diameter d_F. As for the latter criterion, the allowable maximum temperature of cladding material, Zircalloy-2, is 550°C, and thus the design value is set to 358°C. The cladding tube temperature is affected by the heat transfer coefficient at the surface, which is a function of the heat flux q_S, where q_S is approximately proportional to d_F as shown by Eq. (4.130). This means that a limitation exists in d_F also with respect to the cladding tube temperature.

The heat transfer coefficient is, in general, affected by the fuel rod diameter d_0

and the pitch s. A very small pitch, that is, a small clearance between fuel rods, causes a reduction of the local boiling heat transfer coefficient and leads to burnout. The pitch affects the flow resistance in the channel and thus the pumping power. Therefore, the optimum core design should be reached by balancing the nuclear design against the thermohydraulic design.

Core design, as regards the heat transfer problem, is conducted on the basis of the critical heat flux and the critical value of mass flux for a burnout, as described in Sections 4.2 and 4.8.4. In this section, only the design values of mass flux in current reactors are described.

To avoid burnout or excess temperature rise of the fuel rod surface, a sufficient mass flux should be ensured. This situation is similar to that in forced-circulation boilers and once-through boilers. The average value of heat flux in BWRs is about $485\,kW/m^2$ and the local maximum value is about $1260\,kW/m^2$, which are about ten times larger than those in boilers. The mass flux in BWRs is thus considerably higher and the coolant flow rate reaches 48,300 t/h for an evaporation rate of 6430 t/h in the 1100-MWe-class power plant. This value corresponds to a circulation ratio $R = 7.5$, and the average outlet quality becomes $x_e = 0.13$.

The radial distribution of the coolant flow rate is given corresponding to the distribution of power in the core. Table 4.4 shows that the maximum flow rate in the center of the core reaches 68 t/h, which is about twice as large as the flow rate of 32 t/h in the outer region. As the cross-sectional area of the channel is 100.1 cm^2 (Fig. 4.67), the mass fluxes are $1888\,kg/m^2s$ and $888\,kg/m^2s$ and the velocities at the channel inlet are 2.5 m/s and 1.2 m/s, respectively. The pressure drop in the channel is then about 0.21 MPa.

Table 4.5 illustrates the transition of core design for typical BWR types, BWR-1 to ABWRs (Miki and Ohki 1984; Research Committee on Nuclear Reactor Safety 1985a). The power density has increased from 31.2 kW/l for the BWR-1 to 50 kW/l for the current type, BWR-5. The maximum linear fuel rating, however, has reduced from 50–57.4 kW/m for types BWR-1 to BWR-3 to 43.9 kW/m for BWR-5. This has been achieved by making the power distribution uniform. Following a reduction of the fuel rod diameter from 12.1 mm for the BWR-4 to 10.6 mm for the BWR-5, the linear fuel rating also reduced from 60.7 kW/m to 43.9 kW/m, as shown by the relationship of Eq. (4.130), and the number of fuel rods in a fuel rod assembly was increased from 7×7 to 8×8.

4.8.4 Thermal design criteria of reactor core channels

The thermal and hydraulic design of the core is very important for the safety of BWRs, and design instructions have been proposed that are based on a large amount of experimental data. In this section a thermal and hydraulic design method developed at General Electric, denoted by GETAB (General Electric

Table 4.4. *Coolant flow rates in BWRs*

	Average in center region	Maximum	Average in outer region	Average in core
Flow rate per fuel assembly (t/h)	60	68	32	60.8[a]
Coolant mass flux (kg/m²s)	1666	1888	888	1690
Coolant velocity (m/s)	2.2	2.5	1.2	2.3

[a] (Coolant flow rate 46,500 t/h) ÷ (number of assemblies 746).

BWR Thermal Analysis Basis), and a critical heat flux correlation denoted by GEXL (General Electric Critical Quality (x_{cr}) – Boiling Length (L_B) Correlation), which is the basis of GETAB, are introduced briefly (GETAB 1973; Nuclear Reactor Safety Inspection Committee 1976).

(1) GEXL. This is a critical heat flux correlation based on more than 6000 experimental data obtained on full-scale, full-power rod bundles ($7 \times 7, 8 \times 8$) and various axial distributions of heat fluxes. The critical heat flux is defined as the fuel assembly power at which boiling transition (departure from nucleate boiling) is expected to occur. The correlation is expressed in the following form:

$$x_{cr} = f\left(L_B, D_q, G, L, p, R_w\right) \tag{4.131}$$

where, for a particular reactor, x_{cr} is the critical quality, L_B is the boiling length (the length over which boiling takes place), D_q is the hydraulic equivalent diameter, G is the mass flow rate per unit area (mass flux), L is the total heated length, p is the pressure, and R_w is a weighting factor that characterizes the local rod-to-rod peaking pattern for the rod under the most severe conditions. The exact formulation of Eq. (4.131) is revealed only for contractors, and thus is not presented in this book.

In channels of 3.6–3.7 m length and high heat flux, the local boiling transition is affected by the axial heat flux distribution; that is, the boiling transition is affected not only by the local condition but also by the upstream history of heat flux distribution. However, the influence of the heat flux distribution can be compensated by relating the critical quality x_{cr} to the boiling length L_B. The relationship between x_{cr} and L_B is determined independently of the heat flux distribution, for example as shown in Figure 4.77 (GETAB 1973). The values predicted by the GEXL correlation P_{GE} agree with the experimental values P_M for 2700 data for the 7×7 type and 1800 data for the 8×8 type within an error of ±6%. The accuracy is expressed by the ratio of P_{GE} to P_M, referred to as the experimental critical power ratio (ECPR), that is,

Table 4.5. *Development of fuel type in BWR (Miki and Ohki 1984; Research Committee on Nuclear Reactor Safety 1985a)*

	1960	1969	1970	1974	1978	1983	1996
Type	BWR-1	BWR-2	BWR-3	BWR-4	BWR-5	Standard BWR (revised BWR-5)	ABWR
Typical plant	Dresden	Tsuruga-1	Fukushima I-1	Fukushima I-2	Tokai-2	Fukushima II-2	Kashiwazaki-6
Size (MW)	200	650	800	880	1100	1100	1350
Fuel rod lattice	6 × 6	7 × 7	7 × 7	7 × 7	8 × 8	8 × 8	8 × 8
Volume power density (kW/l)	31.2	33.6	41	51	50	50	46
Maximum linear fuel rating (kW/m)	50	57.4	57.4	60.7	43.9	43.9	44.0
Burn-up (MWd/t)	12,000	22,000	21,500	27,500	27,500	29,500	39,500
Fuel pellet material	UO_2	UO_2	UO_2	UO_2 and Gd_2O_3 added UO_2	UO_2 and Gd_2O_3 added UO_2	UO_2 and Gd_2O_3 added UO_2	UO_2 and Gd_2O_3 added UO_2
Diameter (mm)		12.4	12.4	12.1	10.6	10.3	10.4
Length (mm)		22	22	12	11	11	10
Stack length (m)	2.69	3.66	3.66	3.66	3.66 3.71	3.71	3.71
Cladding tube material	Zircaloy-2 Cold working	Zircaloy-2 Stress relief annealing	Zircaloy-2 Stress relief annealing	Zircaloy-2 Recrystallization annealing	Zircaloy-2 Recrystallization annealing	Zircaloy-2 Recrystallization annealing	Zircaloy-2 Recrystallization annealing
OD (mm)	14.4	14.5	14.3	14.3	12.5	12.3	12.3
Thickness (mm)	0.81	0.91	0.81	0.94	0.86	0.86	0.86
Number of fuel rods/assembly	36	49	49	49	63	62	60
Number of water rods	0	0	0	0	1	2	1

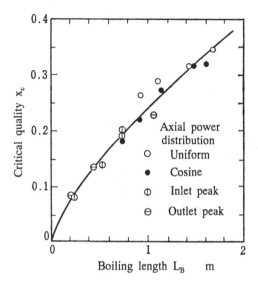

Figure 4.77 Critical quality in round tube (data by Babcock and Wilcox).

$$\text{ECPR} = \frac{P_{\text{GE}}}{P_{\text{M}}} = \frac{\text{critical power predicted by GEXL correlation}}{\text{experimental critical power}} \qquad (4.132)$$

The frequency distribution of ECPR for 4×4 and 8×8 lattices, for example, is shown in Figure 4.78 (Nuclear Reactor Safety Inspection Committee 1976). The average ECPR is 0.9846 and the standard deviation $\sigma = 0.0280$ for the cases shown, and the figures for the 7×7 lattice are 0.9885 and 0.036, respectively. This indicates that the accuracy of the GEXL correlation is very high. The safety margin is then predicted as follows. Assuming the frequency distribution in Figure 4.78 to be a normal probability density distribution shown in Figure 4.79, the region of ECPR where the probability of occurrence of boiling transition is less than 0.1% is greater than 3.09 σ. When the value $3.09 \times 0.028 = 0.0865$ is adopted as the margin for the critical power ratio (CPR), safety will be ensured within 0.1% error.

(2) Critical power ratio. The CPR is used as a thermal criterion, and is defined as

$$\text{CPR} = \frac{P_{\text{cr}}}{P} = \frac{\begin{array}{c}\text{power in fuel assembly at which boiling} \\ \text{transition takes place (critical power)}\end{array}}{\text{actual power in fuel assembly}} \qquad (4.133)$$

Boiling transition occurs at CPR $= 1$; reactor safety is therefore improved by using a higher value of CPR. The occurrence of boiling transition, however, does not necessarily result in damage to the fuel rods, because the heat transfer

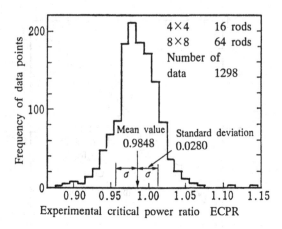

Figure 4.78 An example of frequency distribution of experimental critical power ratio.

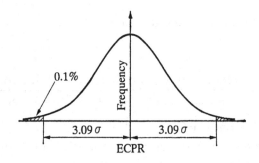

Figure 4.79 CPR margin.

coefficient is very high owing to the high mass flux in the channel. The power and mass flux distributions are not uniform, and consequently the minimum value of CPR in the core is used as the criterion and is referred to as minimum critical power ratio (MCPR).

The CPR is determined by the following procedure, as shown in Figure 4.80. (1) Obtain a relationship between the boiling length L_B and the quality x for the actual power P and the mass flux on the basis of the heat balance (curve A). (2) Obtain a relationship between L_B and x for the same condition by the GEXL correlation (curve B). (3) Obtain L_B versus x relationships for various powers at the given mass flux (curves C and D). (4) Determine an x–L_B curve for a value, P_{cr}, which is tangential to curve B. The critical power P_{cr} is thus determined. (5) Finally, determine the CPR from Eq. (4.133) as P_{cr}/P.

The critical value of MCPR that is referred to as the safety limit minimum critical power ratio (SLMCPR) for normal operation is shown in the first column of

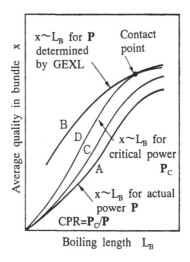

Figure 4.80 Determination of CPR.

Table 4.6. *Critical power ratio*

Type of plant	BWR-2		BWR-3		BWR-4		BWR-5
SLMCPR	1.06		1.06		1.06		1.07
Fuel rod	7 × 7	8 × 8	7 × 7	8 × 8	7 × 7	8 × 8	8 × 8
Maximum ΔMCPR	0.17	0.18	0.20	0.23	0.16–0.18	0.12–0.23	0.19
OLMCPR	1.23	1.24	1.26	1.29	1.22–1.24	1.23–1.28	1.26

Table 4.6. A typical value of MCPR is 1.07 for Type BWR-5. This value was decided on the basis of the probability analysis that the fraction of fuel rods likely to cause boiling transition in relation to the total of fuel rods is less than 0.1%, taking into account errors in the measuring system, production tolerance of fuel and prediction error of GEXL. Figure 4.81 shows the relationship between the fraction of fuel rods f (%) predicted to avoid the boiling transition and MCPR. For $f = 99.9\%$ the value of MCPR is determined as 1.07, as shown in the figure.

(3) Operating limit MCPR (OLMCPR). The operating limits of MCPR for anticipated abnormal transient conditions are provided for various cases as follows (Nuclear Reactor Safety Inspection Committee 1976).

$$OLMCPR = SLMCPR + \Delta MCPR \qquad (4.134)$$

where ΔMCPR is the margin for the transients and is given as follows:

i. One pump trip in water circulation system: 0.07. The decrease of the circulating water results in a decrease in power owing to an increase in void fraction, but the heat flux

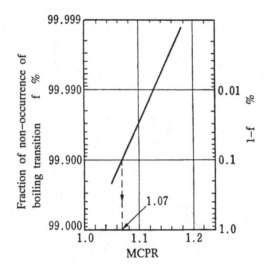

Figure 4.81 Determination of MCPR.

does not decrease owing to the heat capacity of the fuel rods. Consequently the critical power decreases temporarily, and thus MCPR decreases.

ii. Failure in feed-water heater: 0.12. The reduction of feed-water temperature results in an increase in power owing to a decrease in void fraction, which leads to a decrease in MCPR.

iii. Generator load rejection without bypass: 0.19. The main steam valve is closed to protect the turbine and the pressure rises, which results in an increase in power owing to a decrease in void fraction.

iv. Control rod withdrawal: 0.12. Power increases locally in the transients.

These values of ∆MCPR and OLMCPR are listed in Table 4.6. For Type BWR-5, for example, the OLMCPR is 1.26, which is the sum of 1.07 (SLMCPR) and 0.19 (the maximum value of ∆MCPR in case (iii) described above).

4.8.5 Evolution of critical heat flux research and the design criteria of reactor cores

The evolution of research on critical heat flux (CHF) and the design criteria for the reactor core are illustrated in Table 4.7, showing the development of the BWR type. This table suggests the relationship between the prediction method for CHF used in each period and BWR core design at each period.

When the first BWR commercial power plant (Dresden) was built in 1960, only a few very simple correlations of CHF in forced convection boiling, as compared with several correlations of CHF in pool boiling, existed. Data on CHF in a 6 × 6 lattice full-scale model had not yet been obtained. This indicates that the actual

Table 4.7. *Evolution of CHF research and the design criteria*

Year	Type of BWR	Researches on CHF and design criteria
1950	FBR, EBR-1 (100 kWe, 1951)	Pool boiling CHF, McAdams (1949)
		Pool boiling film boiling, Bromeley's equation (1950)
		Pool boiling CHF, Kutateladze's equation (1952)
	BORAX-III Test (1955)	Forced convective boiling CHF, Jens–Lottes' equation (1952)
	Declaration of peaceful utilization of nuclear	
1955	power (USA, 1955)	
1956	BWR(ANL, 3500 kWe, 1956)	Pool boiling CHF, Rohsenow–Griffith's equation (1956)
		Forced convective boiling CHF, Zenkevich, Griffith (1957)
	GCR (Colder Hall, 50 MWe, 1956)	
	PWR (Shipping Port, 1957)	Pool boiling CHF, Zuber's equation (1958)
		Pool boiling, Nishikawa et al. (1958)
1959		
1960	BWR-1 (Dresden, 200 MWe, 1960)	Research Committee on Boiling (JSME) (1961)
	PWR (Yankee Rowe, 175 MWe, 1960)	Jansen–Levy's equation (Concentric annulus, data points 700) (1962)
		JSME "Boiling Heat Transfer" (1964)
		Tong's equation of CHF, W1, W2 (1964)
		CISE-1 Eq. Tong's W3 equation
		Tong "Boiling Heat Transfer and Two-Phase Flow" (1965)
1965	GCR (Tokai, 166 MWe, 1965)	Becker's equation of CHF (1965)
1966		Hench–Levy's equation (2 × 2, 3 × 3 lattice, data points 700) (1966)
		Columbia Univ. (44 lattice, data points 6200) (1968)
		Macbeth's equation of CHF (1968)
		GE (7 × 7, 8 × 8 lattice, Freon loops, data points 6000) (1969)
1969	BWR-2 (Tsuruga-1, 1969)	
1970	BWR-3 (Fukushima I-1, 1970)	Tong–Weisman, "Thermal Analysis of Pressurized Water Reactors" (1970)
		ATLAS loop (GE) (4 × 4, 7 × 7, 8 × 8 array, data points 6000) (1972)
		GEXL (data points 1703) (1973)
	BWR-4 (Fukushima I-2, 1974)	GETAB (1973)
1975		
1976		
	BWR-5 (Tokai, 1978)	Lahey–Moody, "The Thermal-Hydraulics of a Boiling Water Nuclear Reactor" (1977)
1979		
1980		
	Standard BWR (Fukushima II-2, 1983)	

design of the reactor core was conducted using limited information on CHF by designers with excellent engineering sense. In 1962, the Jensen–Levy equation was obtained on the basis of 700 experimental data in an annulus heated by the inner tube, and in 1966 data on 2 × 2 and 3 × 3 lattices were first obtained and the Hench–Levy equation was presented based on these data. At that period Type BWR-2 was developed. After 1966 a large amount of data on 4 × 4, 7 × 7, and 8 × 8 lattices with loops have been obtained at Columbia University, including the GE and ATLAS loops, and the GETAB design method was developed with sufficient accuracy. This period corresponds to the development of Types BWR-3 and BWR-4.

Prior to 1973, when GETAB was established, the critical heat flux ratio defined by Eq. (4.135) was used as a design criterion on the basis of the Hench–Levy equation of critical heat flux, as described as follows:

$$\text{critical heat flux ratio}\left(\text{CHFR}\right) = \frac{\begin{array}{c}\text{heat flux at which boiling}\\ \text{transition occurs at any local}\\ \text{point in fuel rod assembly, } q_{HL}\end{array}}{\text{actual heat flux}} \quad (4.135)$$

q_{HL} is evaluated by Hench–Levy equations (4.46) to (4.49), and is a lower limit line to the 700 data points for 2 × 2 and 3 × 3 lattices, as shown in Figure 4.82. When the heat flux distribution along a fuel rod, q_p, in an actual operation is drawn as illustrated in Figure 4.82, the ratio q_{HL}/q_p takes a minimum value at a certain location. The minimum value of q_{HL}/q_p is used as a design criterion that is designated the minimum critical heat flux ratio (MCHFR):

$$\text{MCHFR} = q_{HL}/q_p \quad (4.136)$$

With increasing MCHFR, the thermal margin for the reactor core increases. Since q_{HL} is the lower limit of experimental data, the case of MCHFR = 1 is the minimum limit for safety. However, the critical heat flux is affected not only by the local quality but also by the pattern of heat flux distribution, which is not included in the Hench–Levy correlation; therefore the location of CHF cannot be accurately predicted. As design criteria, therefore, the following values were adopted: at rated load: MCHFR > 1.9; at overload: MCHFR > 1.0. Since 1973, GETAB has been used for BWR reactor core design.

4.9 Thermal and hydraulic design of pressurized water reactors

4.9.1 System overview of the pressurized water reactor

A typical flow system for the pressurized water reactor (PWR), composed of primary loops at a pressure of 15.4 MPa and secondary loops at a pressure of 6.2

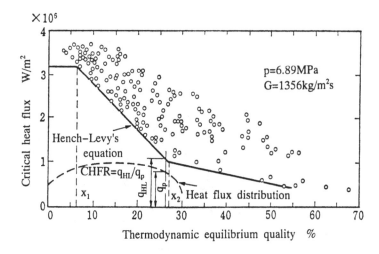

Figure 4.82 CHF and CHFR.

MPa, is shown in Figure 4.83. In the primary loop, water (primary coolant) is circulated by primary coolant pumps from the reactor core channel to the steam generator. The water is heated in the core from 289°C to 323–325°C and supplied to an inverted U-tube heat exchanger in the steam generator. In the secondary loop, the water (secondary coolant) is evaporated in the steam generator and the saturated steam is supplied to the turbine. The exhaust steam is condensed in the condenser and returned to the steam generator through feed-water heaters. The constitution of the secondary loop is similar to the loop in fossil-fuel steam power plants.

In a reactor core in the PWR, boiling does not occur in normal operation and the nuclear reactivity does not change with a change of void fraction. Therefore, thermal and hydraulic analyses in PWR cores are rather simple compared with those in BWR cores, because the flow in the PWR core channel is a single-phase flow. Power control is conducted by changing the primary coolant outlet temperature, which results in a change of the secondary coolant evaporation rate, as shown in Figure 4.84 (Yoshida and Shigemune 1985). The figure illustrates an example of the temperatures in the primary and secondary coolants. The temperature of the primary coolant in the outlet of the core is controlled by a control rod operating system or by a chemical shimming control system. In the chemical shimming control system, which is shown in Figure 4.83, the reactivity in the core can be controlled by changing the boron concentration in the primary coolant over a range of 2000 to 100 ppm; this is done by increasing the concentration by adding H_3BO_3 into the primary coolant, or decreasing it by adding makeup feed water.

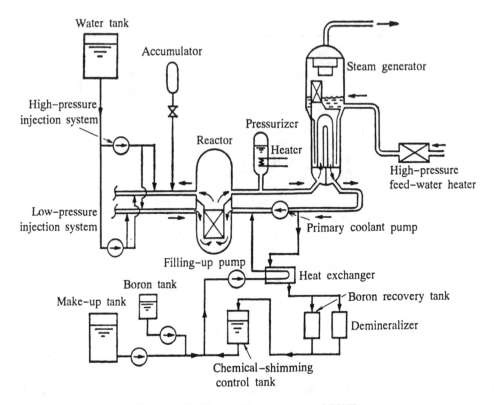

Figure 4.83 Typical flow system of PWR.

An emergency core cooling system (ECCS) against LOCA is installed, furthermore, composed of an accumulator core-cooling system and high-pressure and low-pressure coolant injection systems.

PWR plants consist in general of one reactor and a few steam generators, the number of which, that is, the number of secondary loops, is determined corresponding to the required plant power; for example, two loops in a 600-MWe-class plant, three loops in an 800-MWe-class plant, and four loops in an 1100-MWe-class plant. In each secondary loop, the flow rate of primary coolant is about 1500 t/h (about 2000 m³/h), the power of the circulation pump is 6000 PS, the evaporation rate of secondary coolant is 1600 t/h, and the electric output is about 300 MW.

4.9.2 Reactor core structure and thermal conditions

The flow system in the reactor core in the 1100-MWe-class PWR is shown schematically in Figure 4.85. The feed water supplied by the primary coolant circulation pump enters from the inlet nozzles of the reactor vessel and flows

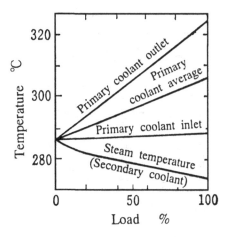

Figure 4.84 Primary coolant temperature in power control program (Yoshida and Shigemune 1985).

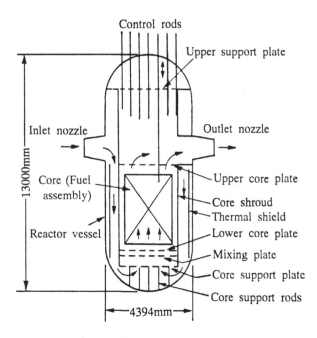

Figure 4.85 Flow in PWR core.

through an annular channel, lower plenum, mixing plate, and fuel assemblies to the outlet nozzles. The inlet and outlet nozzles are installed peripherally at the same height. The dimensions of the reactor vessel in the 1100-MWe-class plant are as follows: height about 13 m, ID about 4.4 m, length of fuel assembly about 3.9 m, and flow rate of primary coolant 60,100 t/h.

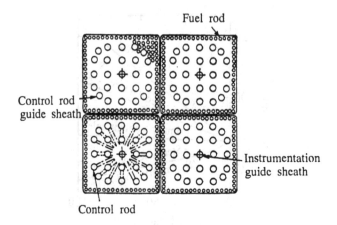

Figure 4.86 Fuel assembly of PWR.

A fuel assembly comprises 264 fuel rods in a 17×17 lattice, one instrumentation guide sheath, and 24 control rod guide sheaths as shown in Figure 4.86. The outside diameter of cladding tubes is 9.5 mm and the pitch is 12.6 mm. As the fuel assemblies are not covered by side plates, the primary coolant is well mixed between fuel assemblies and the cooling performance is therefore high. In fuel assemblies of about 4.06 m length, support plates are installed at nine locations axially, and mixing vanes are installed at seven spacers located in the central part of the core to increase the mixing performance in the coolant and to improve the heat transfer performance (see Figure 4.87 below).

The thermal conditions in a core in a typical PWR are as follows. For a core of equivalent diameter 3.37 m, height of 3.66 m, and thermal power $Q_t = 3411$ MW (electric power = 1100 MWe), the power density $q_V = 3411 \times 10^3/(\pi \times 3.37^2 \times 3.66/4) = 105 \times 10^3 \, \text{kW/m}^3 = 105 \, \text{kW/l}$. This value is about twice as large as the power density in a BWR (50 kW/l). Also, the specific power per unit mass of uranium in the PWR is 38.3 MW/tonne-U, which is considerably higher than the value of 25 MW/tonne-U for the BWR. The average linear fuel rating q_L is a function of the number of fuel rods $N = $ (264/fuel assembly) $\times$ 193 (number of fuel assemblies) = 50,952 and fuel length = 3.66 m/(fuel rod): $q_L = Q_t/3.66N = 3411/3.66 \times 50,952 = 17.9 \, \text{kW/m}$; the maximum linear fuel rating is 43.1 kW/m and the peaking factor is 2.4. This value is approximately equal to the value of 43.9 kW/l in recent BWRs, as shown in Table 4.5. The value of linear fuel rating in the PWR is kept at this level using fuel rods with a smaller diameter of 9.5 mm against 12.3 mm in the BWR, though the volume power density in the PWR is higher than that in the BWR. The average heat flux at the fuel rod surface is $q_s = Q_t/A_F = 60 \times 10^4 \, \text{W/m}^2$ for the total surface area of fuel rods $A_F = 5500 \, \text{m}^2$, and the maximum heat flux is $144 \times 10^4 \, \text{W/m}^2$; this value is about 20% higher

Generation	1st	2nd	3rd	4th	Recent
Plant	Yankee Rowe	Trino	San Onofre	Mihama–1	Ohi–1,2
Side view					
Cross section					

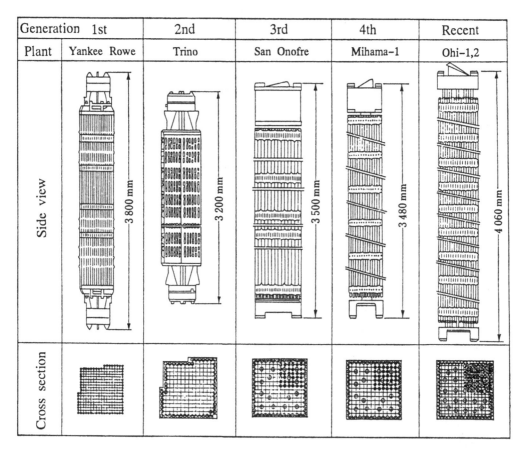

Figure 4.87 Development of fuel assembly in PWR (Morioka 1979; Ogura et al. 1982).

than that in the BWR. It is therefore necessary to give a sufficiently high value to the heat transfer coefficient, and then the rather high mass flux of about 3300 kg/m²s (i.e., velocity = 4.5 m/s) is adopted. The temperatures in a fuel rod can be calculated by the same method as described for the BWR in Section 4.8.2.

The fuel rods of 9.5 mm OD are arranged with a pitch of 12.6 mm and the moderator to fuel ratio V_M/V_F is 1.67, which is at the left side of the peak of the k curve as shown in Figure 4.76. This value of 1.67 is small compared with the design value of 2.5 in the BWR, in which, during normal operation, the value of V_M/V_F is decreased by steam generation from its value in the cold state.

Design changes in the reactor core structure and thermal conditions such as the volume power density, linear fuel rating, and heat flux are illustrated in Table 4.8 and Figure 4.87 (Morioka 1979; Ogura et al. 1982). This table indicates clearly the increases in volume power density, average linear fuel rating, and heat flux in recent reactors. The changes taking place after the fourth period in Japan are

Table 4.8. *Development of PWR (Morioka 1979; Ogura et al. 1982)*

	Generation				
	1st (1958–60)	2nd (1959–62)	3rd (1963–7)	4th (1965–7)	Recent (1974–)
Fuel assembly					
Type	6 × 6	15 × 15	15 × 15	14 × 14, 15 × 15	17 × 17
Number of grid assembly	none		5 (mixing vane)	6	9
Grid assembly material	none	Stainless steel	Inconel 718	Inconel	Inconel
Length (m)	ca. 3.8	ca. 3.3	ca. 3.5	ca. 3.5	ca. 4.0
Fuel					
UO_2 density (%)	93	96	93	93	95
UO_2 pellet diameter (mm)	7.47	8.78	9.79	9.38–9.50	8.2
Cladding tube					
OD (mm)	8.64	9.78	10.4	10.7	9.5
Thickness (mm)	0.533	0.4	0.4	0.6–0.7	0.6
Material	Stainless steel AISI348	Stainless steel AISI304	Stainless steel AISI304	Zircaloy-4	Zircaloy-4
Control rod	Cross-type	Cross-type	Cluster-type	Cluster-type	Cluster-type
Average volume power density (kW/l)	72	64	61	78	105
Average linear fuel rating (kW/m)	10	12.6	14	16.6	17.9
Maximum linear fuel rating (kW/m)	38	41	45	54	43.1
Average heat flux (kW/m²)	370	446	470	543	650
Maximum heat flux (kW/m²)	1400	1450	1510	1760	2000
Minimum departure from nucleate boiling ratio	>2.0	2.4	2.0	1.94	2.0
Hot channel factor E_Q	3.46	3.46	3.25	1.88	1.58
Hot channel factor $E_{\Delta H}$	2.56	2.56	2.11	1.94	2.00

further illustrated in Table 4.9. Typical changes are the type of fuel assembly, that is, from a 14 × 14 lattice to a 17 × 17 lattice, and the increase in average volume power density from 71 to 105 kW/l.

4.9.3 Thermal design criteria for the reactor core

Temperature at center of fuel rod: A criterion of core design is the limitation of temperature of the fuel, UO_2, which has a melting point of over 2800°C. The same applies to BWR design, as mentioned in Section 4.8.2. As for the actual design criteria for the temperature, values of about 2000°C at rated power and 2350°C at maximum linear fuel rating (54 kW/m) have recently been adopted. According

Table 4.9. *Development of typical PWR plants in Japan, by year*

	1970	1972	1975–82	1974–6	1984–5	1979–87
Number of plants	1	1	4	3	4	3
Number of SG	2	2	2	3	3	4
Electric power (MWe)	340	500	559–566	826	870–890	1160–1175
Fuel assembly						
Type	14 × 14	14 × 14	14 × 14	15 × 15	17 × 17	17 × 17
Number	121	121	121	157	157	193
Fuel rod						
OD of cladding tube (mm)	10.7	10.7	10.7	10.7	9.5	9.5
Number of fuel rods/assembly	179	179	179	204	264	264
Average linear fuel rating (kW/m)	15.4	17.9	20.3	20.2	17.1	17.9
Average volume power density (kW/l)	71	83	95	92	100	105
SG						
Type	WH	Type 44	Type 51, 51M	Type 51	Type 51M, 51F	Type 51A, 51FA
Heating surface area (m²)	3380	4127	4784	4784	4784	4784

to this criterion, the average linear fuel rating and maximum linear fuel rating in recent reactors are 17.9 kW/m and 44 kW/m, respectively, as shown in Figure 4.74. Changes in these values are also shown in Tables 4.8 and 4.9.

The temperature at the center of the fuel rod can be predicted as follows. The temperature of a fuel rod changes axially owing to the temperature rise of primary coolant along the channel and the distribution of heat generation along the fuel rod. The axial distribution of the heat generation rate (linear fuel rating) is assumed to be expressed by

$$q_L = q_{L0} \cos \frac{\pi z}{H'} \qquad (4.137)$$

where q_{L0} is the heat generation rate at the center of the core at height $z = 0$, H is the length of the fuel rod, $H' = H + 2\delta$, and δ is an extrapolation distance that takes into account the existence of a reflector. The heat balance in the channel length dz is expressed by

$$q_L dz = GC_p dT \qquad (4.138)$$

where G is the mass flow rate of primary coolant per fuel rod (kg/s) and C_p is the specific heat at constant pressure (J/kg K). The temperature rise of the coolant dT is given by combining Eqs. (4.137) and (4.138),

$$dT = (q_{L0}/GC_p)\cos\left(\frac{\pi z}{H'}\right)dz \qquad (4.139)$$

Integration of this equation from the inlet of the channel $z = -(H/2)$ to z gives the temperature distribution of coolant T_f.

$$T_f - T_{f1} = \int_{-H/2}^{z} dT = \frac{q_{L0}H'}{GC_p\pi}\left(\sin\frac{\pi z}{H'} + \sin\frac{\pi z}{2H'}\right) \qquad (4.140)$$

where T_{f1} is the temperature at the channel inlet.

Next, the surface temperature of the cladding tube T_s is obtained as follows. The heat transferred at the surface is given by Eq. (4.141) under the assumption of a constant heat transfer coefficient along the channel as shown in Figure 4.72.

$$q = h_s\pi d_3(T_s - T_f) \qquad (4.141)$$

Substitution of Eqs. (4.137) and (4.140) into Eq. (4.141) gives the distribution of T_s along the channel:

$$T_s = T_{f1} + q_{L0}\left[\frac{H'}{GC_p\pi}\left(\sin\frac{\pi z}{H'} + \sin\frac{\pi H}{2H'}\right) + \frac{1}{h_s\pi d_3}\cos\frac{\pi z}{H'}\right] \qquad (4.142)$$

The distribution of T_0 is expressed by Eq. (4.143), by rearranging Eqs. (4.120)–(4.122):

$$T_0 = T_{f1} + q_{L0}\left[\left(\frac{1}{4\pi\lambda_{Fm}} + \frac{1}{\pi d_1 h_G} + \frac{\ln(d_3/d_2)}{2\pi\lambda_P}\right)\cos\frac{\pi z}{H'}\right.$$
$$\left. + \frac{H'}{GC_p\pi}\left(\sin\frac{\pi z}{H'} + \sin\frac{\pi H}{2H'}\right) + \frac{1}{\pi d_3 h_s}\cos\frac{\pi z}{H'}\right] \qquad (4.143)$$

where d_1 is the outer diameter of the fuel rod, d_2 and d_3 are the inner diameter and outer diameter of the cladding tube, respectively, h_G is the gap conductance, and λ_P is the thermal conductivity of the cladding tube. The heat conductivity of the fuel material λ_{Fm} is assumed constant, independent of the temperature, and a term $\ln(d_3/d_2)/2\pi\lambda_P$ is approximated by (δ_P/λ_P), where δ_P is the thickness of the cladding tube (see Fig. 4.72).

An example of calculated results for T_f, T_s, and T_0 distributions by Eqs. (4.140), (4.142), and (4.143) for the following typical specification in the 1100 MW class PWR is shown in Figure 4.88. The primary coolant mass flow rate $G_T = 60,100$ t/h, the thermal power $Q_t = 3400$ MW, the number of fuel rods $N = 50,952$, the

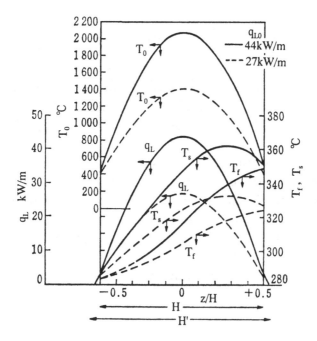

Figure 4.88 Axial temperature distribution in fuel rod.

length of a fuel rod = 3.66 m, the outer diameter of a cladding tube d_3 = 9.5 mm, the pitch s = 12.6 mm, H = 3.66 m, H' = 3.86 m, T_{f1} = 288°C, λ_{Fm} = 2.5 W/m K, λ_P = 17 W/m K, h_G = 5.7 × 10³ W/m² K. According to these values, the following values are adopted: mass flow rate per fuel rod G = G_T/N = 0.31 kg/s, average linear fuel rating q_L = Q_t/NH = 18.2 kW/m, and linear fuel rating at the center height $(Z = 0)$ q_{L0} = 18.2 × 1.48 = 27 kW/m on the assumption that the peaking factor is 1.48; also, the heat transfer coefficient at the cladding tube surface h_s can be obtained from Eq. (4.144).

$$N_u \equiv \frac{h_s d_e}{\lambda} = c\,\text{Re}^{0.8}\,\text{Pr}^{0.4} \qquad (4.144)$$

where Re is the Reynolds number based on the hydraulic equivalent diameter d_e, Pr is the Prandtl number, and the constant c can be obtained from the following Weisman's formula for flow parallel to the tube bundle (Tong and Weisman 1979).

For a square lattice:

$$c = 0.042(s/d) - 0.024 \qquad (4.145)$$

For a triangular lattice:

$$c = 0.026(s/d) - 0.006 \tag{4.146}$$

The constant c is determined from Eq. (4.145) for $d = 9.5\,\text{mm}$ and $S = 12.6\,\text{mm}$ to be 0.03, and $d_e = 4\{s^2 - (\pi/4)d^2\}/d = 0.0118$; thus the value of the heat transfer coefficient is determined by Eq. (4.144) as $h_s = 4.5 \times 10^4\,\text{W/m}^2\text{K}$. Furthermore, the calculated result of temperature distributions for $q_{L0} = 44\,\text{kW/m}$ is shown in Figure 4.88.

The figure illustrates that, for the cosine distribution of q_L, T_f rises along the channel with an S-profile, whereas T_s rises to a maximum at a location between the center and the channel outlet; also that the axial distribution of temperature at the center of the fuel rod T_0 has the maximum value at a location near the center. Thus a design criterion can be attained by obtaining the maximum value of T_0 and confirming it to be lower than the limitation temperature. The location z of maximum temperature T_0 is obtained by a condition of $dT_0/dz = 0$, as

$$z = \frac{H'}{\pi}\tan^{-1}\left[H'\bigg/\left\{GC_p\left(\frac{1}{4\lambda_{Fm}} + \frac{1}{d_1 h_P} + \frac{\ln(d_3/d_2)}{2\lambda_P} + \frac{1}{d_3 h_s}\right)\right\}\right] \tag{4.147}$$

For the conditions mentioned above, the value of z/H is about 0.007 by Eq. (4.147); this means, therefore, that the location of maximum fuel rod temperature is near the axial center of the rod. When the distribution of q_L is axially uniform, T_f, T_s, and T_0 rise linearly along the channel and the location of maximum temperature of T_0 is at the downstream end of the channel.

(2) *Departure from nucleate boiling ratio*: In the initial period of their development, pressurized water reactors were in principle planned so that boiling does not occur in the core. In recent designs of the PWR, however, occurrence of slight boiling, including surface boiling, is allowed. Therefore, a departure from nucleate boiling (DNB) point, which is defined as a transition from nucleate boiling to film boiling with very low heat transfer coefficient, is used as the basis of thermal design of the reactor core. The design criterion is the DNB ratio (DNBR), which is defined as

$$\text{DNBR} = \frac{\text{DNB heat flux predicted by applicable correlation}}{\text{actual local heat flux}} \tag{4.148}$$

The DNBR is identical with CHFR, defined above. The CHFR was used in the past in BWR design, but recently the CPR has been used, as mentioned in Section 4.4.4.

As a correlation for DNB flux, Eq. (4.149) by Tong (1967) has been used in design. This equation is also denoted the (W-3) equation (see Table 4.7):

$$q_{scr} = 3154\big[(2.022 - 0.06238p) + (0.1722 - 0.01427p)$$
$$\times \exp\{(18.177 - 0.5987p)x\}\big]$$
$$\times \big[(0.1484 - 1.596x + 0.1729x|x|)(0.7373G/10^3) + 1.037\big]$$
$$\times (1.157 - 0.869x)\big[0.2664 + 0.8357\exp(-124d_e)\big]$$
$$\times \big[0.8258 + 0.000341(i' - i_i)\big] \quad (kW/m^2) \tag{4.149}$$

where i' is the enthalpy of saturated liquid and i_i is the enthalpy of fluid at the channel inlet. The ranges of application of this equation are as follows: $p = $ 6.9–15.8 MPa, quality $x = -0.15$–0.15, mass flux $G = 1.4 \times 10^3$ to 6.8×10^3 kg/m^{2}s, hydraulic equivalent diameter $d_e = 0.5$–1.78 cm, length of channel $= 0.25$–3.66 m. This formula was obtained in 1964 on the basis of 809 experimental data on a single tube heated with axially uniform heat flux, and is applicable directly to fuel assemblies of 14×14 lattice and 15×15 lattice. For 17×17 lattices the value of $0.88d_e$ should be used instead of d_e (Tong 1967). Furthermore, correction factors for nonuniform axial distribution of the heat generation rate, for tube bundles, for 17×17 lattice, and for spacers have been provided, but these are not given here.

The DNBR shows the safety margin in the design between operating conditions and the CHF (DNB). The minimum value of DNBR in a channel with the most severe condition, that is, minimum departure from nucleate boiling ratio (MDNBR), is used as a criterion for PWR core design. The typical value is MDNBR $= 1.3$ at anticipated abnormal transient conditions; when operating at rated power, MDNBR is about 1.9.

Figure 4.89 illustrates a method to predict the value of MDNBR. Curve a shows the cosine heat flux q_S for an average linear fuel rating $q_L = 17.9$ kW/m (a typical value in a 1100-MWe-class plant); line b is the average heat flux q_{Sav} which is obtained from $q_{Sav} = q_L/D = 17.9/(\pi \times 9.5 \times 10^{-3}) = 600$ kW/m^2; and curve c is the heat flux distribution q_S at the hottest channel on the assumption that $q_{L0} = $ 44 kW/m. For the q_S distribution, the temperature and enthalpy distributions of the primary coolant along the channel are obtained from Eq. (4.140), and then a distribution of DNB (curve d) can be obtained from Eq. (4.149). Here, the qualities $x = (i - i')/(i'' - i')$ have negative values in most of the channel length, since i is smaller than i' in the greatest part of the PWR core. Finally, the DNBR distribution (e) can be obtained as d/c from Eq. (4.148). The figure indicates that the value of MDNBR is determined as about 1.9 at $z/H = 0.19$.

The criterion mentioned above, that DNBR $= 1.3$ at anticipated abnormal transient conditions, was determined taking into account a reliability analysis on Eq. (4.149), which was shown in Figure 4.78. Therefore, when the accuracy of

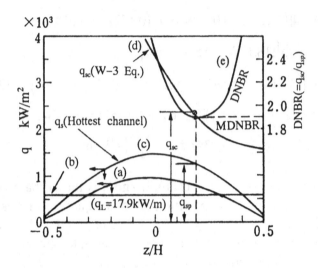

Figure 4.89 Heat flux and DNBR.

DNB correlation is improved, the value of the criterion should be changed. For example, the value of MDNBR is 1.17 for MIRC-1 correlation (Akita 1986) and for the NFI-1 correlation (Nuclear Fuel Industry 1985a,b), with higher accuracy.

4.9.4 Steam generators

A steam generator is an apparatus to generate steam in the secondary loop with lower pressure, using high-pressure, high-temperature water in the primary loop heated in the reactor core. Consequently it is, in principle, a sort of liquid–liquid shell-and-tube-type heat exchanger, though the capacity is very large and higher reliability is needed than for industrial heat exchangers. In the Shippingport PWR plant, opened in 1955 as the first PWR commercial plant, four horizontal steam generators are installed (see Table 4.7). In the Yankee Rowe PWR plant, operated since 1960, and in most of the plants built later, vertical shell and inverted U-tube steam generators are used.

A typical construction of the vertical shell and inverted U-tube steam generator (Type 51 in Table 4.9) is shown schematically in Figure 4.90. The primary coolant, heated in the reactor core at a pressure of 15.4 MPa, enters from the inlet nozzle at the bottom of the steam generator, flows through inverted U-tubes, which generates steam in the secondary coolant, and continues to the outlet nozzle at the bottom of the steam generator. The secondary coolant, supplied from a feed-water heater at a pressure of 6.2 MPa and a temperature of 221°C, is fed to the steam generator through the feed-water nozzles and a feed-water

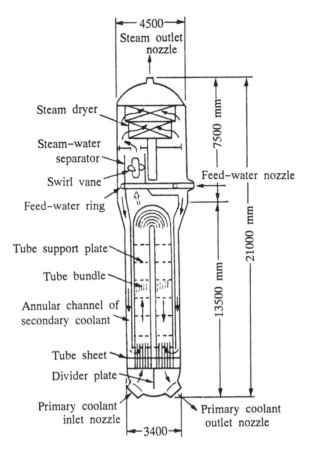

Figure 4.90 Steam generator.

ring. The water flows downward in an annulus between the steam generator shell and a baffle surrounding the inverted U-tubes, and the two-phase mixture flows upward in the channel between the tube bundles, with evaporation. The steam–water mixture flows through the swirl vane steam separator and steam dryer, in which the humidity is reduced to below 0.25% at the rated load, to the steam outlet nozzle at the top of the shell. The water separated in the steam separator is mixed with the feed water and is circulated in the loop composed of the annular downcomer and the tube bundle. The flow is induced by natural circulation with a circulation ratio of 3–5. The circulating flow rate can be calculated by the method described in Section 4.5. Design criteria for water circulation are not, in general, provided for steam generators, because boiling in the secondary coolant is induced by heating in the primary coolant at a temperature of 325°C at the most and then the bundle is not damaged if transient boiling occurs.

The specifications of a steam generator (e.g., Type 51) are as follows. For the primary coolant, pressure = 15.4 MPa, inlet temperature = 325°C, flow rate = 1500 t/h, outer diameter of tube = 22.2 mm, thickness = 1.27 mm, pitch = 32.5 mm, total length of a tube = 21 m, number of tubes = 3383, heating surface area = 4784 m², and sectional area of total tubes = 1.02 m². For the secondary coolant, pressure = 6.2 MPa, inlet temperature = 221°C, outlet temperature = 277.7°C (saturated temperature), evaporation rate = 1600 t/h. The thermal power per unit steam generator Q_t is about 856 MW and the electric power Q_e is about 300 MW; these are obtained from the heat balance based on the values mentioned above.

The characteristic quantities on heat transfer in the steam generator are as follows:

(a) *Heat transfer coefficient* h_i *on the inner surface of the tube*: This value can be obtained from Eq. (4.144) as h_i = 36,000 W/m²K for the inlet velocity u = 5.6 m/s (mass flux G_L = 1500 × 10³/(1.02 × 3600) = 4080 kg/m²s), Pr = 0.864, and λ = 0.5597. The heat conductivity of the tube λ_P/δ is 10,200 m²K/W for the thermal conductivity of Inconel 600 (ASTM B 163) used as the tube material, λ_P = 13 W/m K and the thickness δ = 0.00127 m.

(b) *Heat transfer coefficient* h_o *on the outer surface of the tube*: This is a function of the mass flux, heat flux, and quality as described in Section 4.2, and the value is in the range of 5000–30,000 W/m²K depending on the distribution of heat flux. Since most of the heating surface is covered by the saturated boiling region and only a little part by the subcooled boiling region, the average heat transfer coefficient is on the order of 10^4.

(c) *Overall heat transfer coefficient of the bundle*, k: The fouling factor on the outer surface of the tube (thermal resistance), R_f, is estimated as R_f = 5.3 × 10⁻³ m²K/W (the conductivity = 1/ R_f = 18,900 W/m²K). According to the values mentioned above, the overall heat transfer coefficient k = 1/[(1/h_i) + (δ/λ_P) + (1/R_f) + (1/h_0)] is estimated as k = 3500–7000 W/m²K. In practice, moreover, the value of k should be predicted taking into account the effects of bypass flow and tube support plates. Average values of k can also be predicted by the equation of heat balance Q_t = $kA\Delta T_{ln}$. For the values of Q_t = 856 MW, A = 4784 m², and ΔT_{ln} = 27°C, which is assumed to be given by the logarithmic mean temperature difference for a single channel, the value of k is determined as k = $Q_t/A \Delta T_{ln}$ = 856 × 10⁶/4784 × 27 = 6700 W/m²K.

The flow rate of secondary coolant G_t is determined from the evaporation rate as G_t = $G_s R$ = 1600 × (3–5) = 4800–8000 t/h (G_s = the evaporation rate in t/h, R = the circulation ratio), and for a cross-sectional area of secondary coolant of about 4.5 m² the mass flux G_L is 300–500 kg/m²s and the circulation velocity is 0.39–0.65 m/s. The natural circulating flow of secondary coolant is three-dimensional and very complicated, because the evaporation rate varies in both axial and radial directions and cross-flow between bundles occurs. Therefore, for accurate prediction, a digital computer analysis has recently been conducted using thermal and hydraulic three-dimensional computer codes.

References

Akagawa, K. 1974. *Gas–liquid two-phase flow*. Tokyo: Corona.

Akagawa, K., et al. 1968. Preprint of 43th Meeting of JSME-Kansai, pp. 20–2.

Akagawa, K., et al. 1979. *Bull. MESJ* 14, no. 6: 476–82.

Akagawa, K., et al. 1982. Memoirs of the Faculty of Engineering, Kobe University, no. 29: 189–201.

Akita, Y. 1986. MAPI Technology Report, no. 42: 25–8.

Armand, A.A. 1946, 1947, 1950. *Izvest. Veresoy. Teplotekh. Inst.*, no. 1 (1946), no. 4 (1947), no. 2 (1950).

Bankoff, S.G. 1960. *Trans. ASME, Ser. C* 82: 265.

Barnea, D., et al. 1980. *Int. J. Multiphase Flow* 6: 217–25.

Baroczy, C.Z. 1966. *Chem. Engng. Prog. Symp. Ser.* 62, no. 44: 323.

Bennett, A.W., et al. 1965. AERE-R 4879.

Biasis, L., et al. 1967. *Energia Nuclear* 14, no. 9: 530–6.

Bishop, P.P., et al. 1965. ASME Paper, 65-HT-31.

Chen, J.C. 1966. *Ind. Engng. Proc. Des. Dev.* 5: 322.

Doroshchuk, V.E., and A.C. Konjkov. 1972. *Teploenergetika* 19, no. 3: 72–4.

Doroshchuk, V.E., and B.I. Nigmatulin. 1970. *Teploenergetika* 18, no. 3: 79–80.

Furutera, M., and T. Furukawa. 1985. Proceedings of the Two-Phase Flow Symposium, JSME, 854, no. 6: 5–8.

General Electric BWR Thermal Analysis Basis (GETAB). 1973. Data, correlation and design applications, NEDO-10958.

Griffith, P. 1964. *Developments in heat transfer*. MIT Press.

Handbook of fluid mechanics. 1987. Maruzen.

Handbook of gas–liquid two-phase flow technology. 1989. Tokyo: Corona.

Healzer, J.M., J.E. Hench, E. Janssen, and S. Levy. 1966. APED-5286.

Hetsroni, G. 1982. *Handbook of multiphase systems*. Hemisphere Pub. Co.

Hewitt, G.F., et al. 1970. AERE-R 5590.

Idsinga, W., et al. 1977. *Int. J. Multiphase Flow* 3: 401–3.

Ishigai, S., and K. Akagawa. 1984. Proceedings of the Japan–U.S. Seminar on Two-Phase Flow Dynamics, Toroy, C-1, pp. 1–6.

Ishigai, S., et al. 1969a. *Trans. JSME* 35, no. 280: 2395–400.

Ishigai, S., K. Akagawa, and Y. Takeda. 1969b. *Synthetic steam power engineering*. Tokyo: Corona.

Ishimoto, R., and T. Ohki. 1977. *Thermal and Nuclear Power* 28, no. 9: 826–34.

Isshiki, S., et al. 1967. *J. Atomic Energy Soc. Japan* 9, no. 3: 166.

Iwabuchi, M., et al. 1982. *Trans. JSME, Ser. B* 48, no. 432: 1501–8.

Katto, Y. 1981. *Trans. JSME* 47, no. 413: 139–48.

Kawamura, T., et al. 1980. *Mitsubishi Juko Giho* 17, no. 2: 1–16.

Kono, Y., et al. 1984. *Hitachi Review* 66, no. 4: 247–54.

Kozlov, B.K. 1952. *Zuhr. Tech. Fiz.* 24, no. 4: 656.

Kutatelaze, S.S. 1959. USAEC Report AEC-TR-3770.

Levitan, L.P., and F.L. Lantsman. 1975. *Teploenergetika* 22, no. 1: 80–3.

Macbeth, R.V. 1963. AEEW-R, p. 267.

Masuda, K. 1982. Lecture Notes of 554th Seminar of JSME, pp. 17–31.

Matumoto, S., and T. Ohki. 1978. *Ishikawajima Harima Engineering Review* 18, no. 2: 188–97.

Matumoto, S., et al. 1979. *Thermal and Nuclear Power* 30, no. 9: 925–34.
Miki, M., and A. Ohki. 1984. *Hitachi Review* 66, no. 4: 247–54.
Morioka, N. 1979. *Thermal and Nuclear Power* 32, no. 2: 167–76.
Nakanishi, S. 1986. *Trans. JSME* 52, no. 479: 2682–8.
Nakano, M. 1982. *Thermal and Nuclear Power* 33, no. 11: 1171–88.
Nakano, S., et al. 1972. *J. JSME* 75, no. 637: 235–45.
Nuclear Fuel Industry. 1985a. NEK-8087.
Nuclear Fuel Industry. 1985b. NEK-8088.
Nuclear Reactor Safety Inspection Committee. 1976. *Method of thermal design of reactor core of BWR and procedure of determination of thermal operation criteria.* Tokyo.
Nukiyama, S. 1934. *J. JSME* 34, no. 206: 367.
Ogura, N., et al. 1982. *Mitsubishi Juko Giho* 19, no. 6: 620–41.
Remizov, O.V. 1978. *Teploenergetika* 25, no. 1.
Research Committee on Nuclear Reactor Safety. 1985a. *Fuel behavior in light water reactors.* Tokyo.
Research Committee on Nuclear Reactor Safety. 1985b. *Light water nuclear reactor.* Tokyo.
Rohsenow, W.M., and J.P. Hartnett. 1973. *Handbook of heat transfer.* New York: McGraw-Hill.
Roko, K. 1980. Study on heat transfer performance in sodium heated steam generator. Ph.D. diss., Osaka University.
Roko, K., et al. 1975. JSME Meeting Preprint 754-5, pp. 64–6.
Schrock, V.E., and L.M. Grossman. 1962. *Nucl. Sci. Engng.* 12: 474.
Shiojima, Y., and H. Haneda. 1978. *Mitsubishi Juko Giho* 15, no. 2: 1–16.
Smith, S.L. 1969–70. *Proc. Inst. Mech. Engrs.* 184, pt. 1, no. 36: 647.
Styrikovich, M.A., O.I. Martynova, and Z.L. Miropoljskii. 1969. *Protsessy generatsii para na elektrostantsiyakh* (Energiya). Moscow.
Sudakov, A.V., et al. 1979. *Heat Trans.-Soviet Research* 11, no. 2: 80–8.
Takashima, Y., et al. 1984. *Hitachi Review* 66, no. 4: 305–10.
Thermal and Nuclear Power 32, no. 6 (1981): 603.
Thom, J.R.S. 1964. *Int. J. Heat and Mass Transfer* 7: 709–24.
Tong, L.S. 1967. *J. Nucl. Energy* 6: 21.
Tong, L.S., and J. Weisman. 1979. *Thermal analysis of pressurized water reactors*, 2nd ed. American Nuclear Society.
Ueda, T. 1981. *Gas–liquid two-phase flow.* Tokyo: Yokendo.
Weisman, J. 1979. *Two-phase flow dynamics*, ed. S. Ishigai and A.Z. Bergles. New York: Hemisphere.
Weisman, J., and S.Y. Kang. 1981. *Int. J. Multiphase Flow* 7: 271–91.
Yamashita, J., et al. 1984. *Hitachi Review* 66, no. 4: 319–22.
Yamauchi, S. 1980. Study on fluctuations of flow and wall temperature in evaporator tubes. Ph.D. diss., Osaka University.
Yamauchi, S., and S. Nakanishi. 1985. Proceedings of the JSME, 47th Meeting, no. 854-416, pp. 9–12.
Yamauchi, S., S. Nakanishi, and S. Ishigai. 1981. *Trans. JSME, Ser. B* 47, no. 416: 693–709.
Yanuki, H., et al. 1985. *Ishikawajima Harima Engineering Review* 25, no. 5: 357–65.
Yoshida, T., and K. Shigemune. 1985. *Thermal and Nuclear Power* 36, no. 10: 1102–9.
Zuber, N., et al. 1961. Proceedings of the International Heat Transfer Meeting, Boulder, Colorado, paper no. 27.

5

Flow instability problems in steam-generating tubes

MAMORU OZAWA

Kansai University

5.1 The role of dynamic behavior of thermal hydraulics in steam-generating plant design

In the design of boiler-tube systems and steam generators, the design principles related to thermal hydraulics, including the phase change process, two-phase flow, and heat transfer, are described in the previous chapters. However, such design principles are mainly focused on the static characteristics, which means that the discussion is conducted assuming steady-state behavior or operation of the plant. As can be easily understood in examples such as the sliding-pressure operation of the boilers and the start-up and/or cooldown of the plant, steam power plants cannot be operated without a transient condition; that is, the dynamic behavior of the system is an essential factor for consideration in designing the plant. Automatic control problems are of primary importance in the above-mentioned examples, although discussion on this matter is beyond the scope of this chapter. Instead, this chapter is devoted mainly to unanticipated dynamic behavior, that is, flow instability problems that harm the steady state and/or safety of operation, even in the case of suitable design on the basis of the steady-state characteristics.

Flow instabilities mean large-scale fluctuations of flow in boiling-channel systems that cause large-scale pressure fluctuation, departure from steady-state and safe operation in the heat transfer process, and also the mechanical vibration of tubes. Thus so-called hydrodynamic instabilities, including such as flow fluctuation in the slug flow regime and the flow pattern transition, are not included in this concept, although these phenomena may provide triggers for flow instability. In this sense, the flow instabilities discussed in this chapter are, in principle, one-dimensional macroscopic phenomena, but are not localized in a rather narrow area. In the next section, such flow instabilities are explained phenomenologically, on the basis of a forced-flow boiling experiment that was conducted to measure the pressure drop in a steady-state condition.

5.2 Phenomenological description of flow instabilities

Examples of flow instabilities are reviewed in this section. Figure 5.1 represents
a test facility of a forced-flow boiling channel system having originally a volute-
type water feed pump and an evaporator tube of dimensions of 9.6 mm ID, 16
mm OD, and 42.74 m length. The heat flux was imposed by the Joule heating of
an electric current. When this system was operated by supplying water through
the volute-type pump, the flow rate varied gradually from the predetermined
initial state to increase or decrease to a new steady state of higher or lower flow
condition, as shown in Figure 5.2. The steady-state characteristics of the pressure
drop across the boiling channel were then measured by using the constant-
discharge pump instead of the volute-type pump, and the initial condition was
found to be at point 1 of the negative slope of the pressure drop versus flow rate
curve shown in Figure 5.3. Moreover, the pump characteristics, that is, the pump
head versus flow rate relationship, were found to be described by curve A, which
intersects the pressure drop characteristics of the channel at points 1–3.
Comparing this figure to the dynamic behavior in Figure 5.2, the case of decrease
in flow rate corresponds to the transient from point 1 to the steady-state condition
of lower flow rate at point 2, and the case of increase in the flow rate to the transient state from point 1 to point 3. Thus, the initial condition of point 1 seems to be
a saddle point that is easily transmitted to the other steady states. If the boiling
system were designed to be operated at point 1, the actual flow rate would become
higher or lower than the design point when the constant-discharge pump is not
used. The former may lead to nonboiling single-phase liquid flow at the channel
exit, and the latter may induce a critical heat flux condition. Such an undesirable
phenomenon is referred to as a flow excursion, which will be discussed later.

The above example may suggest that steady-state operation will be achieved
at the positive slope region of the pressure drop characteristics such as point 5 in
Figure 5.3. But this is not always true. Figure 5.4 demonstrates the flow oscillation observed at point 5, that is, the time-averaged state was located at point 5 in
the same boiling system mentioned above. The trace of the flow rates indicates
divergent tendency, and actually the oscillation amplitude grew so as to induce
the flow-reversal condition, that is, a negative value of the flow rate. This type of
instability is not transitional, as in the previous example, but is oscillatory in
nature. When the heat flux is high enough to exceed a certain limit, periodical
dryout is expected to occur and to induce the thermal fatigue of the tube, and
also to induce mechanical vibration in it.

The effect of such flow instabilities is demonstrated by focusing on the critical
heat flux (CHF) problems. Figure 5.5 demonstrates the CHF represented by the
dimensionless coordinate system of the heat flux and the mass flux (Mishima et
al. 1985). The data represented by the solid and empty circles are the CHF for

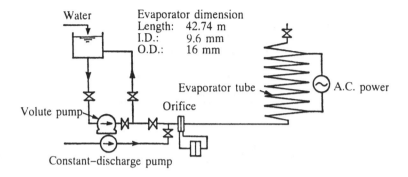

Figure 5.1 Forced-flow boiling system.

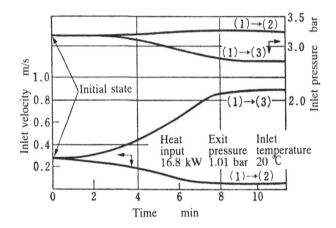

Figure 5.2 Traces of flow rate and inlet pressure.

the stiff system; that is, the mass flow remains constant independent of the system situation. The CHF increases with the increase in the mass flux, and then beyond a certain threshold the CHF becomes almost constant and nearly equal to the CHF in the pool boiling condition, independent of the mass flux. On the other hand, when the system is soft; that is, when the flow rate is a function of the system situation such as the pressure drop characteristics, the CHF reduces drastically from those in the stiff system. This drastic decrease is mainly caused by the flow excursion described above. The mass flux designated at the horizontal axis represents the design base flow rate and is not an actual one; however, the importance of flow instability is well demonstrated in this figure.

Another example of such a CHF problem is shown in Figure 5.6, which describes the reduction of CHF during the oscillatory flow condition (Umekawa et al. 1995). The CHF data, normalized by the steady-state value, are plotted

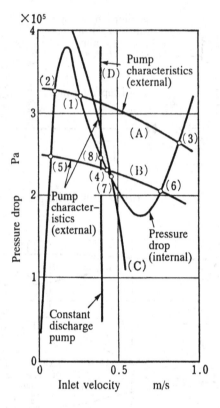

Figure 5.3 Pressure drop and pump characteristics.

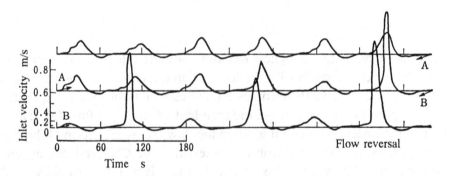

Figure 5.4 Example of flow oscillation in a forced-flow boiling system.

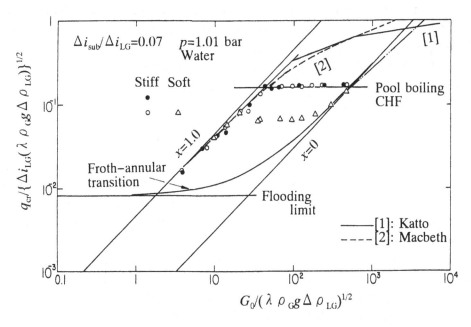

Figure 5.5 Critical heat flux under stable and unstable conditions (Mishima et al. 1985).

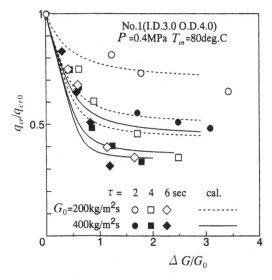

Figure 5.6 Dryout heat flux under oscillatory flow condition (Umekawa et al. 1995).

against the normalized amplitude of the mass flux. When the amplitude increases so as to induce flow reversal, the CHF decreases drastically to about 40–50% of the steady-state value. For a long oscillation period, the CHF has a significantly low value. This example emphasizes that the CHF, especially the dryout in this case, is poorly estimated by only the steady-state condition.

What are the dominant mechanisms of such flow instabilities? How can we detect or predict such flow instabilities in the design stage of systems, and how can we prevent such undesirable phenomena? To answer these questions, the subsequent sections begin, first, with a phenomenological classification of the flow instabilities appearing in boiling channel systems.

5.3 Classification of flow instabilities

Flow instabilities are encountered in various boiling channel systems, and the features of instabilities seem to be of great variety depending on the system configuration and/or the operating conditions. These flow instabilities, however, have several primary features in oscillation periods, amplitudes, and relationships between pressure drop and flow rate. On the basis of these primary features, these flow instabilities have been classified into several types listed in Table 5.1, which was first proposed by Bouré et al. (1971), who promoted extensive work in this field. In the following, their classification is briefly reviewed.

The primary classes of instabilities in Table 5.1 are static instability and dynamic instability. The static type includes flow instabilities whose mechanisms are explained by the static and/or steady-state characteristics of the system, and thus the threshold conditions are predicted only on the basis of the steady-state characteristics of the pressure drop of the boiling channel and the outer system such as a pump. The second type, the dynamic instability, is caused by the dynamic interaction between the flow rate, pressure drop, void fraction, and so on. Thus the stability boundary of this type of instability is predicted on the basis of consideration of the dynamic behavior or time-dependent characteristics of the system.

Prior to further discussion, the instability problem is discussed with reference to mechanical vibration theory. The well-known vibration equation of the second-order system is expressed by

$$a_2 d^2x/dt^2 + a_1 dx/dt + a_0 x = 0 \qquad (5.1)$$

Now the coefficient a_2 is assumed positive and constant, which corresponds to the inertial mass. The second term a_1 corresponds to damping and the third a_0 to spring force. When the system has positive damping, $a_1 > 0$, and positive spring constant, $a_0 > 0$, then the system is statically and dynamically stable. The system with negative damping is dynamically unstable, and the state has an oscillatory feature with growing amplitude. If the system is free from spring force, then the system equation (5.1) reduces simply to the first-order system expressed by the inertial term and the damping term as follows:

$$a_2 d^2x/dt^2 + a_1 x = 0 \qquad (5.2)$$

Table 5.1. *Classification of flow instability (Bouré et al. 1971)*

Class	Type	Mechanism	Characteristics
Static Instabilities			
Fundamental static instabilities	Flow excursion or Ledinegg instability	$(\partial \Delta p / \partial G)_{int} < (\partial \Delta p / \partial G)_{ext}$	Flow undergoes sudden, large amplitude excursion to a new, stable operating condition
	Boiling crisis	Ineffective removal of heat from heated surface	Wall temperature excursion and flow oscillation
Fundamental relaxation instability	Flow pattern transition instability	Bubbly flow has less void but higher Δp than that of annular flow	Cyclic flow pattern transition and flow rate variations
Compound relaxation instability	Bumping, geysering, and chugging	Periodic adjustment of metastable condition, usually due to lack of nucleation sites	Periodical process of superheat and violent evaporation with possible expulsion and refilling
Dynamic Instabilities			
Fundamental dynamic instabilities	Acoustic oscillations	Resonance of pressure wave	High frequency (10–100 Hz) related to time required for pressure wave propagation in system
	Density wave oscillations	Delay and feedback effects in relationship between flow rate, density, and pressure drop	Low frequency (1 Hz) related to transit time of a continuity wave
Compound dynamic instabilities	Thermal oscillations	Interaction of variable heat transfer coefficient with flow dynamics	Occurs in film boiling
	BWR instability	Interaction of void reactivity coupling with flow dynamics and heat transfer	Strong only for a small fuel time constant and under low pressures
	Parallel channel instability	Interaction among small number of parallel channels	Various modes of flow redistribution
Compound dynamic instability as secondary phenomenon	Pressure drop oscillations	Flow excursion initiates dynamic interaction between channel and compressible volume	Very low frequency periodic process (0.1 Hz)

The negative damping factor provides the system instability; that is, a small positive disturbance of x induces a further increase of x. The instability is then transitional in nature.

In any case, the essential condition for the occurrence of instability is negative damping. In the first-order system, the damping factor is, in general, provided by the resistance term, that is, the pressure drop in the present case. On the other hand, in the second-order system, the resistance term is only one of the factors that contribute to damping, as discussed later.

From such a viewpoint of mechanical vibration, the instability itself is, in reality, dynamic behavior, that is, of a time-dependent and/or time-varying nature. Thus, to analyze the instability, it is essential to take into account the unsteady terms, such as the time variation of the velocity and/or the enthalpy in momentum and energy conservation equations, respectively, even in the case of static instability as outlined in Table 5.1. The terms "static" and "dynamic" instabilities in Table 5.1 may not well coincide in meaning with those of vibration theory. The classification in Table 5.1 includes several types that are still not clear in their phenomenological aspect or in their physical mechanism. For relevance to boiler and/or steam power engineering, the classification in Table 5.1 is revised and simplified as shown in Table 5.2, which includes the mechanism as well as typical features. On the basis of Table 5.2, the various flow instabilities are explained briefly in the following. In terms of the force acting on the system, the operating condition is determined as an equilibrium state between a driving term and a resistance term. In the boiling channel system, this driving term is represented by the pressure boundary condition, for example, the pressure drop imposed by pump or headers. The resistance term corresponds to the pressure drop in the boiling channel. The driving term, that is, pressure boundary condition, has a generally negative and in some cases zero gradient against the mass flux. On the other hand, the resistance, that is, the pressure drop in the boiling channel, has a negative and/or positive gradient against the mass flux. Thus, the primary classification in Table 5.2 is made with reference to the negative and positive resistances.

Then the first class of instability is given by the negative resistance of the pressure drop versus flow rate relationship. This class includes three types of instability: flow excursion or Ledinegg instability, flow maldistribution, and pressure drop oscillation. The first type is induced by the interaction with the pump system; the second occurs with many parallel channels connected by two headers or plenums; and the last features the compressible capacity as a spring force. Thus the first two types will be expressed by first-order differential equations and the third by a second-order differential equation.

On the other hand, positive resistance may lead the system to a dynamically and thus statically stable condition. Is this always true? We consider the parallel-

Table 5.2. *Classification of flow instabilities in boiling channel systems*

Category	Pattern	Mechanism	Feature
Negative resistance instability	Flow excursion or Ledinegg instability	Negative damping in first-order system	Transitional
	Flow maldistribution		Appears in parallel-channel system
	Pressure drop oscillation	Dynamic interaction between flow excursion and accumulation mechanism of mass and momentum	Relaxation oscillation with large amplitude and long period
Time-delayed feedback instability	Density wave oscillation	Propagation delay of void wave (kinematic wave) and feedback effect provide negative damping effect	Oscillation period comparable with residence time, appears in positive resistance region
Thermal nonequilibrium instability	Geysering	Insufficient nucleation sites bring about large superheat followed by violent boiling	Relaxation oscillation if liquid refilling mechanism exists

channel system that includes many boiling channels connected by plenums at both inlet and exit. When a certain perturbation is encountered in one of the channels, this perturbation does not propagate over all the channels even in the case of a large fluctuation of the flow rate, and the pressure drop between the plenums retains an almost constant value. Then the flow fluctuation in one of the channels induces fluctuation of the void fraction and thus density fluctuation, which induces pressure drop fluctuation. The pressure drop between the plenums retains a constant value, then the pressure drop fluctuation in the reference channel is compensated by the mass flow fluctuation. Adopting control theory to this situation, the dynamic interaction mentioned above is considered a feedback effect. When this feedback effect is transferred almost simultaneously, the instability does not occur. On the other hand, when a certain suitable amount of time lag, or phase shift exists in the feedback response, the instability may occur as is well known in control theory. The system equation in this case, if expressed simply by the second-order equation, has negative damping owing to such a feedback effect, with time lag even in the case of positive resistance. On the basis of the above-mentioned discussion, the second class is given by the time-delayed feed-back between the pressure drop and the flow rate. This includes the density wave oscillation, which appears in all cases with pump, headers with parallel channels, or compressible capacity.

These instabilities appear even under thermal equilibrium conditions. The heat transfer process, however, is not always steady, but in some cases, such as the start-up of a natural-circulation boiler, a violent boiling following a large superheat of liquid will take place similar to water boiling in a test tube of smooth glass with a closed end. This violent boiling is considered an instability induced by the thermal nonequilibrium condition, which provides the third category in the classification. This class includes, typically, geysering in the case of boiling. In the following sections, each flow instability is described in detail, including a simplified method of analysis.

5.4 Flow excursion and flow maldistribution

In this section, flow excursion instabilities or Ledinegg instability and flow mal-distribution are discussed, with special reference to the fundamental mechanism and simplified stability analysis.

Figure 5.7 shows a model of the boiling channel system, which consists of a single boiling channel and a coolant supply system such as a pump. The pressure drop characteristics of the boiling channel are shown schematically against the mass flow rate in Figure 5.8. The characteristic curve A has no negative resistance, but curve B has. In case of curve A, the pressure drop increases monotonically with the increase in the mass flow rate under constant subcooling at the inlet, constant heat flux, and constant system pressure. On the other hand, in curve B, the pressure drop has a maximum and a minimum, and a certain range of pressure drop corresponds to two or three values of mass flow rate. This negative-resistance characteristic depends on the heat flux, inlet subcooling, pressure, and thus on the components of the pressure drop, such as the frictional, gravitational, and accelerational pressure drops. When the boiling channel is dominated by frictional pressure drop, that is, in a relatively small diameter tube, the normalized pressure drop versus mass flux relationship is as shown in Figure 5.9. The horizontal and the vertical axes are dimensionless values, G/G_0 and $\Delta p/\Delta p_0$, respectively, where Δp represents the pressure drop, G the mass flux, G_0 the mass flux when the exit quality becomes zero under given inlet subcooling and heat flux, and Δp_0 the pressure drop corresponding to G_0. The parameter N_{sub} represents the subcooling number defined as

$$N_{sub} = \Delta \varrho_{LG} \Delta i_{sub} / (\varrho_G \Delta i_{LG}) \tag{5.3}$$

where $\Delta \varrho_{LG} = \varrho_L - \varrho_G, \varrho_G$ and ϱ_L represent densities of the saturated vapor and liquid, Δi_{sub} the subcooled enthalpy at the inlet, and Δi_{LG} the latent heat of vapor-

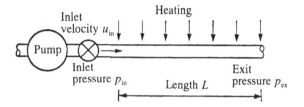

Figure 5.7 Model of boiling channel.

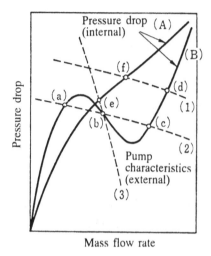

Figure 5.8 Pressure drop and pump characteristics.

ization. When the subcooling number is less than about 7.5, the pressure drop increases monotonically against the mass flux, while a larger subcooling number, $N_{sub} > 7.5$, corresponds to negative resistance against the mass flux. This behavior of the pressure drop in the boiling channel is explained phenomenologically as follows. Suppose that the system has a relatively large frictional component that dominates the total pressure drop in the boiling channel. The coolant at the exit is almost in a liquid state under a relatively high mass flux condition and relatively large inlet subcooling. If the mass flux decreases under the same operating conditions, the pressure drop will decrease accordingly, and the exit velocity be almost equal to that at the inlet. Below a certain limit of the mass flux, net vapor generation takes place and a two-phase mixture is formed in the channel, which results in an increase in the velocity at the exit. The frictional pressure drop is given, in principle, by (density) $\times$ (velocity)^2. The decrease in the mass flux induces a decrease in the density, but results in an increase in the velocity under certain conditions. When this velocity term contribution is larger than that of the density term, the pressure drop increases against the decrease in the mass flux.

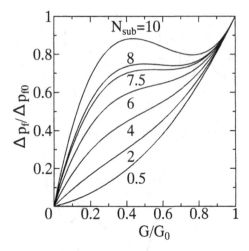

Figure 5.9 Dimensionless characteristics of frictional pressure drop.

This corresponds to negative resistance characteristics. On further decrease in the mass flux under high exit quality, the variation of the void fraction is very small and thus the density variation as well as the velocity variation becomes so small as not to induce a pressure drop increase. Then the pressure drop decreases with the decrease in the mass flux.

Referring to Figure 5.8, the external, relative to the internal, that is, the boiling channel, characteristics given by the pump or any coolant supply system are given by curves 1–3 as examples. The intersection point of these external characteristics and the internal one gives the operating condition. In the model designated in Figure 5.7, the pressure at the inlet of the boiling channel is denoted by p_{in}, the exit pressure by p_{ex}, the pressure drop by Δp, and the velocity at the inlet is given by u_{in}. The dynamic behavior of the boiling channel is lumped by this u_{in} as a single parameter, and the momentum equation of the boiling channel is then given by

$$p_{in} - p_{ex} = d/dt\left(Mu_{in}\right) + \Delta p\left(u_{in}\right) \tag{5.4}$$

where M represents the inertial mass in the boiling channel, defined by

$$M = \int_0^L \varrho dZ \tag{5.5}$$

and is assumed constant against time, where ϱ represents the density of fluid, Z the axial coordinate, and L the channel length. The pump characteristics, that is, the external characteristics Δp_p, are given as a function of u_{in} only as

$$p_{in} - p_{ex} = \Delta p_p\left(u_{in}\right) \tag{5.6}$$

Subtracting Eq. (5.6) from Eq. (5.4), we have

$$Mdu_{in}/dt = \Delta p_p(u_{in}) - \Delta p(u_{in}) \qquad (5.7)$$

On the basis of this equation, the flow behavior is estimated as follows. When the operating condition is located at points a and c–f, respectively, in Figure 5.8, where the pump characteristic curve intersects the boiling channel curve at its positive resistance region, the time derivative term at the left of Eq. (5.7) becomes negative for a larger mass flux than the intersection point, and becomes positive for a smaller mass flux than the intersection point. It means that the state is directed toward the original operating point even if the mass flux increases or decreases to a certain extent. Thus these points are stable operating conditions.

If, on the other hand, the pump characteristics and the pressure drop are given by curves 2 and B, respectively, the state around point b has a tendency for the mass flux to increase further at the right side of point b, and for a further decrease to take place at the left side. Then even a small perturbation from point b induces a large increase or decrease in the mass flux, and the operating condition settles down at points a or c in the positive resistance region. Thus point b is estimated as an unstable condition. These phenomena are referred to as flow excursion instability. When the pump characteristic has only one intersection point with the internal characteristics, as in curve 3, then the excursion does not occur.

General stability analysis is first conducted by using a small perturbation method. Equation (5.7) is linearized by taking account of a small perturbation of the variable u_{in} around the steady-state value, $u_{in} = u_{ino} + \delta u_{in}$; then Eq. (5.7) becomes

$$Md(\delta u_{in})/dt = \Delta p(u_{ino}) - \Delta p(u_{ino})$$
$$+ (d\Delta p_p/du_{ino})\delta u_{in} - (d\Delta p/du_{ino})\delta u_{in} \qquad (5.8)$$

and is Laplace-transformed by using the parameter S,

$$\left\{MS - \left[(d\Delta p_p/du_{ino}) - (d\Delta p/du_{ino})\right]\right\}\bar{u}_{in} = \bar{u}_0 \qquad (5.9)$$

where $\bar{u}_{in}$ represents the Laplace-transformed variable of δu_{in}. The perturbation $\bar{u}_{in}$ is then given by

$$\bar{u}_{in} = \bar{u}_0/\left\{MS - \left[(d\Delta p_p/du_{ino}) - (d\Delta p/du_{ino})\right]\right\} \qquad (5.10)$$

For the infinitesimal small perturbation δu_0 at the initial state, the above equation gives the solution of δu_{in}:

$$\delta u_{in} = (\delta u_0/M)\left\{1 - \exp(-t/\tau_c)\right\}$$
$$\text{in the case of } (d\Delta p_p/du_{ino}) < (d\Delta p/du_{ino}) \qquad (5.11)$$

and

$$\delta u_{in} = \left(\delta u_0/M\right)\left\{\exp\left(t/\tau_c\right) - 1\right\}$$

$$\text{in the case of } \left(d\,\Delta p_p/du_{in0}\right) > \left(d\Delta p/du_{in0}\right) \tag{5.12}$$

where the time constant is given by $\tau_c = M/[(d\Delta p_p/du_{in0}) - (d\Delta p/du_{in0})]$.

The former solution indicates that the velocity perturbation does not develop with time, but it does extend or develop with time in the latter case. Thus the stability criterion is given as:

$$\left(d\Delta p_p/du_{in0}\right) < \left(d\Delta p/du_{in0}\right) \tag{5.13}$$

Now we demonstrate a much simplified stability analysis of this flow excursion, which is an application of Liapunov's direct method. First, a positive-value function is defined as

$$W = \left(1/2\right)M\left(du_{in}/dt\right)^2 \tag{5.14}$$

The time derivative of W gives

$$dW/dt = M\left(du_{in}/dt\right)\left(d^2u_{in}/dt^2\right) \tag{5.15}$$

Equation (5.14) is substituted into Eq. (5.15), then the term d^2u_{in}/dt^2 is eliminated:

$$dW/dt = \left\{\left(d\Delta p_p/du_{in}\right) - \left(d\Delta p/du_{in}\right)\right\}\left(du_{in}/dt\right)^2 \tag{5.16}$$

Following Liapunov's direct stability criterion, the stability of the system requires a negative value of dW/dt, which leads to the same criterion as Eq. (5.13).

This equation indicates that the stability for the Ledinegg instability case is determined by comparing the gradients between the internal and external characteristics at the intersection point. Thus, to stabilize the system, it is necessary to insert additional restriction to increase the frictional pressure drop of single-phase liquid flow to diminish the negative resistance. It is also effective to steepen the pump characteristics to intersect the internal characteristics at only one point.

On the basis of the foregoing discussion, the stability of the flow distribution among the parallel channels is discussed in the following. This is the problem of maldistribution induced by the negative resistance of the boiling channels. The model for reference is shown in Figure 5.10, which is composed of two parallel channels with the same flow characteristics, that is, the pressure drop versus flow rate relationship, and two common plenums or headers. Constant mass flow rate is imposed at the inlet plenum. The mass flow rates in each channel are given by G_1 and G_2, respectively.

Referring to Figure 5.11, a given pressure drop across the two plenums, line A–A in Figure 5.11a, has three intersections 1, 2, and 3 with the pressure drop versus flow rate curve. Then, for this pressure drop Δp_t, the combinations of flow

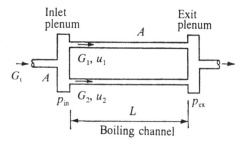

Figure 5.10 Model of parallel-channel system.

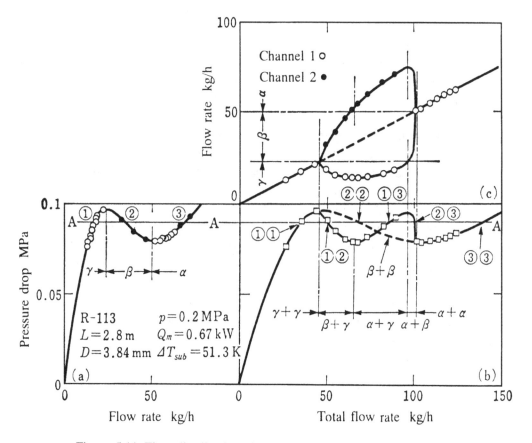

Figure 5.11 Flow distribution characteristics (two-channel system).

distribution become six cases in total: 1–1, 1–2, 1–3, 2–3, 2–2, and 3–3. By plotting these relationships of the pressure drop Δp_t versus the total flow rate $G_t = G_1 + G_2$ and G_1, G_2 versus G_t successively, the solid curves in Figures 5.11b and c are drawn. The symbols plotted in the figures represent the experimental results. We

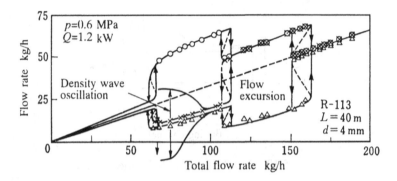

Figure 5.12 Flow distribution characteristics (three-channel system).

then find that the dashed lines have no experimental data on them. These dashed lines correspond to unstable distribution situations that are possible statically but do not appear in actual operation. Increasing the mass flux successively from a very small value of G_t, the flow distribution is, at first, uniform between the two channels, but deviates significantly between channels in the region $G_t = 50\text{--}100$ kg/h in this case, thereafter returning again to uniform flow. Such significant maldistribution induces burnout or dryout of the boiling channel system. A more complicated feature is observed in the three-channel system presented in Figure 5.12 as an example (Akagawa et al. 1971). The flow distribution characteristics show hysteresis. When the number of channels increases further, the distribution becomes extremely complicated. It is interesting to note that Figure 5.12 shows the possibility of occurrence of a different type of instability, that is, density wave oscillation in the channel with low mass flow rates.

Next, simplified analysis is presented on this flow maldistribution. As in the previous analysis, lumped-parameter modeling is applied to the boiling channels with the principal parameters u_1 and u_2 of inlet velocities. Then we have the following two momentum equations:

$$p_{in} - p_{ex} = dM_1 u_1/dt + \Delta p_1(u_1) \tag{5.17}$$

$$p_{in} - p_{ex} = dM_2 u_2/dt + \Delta p_2(u_2) \tag{5.18}$$

where p_{in} and p_{ex} represent the pressures in the plenums at the inlet and exit, respectively, and the pressure drops in both channels are expressed by the functions of respective inlet velocities u_1 and u_2. At the inlet plenum, the following mass conservation relationship holds:

$$\varrho_L A u_{in} = \varrho_L(u_1 + u_2)A = (\text{constant}) \tag{5.19}$$

where A represents the cross-sectional area of the channel, which is assumed to be uniform throughout the system. When the inertial mass M_1 and M_2 are assumed to be constant even during the transient condition, the following equation is obtained:

$$(M_1 + M_2)du_1/dt = \Delta p_2(u_2) - \Delta p_1(u_1) \qquad (5.20)$$

Comparing this equation with Eq. (5.7), the first term $\Delta p_2(u_2)$ can be considered as the pump or external characteristics with respect to channel 1. On the other hand, taking into account the relationship $du_1/dt = -du_2/dt$, the above equation is rewritten as

$$(M_1 + M_2)du_2/dt = \Delta p_1(u_1) - \Delta p_2(u_2) \qquad (5.21)$$

Then the first term $\Delta p_1(u_1)$ on the right side is considered as the pump or external characteristics with respect to channel 2. Thus, the flow maldistribution is found to be caused by the flow excursion or Ledinegg instability between the two channels, where one of the channels plays a role in providing the external characteristics for the other channel. Then the stability analysis mentioned above is directly adapted to this flow maldistribution, and we have the stability criterion

$$d(\Delta p_2)/du_1 - d(\Delta p_1)/du_1 = -d(\Delta p_2)/du_2 - d(\Delta p_1)/du_1 < 0 \qquad (5.22)$$

The pressure boundary condition gives the next relationship,

$$p_{in} - p_{ex} = \Delta p_t = \Delta p_1 = \Delta p_2 \qquad (5.23)$$

Taking into account a small perturbation around the steady-state values of u_1, u_2, and u_{in}, we have the linearized equation of Eq. (5.23):

$$d(\Delta p_t)/du_{in0}\,\delta u_{in} = d(\Delta p_1)/du_{10}\,\delta u_1, \qquad \text{or}$$
$$d(\Delta p_t)/du_{in0}\,\delta u_{in} = d(\Delta p_2)/du_{20}\,\delta u_2 \qquad (5.24)$$

Mass conservation gives

$$\delta u_{in} = \delta u_1 + \delta u_2 \qquad (5.25)$$

Then the next relationship holds between the gradients of the pressure drop against the velocity,

$$1/[d(\Delta p_t)/du_{in0}] = 1/[d(\Delta p_1)/du_{10}] + 1/[d(\Delta p_2)/du_{20}]$$

Substituting this relationship in the above criterion, we then have a simplified criterion for the stability of the flow distribution expressed as

$$[d(\Delta p_t)/du_{in0}][d(\Delta p_1)/du_{10}][d(\Delta p_2)/du_{20}] > 0 \qquad (5.26)$$

This relationship can be easily extended to the three-channel system as

$$\left[d(\Delta p_{\rm t})/du_{\rm in0}\right]\left[d(\Delta p_1)/du_{10}\right]\left[d(\Delta p_2)/du_{20}\right]\left[d(\Delta p_3)/du_{30}\right] > 0 \quad (5.27)$$

This gives quite simplified and easy means for predicting stability without using any computer. If a small computer is available, we can directly solve the momentum equations and the mass conservation equation. Then we have only the stable distributions of solid lines, and the unstable states do not appear. In the hysteresis region, the state bifurcates, depending on the disturbance at the initial state.

In actual plants, the number of parallel channels is tremendous and therefore it is quite difficult to predict the flow distribution. For avoiding such flow maldistribution, each boiling channel should not, at least, have negative resistance characteristics. Then, the problem of flow maldistribution becomes less significant, which does not mean, however, that the problem does not exist. Thus design engineers should be careful to have uniform heat flux distribution, or to ensure uniform flow distribution among the channels so as not to induce dryout or burnout owing to flow maldistribution.

5.5 Pressure drop oscillation

This section focuses on oscillatory flow instability, which is closely related to the negative resistance characteristics of the pressure drop versus flow rate and the compressible capacity upstream of the boiling channel. The inlet mass flux of the boiling channel can vary with time, depending on the pressure difference between the compressible volume and the boiling channel. Thus the compressible capacity accumulates mass and momentum, which is similar to a coiled spring in a mechanical vibration system.

The simplified flow model is shown in Figure 5.13. The system has a single boiling channel with negative resistance as shown in Figure 5.14 and a compressible capacity, that is, an accumulator. Provided that the cross-section of the compressible volume is very large, then a flow fluctuation at the inlet of the boiling channel does not induce any substantial change of the liquid level in the volume. Then the external characteristics, including this compressible volume, become almost horizontal in the pressure drop versus flow rate characteristics. When this horizontal characteristic curve intersects the pressure drop curve at three points, the point of negative resistance is unstable and flow excursion takes place. This becomes a trigger for the pressure drop oscillation discussed in this section.

Suppose that the initial state is located at point A in Figure 5.14. Owing to the positive disturbance of the flow rate, flow excursion occurs toward point B in the positive resistance region. When the inlet flow rate of the system is kept constant, liquid flow in excess of the inlet flow rate proceeds from the capacity. The posi-

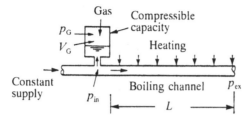

Figure 5.13 Flow model with compressible capacity.

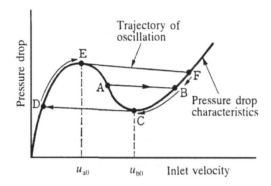

Figure 5.14 Pressure drop oscillation.

tive resistance region is stable, and thus the flow excursion does not expand further; then the flow proceeding from the capacity decreases gradually, with gradual decrease in the pressure in this volume. The state then traces approximately the pressure drop characteristic curve B–C. At point C, further decrease results in an increase in the pressure drop and then the flow rate decreases drastically and catastrophically to point D; that is, flow excursion takes place again. This drastic decrease in the flow rate in the boiling channel is compensated by an increase in the accumulation rate in the volume. With this accumulation of mass, the pressure in the volume increases in turn, which is followed by a gradual decrease in the accumulation rate; the flow rate into the boiling channel, on the contrary, increases gradually. The state in the positive resistance region is also stable and thus moves along the characteristic curve toward point E. Flow excursion takes place again from point E to F. This process is repeated periodically. The oscillation shows typical relaxation oscillation between two steady-state curves D–E and F–C.

This oscillation is induced by two fundamental mechanisms: the triggering due to the negative resistance and the accumulation mechanism, which is exemplified by the model shown in Figure 5.15. The relaxation oscillation is characterized by

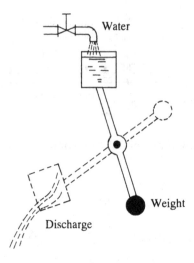

Figure 5.15 Relaxation oscillation (mechanical system) (Magnus 1976).

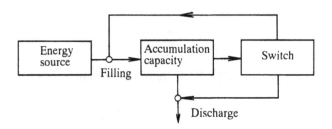

Figure 5.16 Block diagram of relaxation oscillation (Magnus 1976).

(a) the accumulation process of water filling and (b) the discharge process, where the switching is conducted by the weight balance. The block diagram of this process is shown in Figure 5.16.

Here we present a simplified analysis of pressure drop oscillation using the lumped-parameter model similar to the case of Ledinegg instability. In this pressure drop oscillation the stability criterion is quite simple, because the external characteristic given by the liquid supply system, including the capacity, is horizontal on the pressure drop versus mass flux plane. Thus the existence of negative resistance in the pressure drop versus flow rate relationship becomes a stability criterion. In this part, we present a simplified analysis and deduce the relevant mechanism of the pressure drop oscillation. In general, pressure drop oscillation has a long period, depending, of course, on the amount of the compressible volume. The transit time or residence time in the boiling channel is usually so small as to be neglected with respect to the oscillation period. This hypothesis makes it possible to apply lumped-parameter modeling.

Provided that the channel cross-sections of the whole system have the same value A, and that the inlet velocity u_0 of the system and the pressure p_{ex} at the channel exit are constant throughout the transients, the following mass and momentum conservation equations for the channel and compressible capacity are obtained:

Mass: $$u_0 A\, \varrho_L = u_c A\, \varrho_L + u_{in} A\, \varrho_L \tag{5.28}$$

Momentum:

for capacity $$p_{in} - p_c = d\left(M_c u_c\right)/dt \tag{5.29}$$

for channel $$p_{in} - p_{ex} = d\left(M u_{in}\right)/dt + \Delta p\left(u_{in}\right) \tag{5.30}$$

where p_c represents the pressure in the compressible volume, M_c and M the inertial mass in the channel to the volume and the boiling channel, respectively, which are assumed to be constant for simplicity, and u_c and u_{in} represent the inflow velocities to the capacity and the boiling channel, respectively.

The volume change of the compressible capacity is determined by the total amount of inflow from the primary channel:

$$dV_c/dt = -A u_c \tag{5.31}$$

If the gas that gives the compressibility undergoes an isothermal state change, we then have $p_c V_c = $ constant, thus the inflow velocity into the volume u_c is given as

$$u_c = \left(V_c/A p_c\right) dp_c/dt = C dp_c/dt \tag{5.32}$$

The parameter C is assumed to be constant against time, which corresponds to a capacitance. As the pressure drop in the boiling channel is expressed on the basis of the lumped parameter of u_{in}, then the time derivative of the pressure drop is expressed by

$$d(\Delta p)/dt = \left[d(\Delta p)/du_{in}\right] du_{in}/dt \tag{5.33}$$

The above equation system results in the following second-order differential equation:

$$\left(M + M_c\right) d^2\left(\delta u_{in}\right)/dt^2 + \left[d(\Delta p)/du_{in}\right] d\left(\delta u_{in}\right)/dt + \left(1/C\right)\delta u_{in} = 0 \tag{5.34}$$

where $\delta u_{in} = u_{in} - u_0$ and is the deviation from the steady-state value. When all of the above-mentioned coefficients of the δu_{in} terms are constant, the equation is a linear ordinary differential equation. If all the coefficients are positive constant – in reality the first and third terms are always positive constant – then the system is statically and dynamically stable. On the other hand, when the second term is negative constant, which corresponds to the operating condition of u_0

located in the negative resistance region, then the system represents sustained oscillation with developing amplitude. This is a linear stability analysis.

Next, nonlinear analysis is conducted, assuming the relationship between the pressure drop and the inlet velocity to be a third-order polynomial (Ozawa et al. 1979a). The derivative of the pressure drop with respect to the inlet velocity is expressed by

$$d(\Delta p)/du_{in} = \xi(u_{in} - u_{a0})(u_{in} - u_{b0}) \tag{5.35}$$

where ξ is a positive constant, and u_{a0} and u_{b0} represent the velocities corresponding to maximum and minimum pressure drops, respectively, as shown in Figure 5.14. This equation is substituted into Eq. (5.34), and we then have the next normalized equation with dimensionless variables:

$$d^2y/d\tau^2 - \varepsilon(1 - 2\beta y - y^2)dy/d\tau + y = 0 \tag{5.36}$$

where

$$y = (u_{in} - u_0)/[(u_0 - u_{a0})(u_{b0} - u_0)]^{0.5}$$
$$\tau = t/[C(M + M_C)]^{0.5} \tag{5.37}$$

and these variables represent the normalized velocity at the inlet and the normalized time, respectively. The dimensionless parameters are defined by

$$\varepsilon = [C/(M + M_c)]^{0.5}\xi(u_0 - u_{a0})(u_{b0} - u_0)$$
$$\beta = [u_0 - (u_{a0} + u_{b0})/2]/[(u_0 - u_{a0})(u_{b0} - u_0)]^{0.5} \tag{5.38}$$

The above equation (5.36) is the well-known van der Pol equation, which represents limit cycle oscillation (van der Pol 1926). The oscillation trace of y shows a sinusoidal curve for small values of ε. On the other hand, the oscillation becomes a relaxation oscillation for relatively large values of ε, as shown in Figure 5.17.

One of the experimental examples of pressure drop oscillation is shown in Figure 5.18. The upper trace represents the mass flow rate and the lower the pressure drop in the boiling channel. The processes E–F and C–D correspond to the excursions from lower to higher and from higher to lower flow rates, respectively, in the figure. The processes F–C and especially D–E undergo relatively slow transition, which corresponds to the process when the state moves approximately along the steady-state characteristic curve of the pressure drop versus mass flow rate in Figure 5.14. In this case, the calculated value of ε becomes greater than 30, which is high enough to become relaxation oscillation.

On the basis of the van der Pol equation, the oscillation period is approximately proportional to ε at relatively large values of ε; then

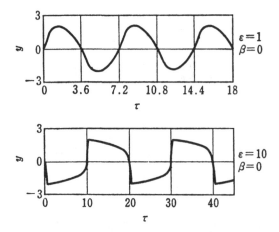

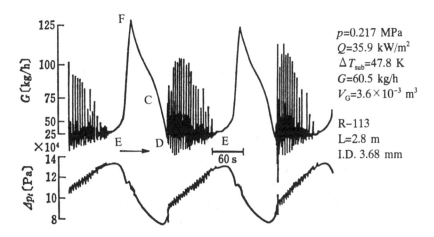

Figure 5.17 Oscillation trace of van der Pol equation (van der Pol 1926).

Figure 5.18 Oscillation trace of pressure drop oscillation.

$$T/\left[C(M + M_c)\right]^{0.5} \propto \varepsilon \tag{5.39}$$

This is rewritten as

$$T \propto \left[- d(\Delta p)/du_0\right]V_c/(Ap_c) \tag{5.40}$$

that is, the oscillation period is proportional to the compressible volume, which is well confirmed by experiment, as shown in Figure 5.19. It is quite interesting to note that the actual system may suffer from not only pressure drop oscillation but also density wave oscillation, as shown in Figure 5.18, where density wave oscillation is also observed during the transient process from D to E. To prevent

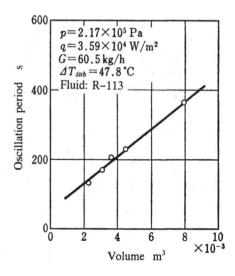

Figure 5.19 Oscillation period of pressure drop oscillation.

such pressure drop oscillation, it will be necessary to remove either the upstream compressibility or the negative resistance characteristics.

5.6 Geysering

The instability known as geysering is closely related to heat transfer, especially to the boiling mechanism. Geysering is a violent boiling occurring periodically in a vertical and/or inclined column or channel with an upper plenum or drum. The phenomenological explanation is shown in Figure 5.20. The geysering process is described on the basis of this figure.

At the first stage of the process (a), the liquid in the channel is heated to have a certain level of superheat mainly owing to insufficient nucleation sites. Boiling then initiates. The generated vapor pushes the liquid up above the vapor column to the drum (b). When the drum capacity is fairly small, this liquid ejection induces a pressure increase in the drum, which suppresses the violent boiling. On the other hand, when the drum capacity is large enough, the drum pressure retains an almost constant value in spite of the inflow of liquid from the channel. Then the static head in the channel decreases, which enhances the self-evaporation or flashing, and the mixture of liquid and vapor is rapidly blown up into the drum (c). The tube wall will then dry out under the relatively high heat flux. Vapor generation then ceases with time and the vapor velocity decreases below the so-called countercurrent flow limitation (CCFL condition), which is followed by flow reversal from the drum (refilling). Finally, the channel is refilled by liquid to return to

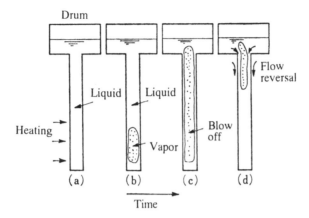

Figure 5.20 Geysering process.

the initial stage. In certain cases, a large pressure wave is generated owing to the collapse of the vapor during the refilling process. These processes are repeated periodically.

Such processes are well demonstrated in the recording trace of the pressure drop shown in Figure 5.21. Under very low mass flux, in this case 0.01 m/s, the successive increase in heat flux induces geysering, which is clearly observed in the large pressure drop fluctuation. The velocity is very low and thus the pressure drop corresponds almost to the gravitational component, that is, this large fluctuation corresponds to the void fraction fluctuation. During the period τ_1 the liquid is superheated. The period τ_2 corresponds to the blowing-up of the liquid or two-phase mixture, and the third period τ_3 corresponds to the refilling process. The period of geysering increases, in general, with an increase in heat flux, but ceases beyond a certain limit of heat flux. The stability diagram is shown in Figure 5.22 in the heat flux versus inlet velocity curve (Ozawa et al. 1979b). As mentioned above, geysering is closely related to violent boiling owing to the liquid super-heat. Thus, to avoid geysering, it is effective to increase the inlet velocity and/or the heat flux. In a natural-circulation system, this geysering is accompanied by periodical enhancement of the circulation rate. Thus the geysering appears possibly during start-up with relatively low heat flux and low flow rate.

5.7 Density wave oscillation

It is essential to have a certain accumulation mechanism or flexibility in the supply system of the coolant to allow for the occurrence of sustained oscillation in boiling channels. These accumulation mechanisms bring about, at the same time, feedback between the pressure drop and the flow rate. In this sense, the

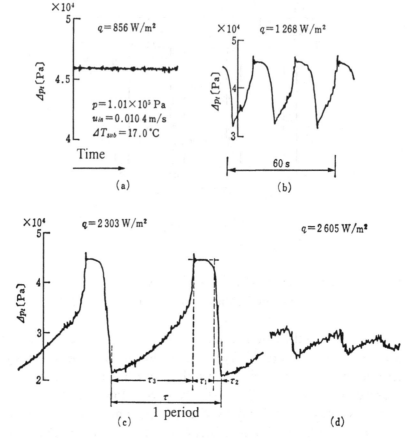

Figure 5.21 Fluctuation of pressure difference during geysering.

overview of density wave oscillation seems to be similar to that of pressure drop oscillation. The main difference exists in the fact that density wave oscillation is induced by negative damping owing to feedback and the time delay, even in the positive resistance region of the pressure drop versus flow rate characteristics, while in the pressure drop oscillation the negative resistance plays a direct part in the negative damping.

The flow models in Figure 5.23 represent examples where density wave oscillation occurs. The first one (a) is a parallel-channel system with an extremely large number of channels; the second (b) has a large bypass slightly heated or unheated. These two systems are typical examples where the pressure drop across the channel is maintained constant even during the transient states. The natural-circulation system gives the same situation as (b), and the downcomer that makes constant pressure drop corresponds to the large bypass, while the flow direction is different from case (b). In case (c), the pressure drop fluctuation is fed back as

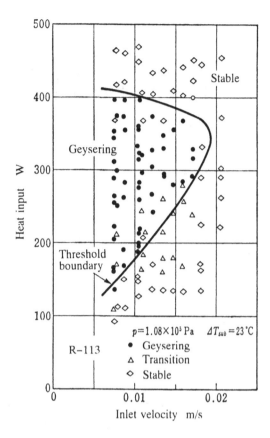

Figure 5.22 Threshold boundary of geysering.

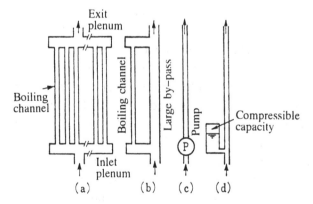

Figure 5.23 Fundamental flow model for density wave oscillation.

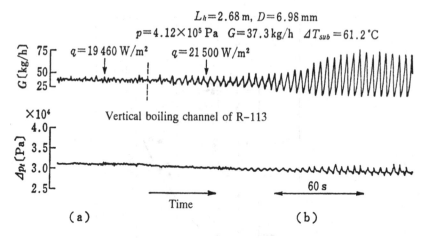

Figure 5.24 Example of density wave oscillation.

flow rate fluctuation through the pump characteristics. The last model (d) has the compressible volume upstream of the boiling channel, as in the case of pressure drop oscillation.

5.7.1 General feature of density wave oscillation

In this section, the general feature of density wave oscillation is discussed on the basis of the experimental experience. One of the typical examples of density wave oscillation observed in laboratory experiments is shown in Figure 5.24. The experiment was conducted using a forced-flow boiling system of type d in Figure 5.23 to detect the threshold condition of the oscillation by successive increases in the heat flux under constant pressure, mass flux, and inlet subcooling. The upper trace represents the mass flow rate, and the lower the pressure drop across the boiling channel. For relatively low heat flux, the mass flow rate is almost constant but with small fluctuations. These are induced by the fluctuation of the two-phase flow itself. If the heat flux is increased beyond a certain limit (state b), an oscillation with a certain definite frequency appears and then dominates the fluctuations with developing amplitude. This oscillation finally reaches the sustained limit cycle oscillation. Further increase in the heat flux results in an increase in the oscillation amplitude, but shows relaxation-like oscillation accompanying the flow reversal; that is, the oscillation trace makes a transition from harmonic-like to relaxation-like oscillations. On the other hand, the oscillation period is a decreasing function of the heat flux.

The oscillation period of such a density wave oscillation is plotted against the residence time in the boiling channel in Figure 5.25, where the residence time is

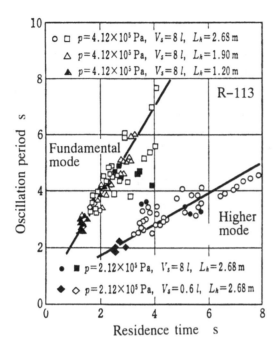

Figure 5.25 Oscillation period of density wave oscillation.

calculated by applying the drift–flux model discussed later. The parameter V_S in the figure represents the volume of compressible capacity upstream of the boiling channel. A typical feature of this oscillation is that the oscillation period is of the same order as the residence time, and this feature is quite different from pressure drop oscillation. The experimental relationship is, however, mainly classified into two groups in this case; that is, one group of data has a period twice the residence time, while the other has a period about 30% less than the residence time. The first one is referred to as fundamental-mode oscillation, and the latter as higher-mode oscillation. The oscillation period seems to be independent of the volume of the compressible capacity, but the threshold condition depends on the volume below a certain critical volume determined by the system. Thus the experimental work needs some examination of the volume of the capacity when system d in Figure 5.23 is used. The threshold condition is exemplified in Figure 5.26, using inlet subcooling and heat flux as coordinates. These data demonstrate that the threshold condition is approximately expressed by a constant exit-quality line. But this may be limited only when the dominant pressure drop is a frictional component in the boiling channel. Further discussion is presented later on the stability effect of various parameters. When the inlet subcooling is high, that is, the inlet temperature of the liquid is low, then the higher-mode oscillation appears at first but changes to the fundamental mode with a further increase in heat flux.

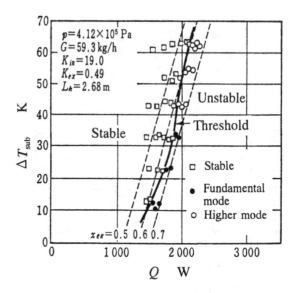

Figure 5.26 Stability boundary in a forced-flow boiling-channel system.

On the other hand, for low inlet subcooling, fundamental-mode oscillation takes place. The mechanism of the appearance of this oscillation mode depends mainly on the existence of the standing wave of enthalpy in the subcooled region, which is well discussed by Yadigaroglu and Bergles (1972).

Density wave oscillation is also observed in the natural-circulation system because of the existence of constant pressure drop across the boiling channel, as discussed with respect to system b in Figure 5.23. In the natural-circulation system, the mass flow rate depends on the heat input, and thus it may not be reasonable to use the same coordinate system as in the forced-flow system. The threshold condition in a natural-circulation system is therefore shown in Figure 5.27 using the heat input versus mass flow rate coordinate system. Under constant pressure and inlet subcooling, the mass flow rate increases with an increase in the heat input, attains a maximum, and then decreases. This characteristic is typical of a natural-circulation system. In this system, flow oscillation is observed in two regions. The low heat input region on the left is mainly observed at the start-up of the system, that is, periodical alternation between bubble generation and cooling down induced by the natural circulation. On the other hand, the high heat input region corresponds to density wave oscillation with the same period–residence time relationship as in Figure 5.25.

At the beginning of this section, it was indicated that density wave oscillation is, in general, observed in the positive resistance region. Is this always true? In the positive resistance region, the state is definite and thus the flow rate is

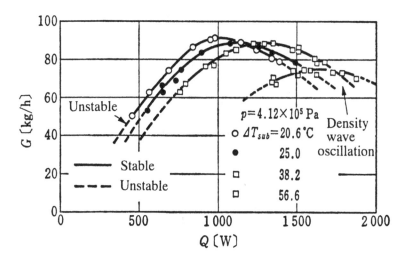

Figure 5.27 Flow rate–heat input relationship in a natural-circulation system.

uniquely determined for a given pressure drop across the channel. Then instability due to the flow excursion mechanism does not occur, and only density wave oscillation can appear. In the negative resistance region, we cannot deny the possibility of such density wave oscillation. However, flow excursion and/or pressure drop oscillation is of prime importance in system dynamics, and density wave oscillation is not apparent in itself. Such density wave oscillation is described in the next section, using phenomenological modeling.

5.7.2 *Mechanism of oscillation*

Here we discuss the dynamic behavior of the flow in the simplified model shown in Figure 5.28. The model is the single-channel boiling system with a large bypass parallel to the boiling channel. This bypass represents a large number of parallel channels, which brings about the boundary condition of a constant pressure drop across the boiling channel, as mentioned before. The single-phase liquid flows through this large bypass with relatively low velocity, and thus the bypass provides an almost constant pressure drop expressed only by the gravitational component.

The boiling channel is expressed as a lumped-parameter system with flow restrictions in the subcooled region and at the channel exit, which provide the pressure drops, a simple riser that provides the residence time, and a vaporizer that produces the two-phase mixture. The restriction in the subcooled region represents the single-phase pressure drop (frictional pressure drop), and that at the exit the two-phase pressure drop. Now let us consider the sequence of the process

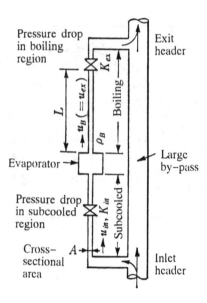

Figure 5.28 Simplified model of boiling-channel system.

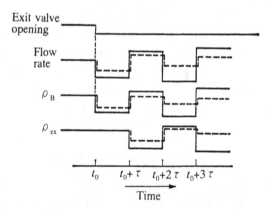

Figure 5.29 History of various parameters after perturbation of exit valve opening.

during the transients. At the initial state, the exit valve opening is slightly closed at $t = t_0$ (see Fig. 5.29). The pressure drop across the boiling channel retains a constant value, and thus the small change in the exit valve opening induces a small decrease in the mass flow rate flowing into the boiling channel. Under constant heat flux and inlet subcooling, positive enthalpy perturbation arises, which is followed by a decrease in the density owing to the increase in the evaporation rate. The density ϱ_B of the fluid flowing out of the vaporizer thus decreases as shown in Figure 5.29 at t_0. The perturbation of this density propagates through the riser, taking time τ ($= L/u_{ex}$, where L represents the length of the riser and u_{ex} the mixture velocity in the riser), and reaches the exit restriction. Then the exit density ϱ_{ex} decreases from its initial state to the value ϱ_B at time $t_0 + \tau$. This density perturbation induces a decrease in the two-phase pressure drop, which is compensated by an increase in the inflow mass flux. When the influence of the disturbance in the pressure drop accelerates the disturbance in the mass flux, successive processes develop to sustain the oscillation. The perturbation of ϱ_B is essentially 180° out of phase from that of ϱ_{ex}, while the flow rate is also out of phase with the two-phase pressure drop. Experimental results support this tendency but are valid only for the system in which the frictional component dominates the pressure drop in the boiling channel. On the other hand, if the disturbance weakens, the oscillation ceases with time.

Imagine a situation in which the residence time is so short as to be neglected. Then the initial disturbance in the density ϱ_B results immediately in a decrease in ϱ_{ex}. The two-phase pressure drop then decreases at the same time, but this is compensated immediately by an increase in the mass flux and the pressure drop in the single-phase region. The oscillation is easy to damp out owing to the positive resistance. Thus it is essential to have a certain time lag in response between the disturbances of the density and the pressure drop. In the boiling channel system this time lag is, in general, produced by a delay in propagation of the void fraction perturbation. This is referred to as the void wave, which propagates with the flow through the boiling channel. Thus the void wave is a kinematic wave in this sense. The oscillation period is then fundamentally expressed as twice the residence time, as in the fundamental mode shown in Figure 5.25. The term "density wave" may cause some misunderstanding, that is, complication with a kinematic wave and a dynamic wave. The dynamic wave is a density disturbance due to compressibility. As to flow instability relating to this dynamic wave, there exists a certain class of instability termed acoustic oscillation, listed in Table 5.1, which is mainly caused by a delay in propagation of the pressure wave in the channel. A detailed description of this instability is beyond the scope of this book, and those who are interested in it are referred to Bouré et al. (1971).

Now let us carry out a simplified analysis of the density wave oscillation based on Figure 5.28. In the formulation, the gravitational and accelerational pressure drops are not taken into account but only the frictional terms or flow resistance terms are included. The pressure drop in the single-phase region is expressed as the lumped-parameter model by

$$\Delta p_{in} = M du_{in}/dt + K_{in}\varrho_L u_{in}^2 \tag{5.41}$$

where M represents the inertial mass of the single-phase region, u_{in} the inlet velocity, and K_{in} the resistance coefficient. The two-phase pressure drop is expressed by

$$\Delta p_{ex} = K_{ex} \varrho_{ex} u_{ex}^2 \tag{5.42}$$

where the inertial mass and thus the inertial term in the two-phase region is assumed to be so small compared with the resistance term as to be neglected. The variables with the subscript "ex" represent the values at the exit of the riser section.

In the vaporizer upstream of the riser, the outflow enthalpy i_B is given in terms of the heat input Q_B, the mass flow rate m_{in} ($= A\varrho_L u_{in}$), and the inlet enthalpy i_{in} as

$$i_B = i_{in} + Q_B/m_{in} \tag{5.43}$$

where A represents the cross-sectional area of the channel. Then the density ϱ_B of the outflow from the vaporizer is expressed as a function of the flow quality x_B ($= (i_B - i_L)/\Delta i_{LG}$) on the basis of the homogeneous flow assumption,

$$1/\varrho_B = 1/\varrho_L + (i_B - i_L)/\Delta i_{LG}(1/\varrho_G - 1/\varrho_L) \tag{5.44}$$

where Δi_{LG} represents the latent heat of vaporization and i_L the saturated liquid enthalpy. As the mass accumulation is assumed to be neglected in the vaporizer, the mass continuity relationship gives

$$\varrho_L u_{in} = \varrho_B u_B \tag{5.45}$$

Then the outflow velocity u_B of the two-phase mixture is given by

$$u_B = u_{in}\left\{1 + (\Delta\varrho_{LG}/\Delta i_{LG}\varrho_G)(Q_B/m_{in} - \Delta i_{sub})\right\} \tag{5.46}$$

The fundamental equation system has already been presented at this stage. The next stage is how to solve the equation system to obtain the threshold condition and/or stability criterion. One approach is to apply the small perturbation method; that is, a small perturbation is imposed upon the steady-state value to linearize the equation system:

$$u_{in} = u_{in0} + \delta u_{in}, \quad u_{ex} = u_{ex0} + \delta u_{ex}, \quad \varrho_{ex} = \varrho_{ex0} + \delta\varrho_{ex}$$
$$\varrho_B = \varrho_{B0} + \delta\varrho_B, \quad m_{in} = m_{in0} + \delta m_{in}, \quad \Delta p_{in} = \Delta p_{in0} + \delta(\Delta p_{in}),$$
$$\Delta p_{ex} = \Delta p_{ex0} + \delta(\Delta p_{ex}).$$

These variables are substituted into the above equations, and the steady-state variables are eliminated.

$$\delta(\Delta p_{in}) = Md(\delta u_{in})/dt + 2K_{in}\varrho_L u_{in0}\delta u_{in} \tag{5.47}$$

$$\delta(\Delta p_{ex}) = 2K_{ex}\varrho_{ex0}u_{ex0}\delta u_{ex} + K_{ex}u_{ex0}^2\delta\varrho_{ex} \tag{5.48}$$

$$\delta\varrho_B = \left[\Delta\varrho_{LG}Q_B/(\Delta i_{LG}\varrho_G m_{in0})\right]\left[\varrho_{B0}^2/(u_{in0}\varrho_L)\right]\delta u_{in} \tag{5.49}$$

$$\delta u_B = \left[1 - \Delta\varrho_{LG}\Delta i_{sub}\big/\left(\Delta i_{LG}\varrho_G\right)\right]\delta u_{in} \tag{5.50}$$

Now the dimensionless parameters, the subcooling number N_{sub} and the phase change number N_{pch}, which are defined by the next equations, are introduced into Eqs. (5.49) and (5.50).

$$\delta\varrho_B = N_{pch}\,\varrho_{B0}{}^2\big/\left(\varrho_L u_{in0}\right)\delta u_{in} \tag{5.51}$$

$$\delta u_B = \left(1 - N_{sub}\right)\delta u_{in} \tag{5.52}$$

where $N_{pch} = \Delta\varrho_{LG}Q_B/(\Delta i_{LG}\varrho_G m_{in0})$ and $N_{sub} = \Delta\varrho_{LG}\Delta i_{sub}/(\Delta i_{LG}\varrho_G)$, and the first one represents the vaporization rate and the latter the amount of the subcooling.

The density perturbation $\delta\varrho_B$ propagates through the riser section and thus, at the riser exit, $\delta\varrho_{ex}$ is expressed using the delayed step function $U(t - \tau)$,

$$\delta\varrho_{ex} = \delta\varrho_B U\left(t - \tau\right) \tag{5.53}$$

The parameter τ corresponds to the residence time of the two-phase mixture in the riser section. The velocity of the two-phase mixture is considered to be uniform in the riser, that is,

$$\delta u_B = \delta u_{ex}, \tag{5.54}$$

and the constant pressure drop is imposed across the boiling channel,

$$\delta\left(\Delta p_{in}\right) + \delta\left(\Delta p_{ex}\right) = 0 \tag{5.55}$$

which are also applied to Eqs. (5.47) and (5.48) above. Then we have

$$Md\left(\delta u_{in}\right)\big/dt + 2\varrho_L u_{in0}\left[K_{in} - K_{ex}\left(N_{sub} - 1\right)\right]\delta u_{in}$$
$$+ K_{ex}\varrho_L u_{in0}N_{pch}\delta u_{in}U\left(t - \tau\right) = 0 \tag{5.56}$$

Furthermore, this equation is Laplace-transformed using the parameter S. Then,

$$M\bar{u}_{in}S + 2\varrho_L u_{in0}\left[K_{in} - K_{ex}\left(N_{sub} - 1\right)\right]\bar{u}_{in}$$
$$+ \varrho_L K_{ex}u_{in0}N_{pch}\bar{u}_{in}\exp\left(-\tau S\right) = 0 \tag{5.57}$$

where $\bar{u}_{in}$ is the Laplace-transformed valuable of δu_{in}. The exponential term is approximated by applying the first-order Pade approximation:

$$\exp\left(-\tau S\right) \cong 1\big/\left(1 + \tau S\right) \tag{5.58}$$

Then, finally, we have the next second-order equation to express the dynamic behavior of the present simplified model:

$$MS^2 + \left\{M/\tau + 2\varrho_L u_{in0}\left[K_{in} - K_{ex}\left(N_{sub} - 1\right)\right]\right\}S +$$
$$\left\{2\left[K_{in} - K_{ex}\left(N_{sub} - 1\right)\right] + K_{ex}N_{pch}\right\}\varrho_L u_{in0}/\tau = 0 \tag{5.59}$$

This equation is equivalent to that of a mechanical system with a coiled spring and a dashpot (damper) given by Eq. (5.1):

$$MS^2 + CS + K = 0 \tag{5.60}$$

As discussed in the previous section, when the parameter C of the second term is negative, the perturbation develops with time and results in oscillatory instability. In the region $N_{sub} < 1$, the parameters C and K, corresponding to the second and third terms in Eq. (5.59), become positive, which results in a dynamically stable state. In the region $N_{sub} > 1$, the increase in K_{ex} may lead to a negative value of the second term C, and the increase in K_{in} may lead to a positive one. The restriction K_{in} is a stabilizing factor: that is, by increasing the inlet restriction we can stabilize the flow; on the contrary, K_{ex} is a destabilizing factor. The stability boundary is given by

$$1 < N_{sub} < K_{in}/K_{ex} + 1 \quad \text{or} \quad N_{sub} < 1 \tag{5.61}$$

Such simplified analysis makes it possible to have a certain analogy with a mechanical vibration system, and it may also be possible to understand density wave oscillation phenomenologically. When we are interested in a more detailed analysis of this instability, we refer to the respective sections appearing later.

5.7.3 Block diagram and the role of the pressure-boundary condition

In the previous section, it was pointed out that two important factors for the occurrence of density wave oscillation are the time delay of the perturbation propagation along the channel and the pressure drop boundary condition that enables feedback to the flow perturbation. Now, in this section, the latter pressure drop boundary is explained on the basis of the flow models shown in Figure 5.23.

In engineering control theory, block diagrams are often used to understand the dynamic interrelationship between the elements. We apply a similar technique to the discussion on the pressure drop boundary conditions. Equations (5.47) and (5.48) are Laplace-transformed, and the transfer functions between the pressure drop and the perturbation of the inlet velocity u_{in} are expressed by

$$\left. \begin{array}{l} \overline{\Delta p}_{in} = G_{in}(S)\overline{u}_{in}, \\ \overline{\Delta p}_{ex} = G_{ex}(S)\overline{u}_{in} \end{array} \right\} \tag{5.62}$$

As for the pressure drop across the boiling channel, the disturbance term $\bar{p}_{dis}$ is added for convenience in constructing the block diagram:

$$\overline{\Delta p}_t + \bar{p}_{dis} = \overline{\Delta p}_{in} + \overline{\Delta p}_{ex} \tag{5.63}$$

First we consider the case of the parallel-channel system and/or the system with a large bypass parallel to the single boiling channel. The pressure drop boundary condition is given by

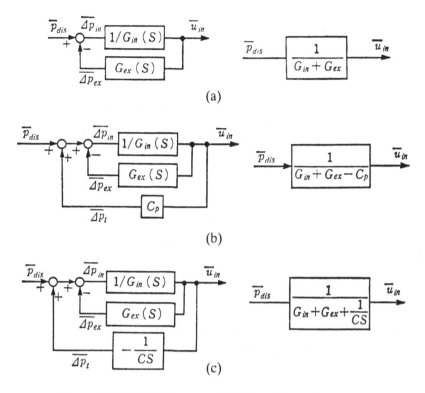

Figure 5.30 Block diagram of boiling-channel system.

$$\overline{\Delta p}_t = 0 \tag{5.64}$$

This pressure drop boundary condition does not limit the flow direction. Thus the natural-circulation system results in the same relationship as Eq. (5.64).

When the block diagram is drawn using these equations, there is no restriction on the selection of a feed-forward element and a feedback element. Here the input is selected as the pressure drop disturbance $\bar{p}_{dis}$ and the output is set to the flow rate perturbation $\bar{u}_{in}$, for convenience of discussion. The diagram is shown in Figure 5.30a (left), where the input for the transfer function $1/G_{in}(S)$ is the disturbance $\overline{\Delta p}_{in}$ and then the output is the inlet velocity disturbance $\bar{u}_{in}$. This $\bar{u}_{in}$ is fed back through the transfer function $G_{ex}(S)$ to give the two-phase pressure drop disturbance $\overline{\Delta p}_{ex}$. This pressure drop disturbance is subtracted from the pressure disturbance $\bar{p}_{dis}$ to give the disturbance $\overline{\Delta p}_{in}$. This feedback circuit is rewritten as shown in Figure 5.30a (right) as a feed-forward system. The following characteristic equation is then examined to see whether or not the system is stable, independent of the disturbance $\bar{p}_{dis}$:

$$G_{in}(S) + G_{ex}(S) = 0 \tag{5.65}$$

Next, the system with a pump is discussed. As in the first model, the single-channel system is considered here. The general feature of the pump characteristics is expressed, after linearizing, by

$$\overline{\Delta p}_t = C_p \bar{u}_{in} \qquad (5.66)$$

where C_p is, in general, negative. On the basis of this equation and Eqs. (5.62) and (5.63), the block diagram is drawn as shown in Figure 5.30b, where the pump characteristics act as the feedback element. The characteristic equation is expressed in this case by

$$G_{in}(S) + G_{ex}(S) - C_p = 0 \qquad (5.67)$$

Considering the limiting case of the value of C_p being nearly zero, $C_p \cong 0$, the block diagram is the same as in the case of Figure 5.30a with the constant pressure drop condition.

The last system is one with compressible capacity but with a single channel. The dynamic behavior of the boiling channel is expressed by Eqs. (5.62) and (5.63), while the compressible capacity is expressed by Eqs. (5.28), (5.29), and (5.32). The inertia term of the piping element between the compressible capacity and the main feed piping is here neglected for simplicity, and the linearized equation system is derived, using the small perturbation method, as

$$\bar{u}_{in} + \bar{u}_c = 0, \qquad \overline{p}_{in} - \bar{p}_c = 0, \qquad \bar{u}_c = C\bar{p}_c S \qquad (5.68)$$

The imposed disturbance $\overline{p}_{in}$ is equal to that of the total pressure drop $\overline{\Delta p}_t$ under the condition of constant exit pressure; we then have

$$\overline{\Delta p}_t = \left[-1/(CS) \right]\bar{u}_{in} \qquad (5.69)$$

On the basis of equations (5.62), (5.63), and (5.69), the block diagram is represented as shown in Figure 5.30c. The characteristic equation is then expressed by

$$G_{in}(S) + G_{ex}(S) + 1/(CS) = 0 \qquad (5.70)$$

In the limiting case of extremely large capacity, the parameter C becomes so large that the last term in Eq. (5.70) can be neglected, which gives the same relationship as shown in Figure 5.30a. This suggests that the volume of the compressible capacity has an effect on the stability of the flow.

Boiling two-phase flow systems consist usually of a certain combination of parallel channels and an external coolant-feed system such as a pump, as shown in Figure 5.31a. With respect to the parallel channels between headers, the boundary condition of the constant pressure drop holds. Then, among the parallel channels, a certain criterion for stability exists. Next, another pressure boundary condition holds between the boiling channels and the external pump system, and another stability criterion exists for this case. Thus, in the total system, two modes

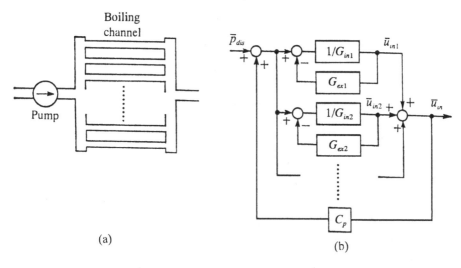

(a)

(b)

Figure 5.31 Block diagram of parallel-channel system with liquid feed pump.

of flow instability will appear, depending on the operating parameters and/or the system configuration. One of the examples of the block diagram is shown in Figure 5.31b. Those who wish to analyze flow stability should identify the frame of reference for consideration as well as the boundary condition, for example, the pressure and the flow rate.

5.7.4 Simplified stability criterion

The pressure drop has three components, frictional, accelerational, and gravitational. The influence of these components, especially their dynamic behavior, differs between them, which results in differences in the stability boundary. Provided that only the frictional component dominates the dynamics of the boiling channel, the simplified stability criterion has been formulated. Here some examples are presented.

The first criterion, by Ishii and Zuber (1970), is applicable to a system with constant heat flux, inlet, and exit restrictions and without the superheated vapor region; that is, the boiling channel is composed of subcooled and two-phase regions. The criterion is expressed as a function of the phase change number N_{pch}, the subcooling number N_{sub}, the inlet and exit restrictions, and the friction number Λ_m, as follows:

$$N_{pch} - N_{sub} = x_{eq}\, \Delta\varrho_{LG}/\varrho_G$$
$$< \left(2K_{in} + 2\Lambda_m + 2K_{ex}\right)\Big/\left(1 + \Lambda_m/2 + K_{ex}\right) \tag{5.71}$$

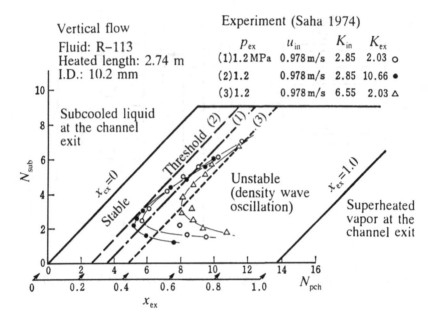

Figure 5.32 Stability map by Ishii and Zuber (1970).

where Λ_m represents the friction number defined by $\Lambda_m = \lambda_m L/(2d)$, L the length of the boiling channel, d the tube diameter, and λ_m the two-phase friction factor. For the value λ_{in} it is recommended that a value twice that of the single-phase value be chosen. Equation (5.71) suggests that the flow stability, that is, stability against density wave oscillation, is determined by the phase change number N_{pch} and the subcooling number N_{sub}, inlet and exit restrictions, and the friction number. With given values of K_{in}, K_{ex}, and Λ_m, it will be convenient to use N_{pch} and N_{sub} for the stability map. Figure 5.32 is the stability map thus obtained that includes the constant-quality line at the exit, that is, $N_{pch} - N_{sub} = $ constant. The plotted data indicate that the threshold condition expressed by the straight lines almost coincides with the experimental results. The experimental data, however, show a large discrepancy from the criterion at low values of N_{sub}; that is, the threshold boundary becomes an almost horizontal line with a constant value of N_{sub}. This corresponds well to Eq. (5.61).

On the other hand, when the system has a superheated vapor region but no inlet or exit restriction, the simplified stability criterion gives (Nakanishi et al. 1986)

$$\varrho_G N_{pch}/\Delta\varrho_{LG} < \left(0.75 + 1.4 N_{sub}\varrho_G/\Delta\varrho_{LG}\right)/\left(1 - 1/\Lambda\right) \qquad (5.72)$$

where Λ is also the friction number defined in the single-phase flow of liquid. In cases with a relatively small subcooling number, it is not suitable to predict the stability boundary with a method similar to Ishii's. The application is limited in the range $\Delta i_{sub}/\Delta i_{LG} > 0.2$ and $\varrho_L/\varrho_G > 30$.

These simplified stability criteria are derived on the basis of a stability analysis conducted using a distributed-parameter model of the boiling system. Detailed discussion of the distributed-parameter system is presented in the next section. Although the application is limited to a certain range of system configurations and/or operating parameters, the predictions using these criteria are extremely easy to make and it is also easy to understand the important parameters.

5.8 Linear stability analysis of density wave oscillation

In the previous sections, simplified modeling and stability analysis are discussed with special emphasis on the phenomenological description. In the following sections, more detailed analysis, that is, distributed-parameter analysis, will be described to demonstrate the method of analysis of density wave oscillation. Prior to the discussion, various stability analyses are reviewed, including nonlinear stability analysis.

Stability analysis is classified into two classes, linear and nonlinear. The former uses the conservation equations linearized by using a small perturbation method and then the Laplace transformation, which is followed by the stability analysis in a frequency domain. This analysis has the merit of understanding the effects of parameters on system stability in the form of functional relationships. The stability is determined whether or not a small perturbation develops with time. In an actual system, limit cycle oscillation will soon appear after the onset of the oscillation, which cannot be expressed, in principle, in this linear stability analysis.

Nonlinear stability analysis needs essentially numerical computations. One approach is to use a finite-difference method to solve a partial differential equation system of conservation relationships. The other approach is to apply lumped-parameter modeling, which leads to ordinary differential equations. The latter modeling suffers from the problem of less detailed information on the distributed-parameter effects but has a merit in the computation process. For the formulation of this lumped-parameter nonlinear analysis, it is necessary to make an assumption on the distribution of various parameters along a boiling channel, such as velocity and density, to integrate the conservation equations. Thus, this approach is, in some sense, semi-analytical stability analysis, and therefore the treatment of equations is quite similar to linear stability analysis as explained in this section. Nonlinear stability analysis is conducted in the time domain and thus enables us to obtain the oscillation trace of the limit cycle oscillation. In this book we describe first linear stability analysis and then lumped-parameter stability analysis.

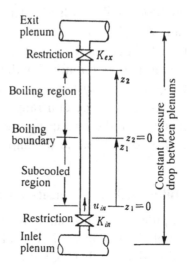

Figure 5.33 Flow model of boiling-channel system.

5.8.1 Model description

Analysis of the dynamic behavior of a two-phase flow system needs mass, momentum, and energy conservation equations of two-phase flow with various constitutive equations and the equation of state. In the boiling channel system, the energy equation reduces to the thermal energy equations owing to the fact that the mechanical energy is so small compared with the thermal energy as to be neglected. The solution of this equation system is described by the following sequence. First, the mass and energy conservation equations are solved to give the dynamic behavior of the velocity and the density. Then these variables are inserted into the momentum equation to obtain the transfer function between the pressure drop and the flow rate.

The analytical model is exemplified in Figure 5.33, which consists of a single vertical boiling channel connected between two plenums with constant pressure drop boundary condition. The flow is one-dimensional and only one velocity is defined in the cross-section of the channel. Furthermore, subcooled boiling is ignored and thus thermodynamic equilibrium is assumed. More detailed analysis emphasizes the importance of subcooled boiling, which is not essential to the present demonstration of the method of analysis. The heat capacity of the tube wall has, in fact, a certain influence on the stability boundary or the threshold condition, but is ignored in this analysis for simplicity. Moreover, in the general analysis, the thermophysical properties are assumed constant; that is, the effect of the pressure on them is ignored in the analysis. The boiling channel is then divided into two parts, the subcooled liquid region and the boiling region. Both these regions are dynamically connected by using jump conditions. In the subcooled liquid region, the specific volume and thus the density is assumed constant and equal to that of the saturated liquid. In the boiling region, two-phase flow is modeled using the drift–flux model, which includes the homogeneous flow model as a limiting case. The heat flux is assumed uniform and constant along the channel. On the basis of these assumptions, the formulation is made in the following sections.

5.8.2 Dynamic behavior in the single-phase region

In this section, the formulation of the dynamic behavior of the subcooled region is discussed. First, the conservation equations are listed, linearized, and Laplace-transformed, and then the dynamic behavior of the velocity and the enthalpy are derived.

Mass and energy conservation equations are expressed, respectively, by

$$\partial(\varrho_L u_1)/\partial z = 0 \tag{5.73}$$

$$\partial(\varrho_L e)/\partial t + \partial(\varrho_L u_1 i)/\partial z = q_v \tag{5.74}$$

where u_1 represents the velocity, e the internal energy, i the enthalpy, and q_v the volume heat flux of fluid. In such a boiling channel system, the internal energy is assumed to be equal to the enthalpy, because the mechanical work is so small compared with the internal energy. Then the energy conservation equation is rewritten as

$$\frac{\partial \varrho_L i}{\partial t} + \frac{\partial \varrho_L u_1 i}{\partial z} = q_v \tag{5.75}$$

Next, these two conservation equations (5.73) and (5.75) are linearized by introducing a small perturbation; that is, the velocity $u_1 = u_{10} + \delta u_1$ and the enthalpy $i = i_0 + \delta i$ are substituted into these equations. The higher-order terms are neglected and the conservation equations then become

$$\partial(\delta u_1)/\partial z = 0 \tag{5.76}$$

$$\partial(\delta i)/\partial t + \partial/\partial z(u_{10}\delta i + \delta u_1 i_0) = 0 \tag{5.77}$$

Then, Laplace transformation of these equations gives the following linearized equations,

$$\bar{u}_1 = \bar{u}_{in} \tag{5.78}$$

$$d\bar{i}/dz + S\bar{i}/u_{10} = -(q_v/\varrho_L u_{10})(\bar{u}_1/u_{10}) \tag{5.79}$$

where the relationship $di_0/dz = q_v/\varrho_L u_{10}$ is used, and $\bar{u}_{in}$ represents the inlet velocity perturbation. Equation (5.79) is easily solved and the perturbation of the enthalpy at the subcooled region exit is given by

$$\bar{i}_1 = -\frac{1 - \exp(-\tau_1 S)}{S} \cdot \frac{q_v}{\varrho_L u_{10}} \cdot \bar{u}_{in} \tag{5.80}$$

where τ_1 represents the residence time in the subcooled region and is given by $\tau_1 = z_{10}/u_{10}$. Thus the transfer functions of velocity and the enthalpy are expressed by Eqs. (5.78) and (5.80) against the perturbation of the inlet velocity.

5.8.3 Dynamic behavior in the two-phase region

In the boiling region, two-phase flow modeling is needed to express the dynamics. The fundamental approach is a homogeneous flow model that leads to an equation system similar to that of a single-phase flow. Another well-known modeling technique is to apply the drift–flux model, which takes into account the relative velocity between phases. These two models are referred to as the mixture model. In this section, we formulate the two-phase flow dynamics on the basis of the drift–flux model that includes the homogeneous flow model as a limiting case. Thus the formulation brings about a slight complication compared with homogeneous flow modeling but does not harm the understanding and the principles of this book.

Mass and energy conservation equations are listed as follows:

$$\frac{\partial}{\partial t}\left[\varrho_L(1 - \alpha) + \varrho_G \alpha\right] + \frac{\partial}{\partial z}\left[\varrho_L(1 - \alpha)u_L + \varrho_G \alpha u_G\right] = 0 \tag{5.81}$$

$$\frac{\partial}{\partial t}\left[\varrho_L(1 - \alpha)e_L + \varrho_G \alpha e_G\right] + \frac{\partial}{\partial z}\left[\varrho_L(1 - \alpha)u_L i_L + \varrho_G \alpha u_G i_G\right] = q_v \tag{5.82}$$

where α represents the void fraction, u_L and u_G the velocities of liquid and vapor phases, and e_L and e_G the internal energies of two phases. These internal energies are approximated by enthalpies i_L and i_G, respectively, as in the subcooled region.

Following the drift–flux model, the volumetric flux and the vapor phase velocity are expressed by

$$J = (1 - \alpha)u_L + \alpha u_G \tag{5.83}$$

$$u_G = C_0 J + V_{GJ} \tag{5.84}$$

where C_0 is referred to as the distribution parameter that is a function of the phase and velocity distributions across the channel cross-section, and V_{GJ} represents the drift velocity. Equation (5.83) is interpreted such that the vapor phase velocity is larger by the amount $(C_0 - 1)J + V_{GJ}$ than the total volumetric flux J. A detailed discussion on this modeling can be found in Zuber and Findlay (1965). In case of $C_0 = 1$ and $V_{GJ} = 0$, the flow model reduces to the homogeneous flow model, and the vapor velocity coincides with the volumetric flux. The parameters C_0 and V_{GJ} are, in general, functions of the flow regime, for example, $C_0 = 1.2$ and $V_{GJ} = 0.35\sqrt{gd\Delta\varrho_{LG}/\varrho_L}$ for slug flow.

Mass flux and thermodynamic equilibrium quality are expressed by

$$G_L = \varrho_L(1 - \alpha)u_L \tag{5.85}$$

$$G_G = \varrho_G \alpha u_G \tag{5.86}$$

$$G_t = G_L + G_G \tag{5.87}$$

$$x = G_G/G_t \tag{5.88}$$

By manipulating this equation system, the dynamics of the boiling region is formulated as follows. Equation (5.81) multiplied by i_G or i_L and Eq. (5.82) lead to the next void propagation equations

$$\frac{\partial \alpha}{\partial t} + \frac{\partial}{\partial z}(\alpha u_G) = \frac{q_0 \varrho_L}{\Delta \varrho_{LG}} \tag{5.89}$$

$$\frac{\partial}{\partial t}(1 - \alpha) + \frac{\partial}{\partial z}\{(1 - \alpha)u_L\} = -\frac{q_0 \varrho_G}{\Delta \varrho_{LG}} \tag{5.90}$$

where $q_0 = \Delta \varrho_{LG} q_v/(\varrho_L \varrho_G \Delta i_{LG})$, which has a dimension of $1/s$. The right-hand term represents a source term due to the phase change. Eliminating this source term, we have the next simplified equation of the volumetric flux distribution,

$$\partial J/\partial z = q_0 \tag{5.91}$$

which is integrated to give

$$J = J_{in} + q_0 z_2 \tag{5.92}$$

The coordinate z_2 represents the axial distance from the boiling boundary. Equation (5.92) indicates that the volumetric flux at a certain position is a linear function of J_{in} under given pressure and heat flux. This relationship makes the analysis an easy one.

(a) *Steady state*: On the basis of Eqs. (5.83), (5.84), and (5.92), the steady-state values of the volumetric flux and phase velocities are expressed by

$$J_0 = J_{in0} + q_0 z_2 \tag{5.93}$$

$$u_{G0} = C_0 J_{in} + V_{GJ} + C_0 q_0 z_2 = u_{Gin0} + C_0 q_0 z_2 \tag{5.94}$$

$$u_{L0} = \{(J_{in0} + q_0 z_2)(1 - C_0 \alpha_0) - \alpha_0 V_{GJ}\}/(1 - \alpha_0) \tag{5.95}$$

When two-phase flow is expressed by the homogeneous flow model, the velocity and the specific volume show a linear distribution, as shown in Figure 5.38. The linear distribution of the enthalpy is based on the assumption in the present analysis. When the drift–flux model is used, the volumetric flux and the gas phase velocity are also linear in the boiling region, but this is not the case for the liquid phase velocity or the specific volume.

Now, we define the residence time of the vapor phase from the boiling boundary to a certain position z_2 by

$$\tau_2 = \int_0^{z_2} \frac{dz}{u_{G0}} \tag{5.96}$$

which corresponds to the residence time of the void wave. Substitution of Eq. (5.94) into Eq. (5.96), followed by integration, gives the residence time as

$$\tau_2 = \frac{1}{C_0 q_0} \ln \frac{u_{G0}}{u_{Gin0}} \quad \text{or} \quad u_{G0} = u_{Gin0} \exp(C_0 q_0 \tau_2) \tag{5.97}$$

Returning to the void propagation equation (5.89), under the steady-state condition, it results in

$$\frac{d\alpha_0 u_{G0}}{dz} = \frac{\varrho_L}{\Delta\varrho_{LG}} q_0 \tag{5.98}$$

Equations (5.96) and (5.97) are substituted into Eq. (5.98), and integration is carried out to give the void fraction distribution expressed by

$$\alpha_0 = \frac{\varrho_L}{C_0 \Delta\varrho_{LG}} \left\{ 1 - \exp(-C_0 q_0 \tau_2) \right\} \tag{5.99}$$

This means that the void fraction distribution is an exponential function of the residence time under the steady-state condition. By using these steady-state values of the void fraction and phase velocities, the mass flux and quality distributions are given by

$$G_{L0} = \varrho_L (1 - \alpha_0) u_{L0}, \quad G_{G0} = \varrho_G \alpha_0 u_{G0},$$
$$x_0 = \varrho_G \alpha u_{G0} / (G_{L0} + G_{G0}) \tag{5.100}$$

(b) *Dynamic behavior:* At the next step, the dynamic behavior of these variables are formulated as follows. As in the case of the subcooled region, small perturbations of the volumetric fluxes J and J_{in}, phase velocities u_L and u_G, and void fraction α from the steady-state values are introduced and are substituted into Eqs. (5.83), (5.84), and (5.92). Then we have the following linear equations of the volumetric flux and phase velocities after the Laplace transformation:

$$\bar{J} = \bar{J}_{in}, \quad \bar{u}_G = C_0 \bar{J},$$
$$\bar{u}_L = \left\{ \bar{J}(1 - C_0 \alpha_0) + \bar{\alpha}(u_{L0} - u_{G0}) \right\} / (1 - \alpha_0) \tag{5.101}$$

The perturbation equation of the void fraction is obtained by applying the small perturbation method and the Laplace transformation to Eq. (5.89),

$$\bar{\alpha} S + \frac{d}{dz}(u_{G0}\bar{\alpha}) = -C_0 \bar{J}_{in} \frac{d\alpha_0}{dz} \tag{5.102}$$

which is further rewritten by using the relationship between the axial coordinate z_2 and the residence time τ_2,

$$\left(u_{G0}\bar{\alpha}\right)S + \frac{d}{d\tau_2}\left(u_{G0}\bar{\alpha}\right) = -C_0\bar{J}_{in}\frac{d\alpha_0}{d\tau_2} \tag{5.103}$$

This is easily integrated to give the dynamics of the void fraction

$$u_{G0}\bar{\alpha} = u_{Gin0}\exp\left(-S\tau_2\right)\bar{\alpha}_{in} - \frac{\varrho_L C_0 q_0}{\Delta\varrho_{LG}}\frac{\exp\left(-C_0 q_0 \tau_2\right) - \exp\left(-S\tau_2\right)}{S - C_0 q_0}\bar{J}_{in} \tag{5.104}$$

Similar manipulation is adopted to give the dynamics of the mass flux, Eq. (5.105), and quality, Eq. (5.106):

$$\bar{G}_t = \frac{S\exp\left(-C_0 q_0 \tau_2\right) - C_0 q_0 \exp\left(-S\tau_2\right)}{S - C_0 q_0}\varrho_L \bar{J}_{in} - \exp\left(-S\tau_2\right)u_{Gin0}\Delta\varrho_{LG}\bar{\alpha}_{in} \tag{5.105}$$

$$\begin{aligned}
\bar{x} = \frac{\varrho_G}{\left(G_{t0}\right)}\Bigg[& G_{t0}\frac{\varrho_L}{\Delta\varrho_{LG}}\left(1 - \exp\left[-C_0 q_0 \tau_2\right]\right)\bar{J}_{in} \\
& + G_{t0}\left(u_{Gin0}\exp\left[-S\tau_2\right]\bar{\alpha}_{in} - \frac{\varrho_L C_0 q_0}{\Delta\varrho_{LG}}\frac{\exp\left[-C_0 q_0 \tau_2\right] - \exp\left[-S\tau_2\right]}{S - C_0 q_0}\bar{J}_{in}\right) \\
& - \frac{\varrho_L u_{Gin0}}{C_0 \Delta\varrho_{LG}}\left\{\frac{\varrho_L S}{S - C_0 q_0}\left(1 - \exp\left[-C_0 q_0 \tau_2\right]\right)\bar{J}_{in}\right. \\
& \left.- \left(\Delta\varrho_{LG}u_{Gin0}\bar{\alpha}_{in} + \frac{\varrho_L C_0 q_0}{S - C_0 q_0}\bar{J}_{in}\right)\left(\exp\left[\left(C_0 q_0 - S\right)\tau_2\right] - \exp\left[-S\tau_2\right]\right)\right\}\Bigg]
\end{aligned}$$

$$\tag{5.106}$$

Thus, the important variables for the boiling region, that is, the volumetric flux, the phase velocity, the mass flux, the void fraction, and the quality, have been expressed by the functions of the perturbation terms of the volumetric flux J_{in} and the void fraction $\bar{\alpha}_{in}$ at the boiling boundary. In other words, transfer functions have been given between the variables mentioned above and the input variables J_{in} and $\bar{\alpha}_{in}$.

5.8.4 Boiling-boundary dynamics

The next important formulation is how to get the jump conditions to relate to two reference regions at the boiling boundary. The reader will encounter the same problem in the lumped-parameter nonlinear analysis discussed later. The resulting jump condition will be the same one, and therefore in this section we explain phenomenological modeling. A more elegant solution can be obtained by using Leibnitz's rule.

Let us suppose a transient condition of the velocity fluctuation at the boiling channel inlet. In this flow fluctuation, the subcooled length and therefore the boiling length change with time. As discussed in the previous section, the transfer functions of the boiling regions are expressed in terms of the input variables J_{in} and $\bar{\alpha}_{in}$. Thus the jump conditions need the relationship between the perturbation of the subcooled and thus boiling lengths and the inlet velocity fluctuation $\bar{u}_{in}$, and the relationship of J_{in} and $\bar{\alpha}_{in}$ to $\bar{u}_{in}$. The fundamental principle to identify the boiling boundary is that the subcooled liquid enthalpy reaches the saturation enthalpy. The energy equation (5.74) in the subcooled region is rewritten, using Eq. (5.73), as

$$\frac{\partial i}{\partial t} + u_1 \frac{\partial i}{\partial z} = \frac{Di}{Dt} = \frac{q_v}{\varrho_L} \tag{5.107}$$

The interpretation of this relationship is that the fluid parcel moving with velocity u_1 has an enthalpy increase of q_v/ϱ_L. Now, the steady-state position of the boiling boundary is expressed by z_{10}. Suppose that this boiling boundary moves from z_{10} to $z_{10} + \delta z_1$. This small movement δz_1 corresponds to the short time period $\delta z_1/u_1$ of the parcel movement. Then the enthalpy before the period $\delta z_1/u_1$ is given by $i_L - (q_v/\varrho_L)(\delta z_1/u_1)$. This means that the enthalpy perturbation at the position z_{10} is expressed by

$$\delta i_1 = i_1 - i_L = -\left(q_v/\varrho_L\right)\left(\delta z_1/u_1\right) \tag{5.108}$$

Thus the boiling-boundary movement is expressed as a function of the enthalpy perturbation at the subcooled region exit:

$$\bar{z}_1 = -\left(\varrho_L u_{10}/q_v\right)\bar{i}_1 \tag{5.109}$$

This equation indicates that the subcooled length increases against the enthalpy decrease and also against the increase in the inlet velocity as expressed in Eq. (5.80).

The velocity relationship at the boiling boundary is derived as follows: Providing short intervals of $\pm\delta z_1$ around z_{10}, the continuity relationship is expressed by

$$u_1\big|_{-\delta z_1} \cong J\big|_{+\delta z_1} \cong J_{in} + q_0\delta z_1 \tag{5.110}$$

This relationship is linearized and Laplace-transformed to give the next relationship

$$\bar{J}_{in} = \bar{u}_1 - q_0\bar{z}_1 \tag{5.111}$$

This is well demonstrated in Figure 5.34a. Similarly to the above discussion, the void fraction perturbation is expressed by

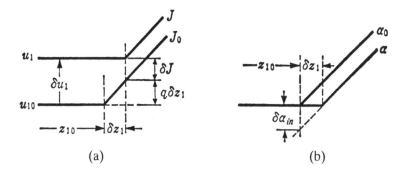

Figure 5.34 Jump condition at boiling boundary. (a) Velocity; (b) void fraction.

$$\bar{\alpha}_{in} = -\bar{z}_1 (d\alpha_0/dz)_{z_{10}} = -\frac{\varrho_L q_0}{\Delta\varrho_{LG} u_{Gin0}} \bar{z}_1 \qquad (5.112)$$

which is also shown in Figure 5.34b.

Now we have the jump conditions of the boiling length, velocity, and void fraction at the boiling boundary. Next, we will solve the momentum equations using various equations derived above.

5.8.5 Dynamic behavior of pressure drop and stability analysis

The momentum equation represents the pressure balance throughout the boiling channel system. The pressure drop in the boiling channel is then expressed by the sum of the gravitational, accelerational, and frictional terms including the flow restriction, for example, for the subcooled region:

Gravitational:
$$\left(-\frac{\partial p}{\partial z}\right)_{grl} = \varrho_L g \qquad (5.113)$$

Accelerational:
$$\left(-\frac{\partial p}{\partial z}\right)_{al} = \varrho_L u_1 \partial u_1/\partial z + \varrho_L \partial u_1/\partial t \qquad (5.114)$$

Frictional:
$$\left(-\frac{\partial p}{\partial z}\right)_{fl} = (\lambda/2d)\varrho_L u_1^2 \qquad (5.115)$$

Inlet restriction: $\Delta p_{Kin} = K_{in}\varrho_L u_1^2 \qquad (5.116)$

These pressure drop equations are linearized by taking into account small perturbations of the pressure drop, the velocity, and the subcooled length; we then have

$$\delta(\Delta p_{grl}) = \int_0^{z_{10}+\delta z_1} \varrho_L g\, dz - \int_0^{z_{20}} \varrho_L g\, dz = \varrho_L g\delta z_1$$

for the gravitational term, which is Laplace-transformed to give

$$\overline{\Delta p}_{gr1} = \varrho_L g \overline{z}_1 \tag{5.117}$$

A similar manipulation is applied to the other equations (5.114)–(5.116):

$$\overline{\Delta p}_{a1} = S\tau_1 \varrho_L u_{10} \overline{u}_1 \tag{5.118}$$

$$\overline{\Delta p}_{f1} = (\lambda/d)\varrho_L u_{10} z_{10} \overline{u}_1 + (\lambda/2d)\varrho_L (u_{10})^2 \overline{z}_1 \tag{5.119}$$

$$\overline{\Delta p}_{Kin} = 2K_{in}\varrho_L u_{10} \overline{u}_1 \tag{5.120}$$

In the boiling region, each pressure drop term is expressed by

Gravitational: $$\left(-\frac{\partial p}{\partial z}\right)_{gr2} = g\left[\varrho_L(1 - a) + \varrho_G a\right] \tag{5.121}$$

Accelerational: $$\left(-\frac{\partial p}{\partial z}\right)_{a2} = \partial/\partial z\left[\varrho_L(1 - a)u_L^2 + \varrho_G a u_G^2\right]$$
$$+\partial/\partial t\left[\varrho_L(1 - a)u_L + \varrho_G a u_G\right] \tag{5.122}$$

Frictional: $$\left(-\frac{\partial p}{\partial z}\right)_{f2} = (\lambda/2d)(\phi^2)G_t^2/\varrho_L \tag{5.123}$$

Exit restriction: $$\Delta p_{Kex} = K_{ex}(\phi_{ex}^2)G_t^2/\varrho_L \tag{5.124}$$

where ϕ^2 represents the friction multiplier given by, for example, Martinelli and Nelson's (1948) or Thom's (1964) correlations, and is a function of the flow quality. For convenience of integration, the next type of approximation is useful for this friction multiplier:

$$\phi^2 = 1 + ax + bx^2 \tag{5.125}$$

where coefficients a and b are functions of the pressure and thermophysical properties.

Equations (5.121)–(5.123) are integrated along the coordinate z. Taking into account the boiling-boundary movement δz_1, the gravitational term (5.121) is integrated from δz_1 to z_2:

$$\delta(\Delta p_{gr2}) = \int_{\delta z_1}^{z_2} g\left[\varrho_L - \Delta\varrho_{LG}(a_0 + \delta a)\right]dz - \int_0^{z_2} g\left[\varrho_L - \Delta\varrho_{LG}a_0\right]dz$$
$$\cong \int_0^{z_2} g\left[\varrho_L - \Delta\varrho_{LG}a_0 - \Delta\varrho_{LG}\delta a\right]dz - \int_0^{z_2} g\left[\varrho_L - \Delta\varrho_{LG}a_0\right]dz$$
$$- \int_0^{\delta z_1} g\left[\varrho_L - \Delta\varrho_{LG}a_0\right]dz \tag{5.126}$$

where the void fraction is approximately zero in the region from $z_2 = 0$ to δz_1 of the small perturbation, and thus the above equation becomes

$$\delta\left(\Delta p_{gr2}\right) = -g\Delta\varrho_{LG}\int_0^{z_2} \delta\alpha\,dz - g\varrho_L\,\delta z_1 \tag{5.127}$$

The equation is rewritten as an integration with respect to τ_2, and is Laplace-transformed; then

$$\overline{\Delta p}_{gr2} = -g\Delta\varrho_{LG}\int_0^{\tau_2} \overline{\alpha}u_{G0}\,d\tau_2 - g\varrho_L\overline{z}_1 \tag{5.128}$$

Substitution of the dynamic responses, Eqs. (5.104) and (5.109), of the void fraction and the subcooled length into Eq. (5.128) gives the dynamic response of the gravitational pressure drop against the inlet velocity perturbation $\overline{u}_{in}$.

The accelerational pressure drop is expressed as a function of the mass flux:

$$\delta\left(\Delta p_{a2}\right) = \int_{\delta z_1}^{z_2} \frac{\partial}{\partial z}\big[G_{L0}u_{L0} + G_{L0}\delta u_L + \delta G_L u_{L0} + G_{G0}u_{G0} + G_{G0}\delta u_G$$

$$+ \delta G_G u_{G0}\big]dz - \int_0^{z_2} \frac{\partial}{\partial z}\left(G_{L0}u_{L0} + G_{G0}u_{G0}\right)dz + \int_0^{z_2}\left(\partial(\delta G_t)/\partial t\right)dz \tag{5.129}$$

This equation is Laplace-transformed and integrated; we then have

$$\overline{\Delta p}_{a2} = \big[G_{L0}\overline{u}_L + \overline{G}_L u_{L0} + G_{G0}\overline{u}_G + \overline{G}_G u_{G0}\big]_0^{z_2}$$

$$- \overline{z}_1\left[\frac{d}{dz}\left(G_{L0}u_{L0} + G_{G0}u_{G0}\right)\right]_{z=0} + S\int_0^{\tau_2}\overline{G}_t u_{G0}\,d\tau_2 \tag{5.130}$$

Substitution of the dynamic responses, $\overline{u}_L$, $\overline{u}_G$, $\overline{G}_L$, $\overline{G}_G$, and $\overline{G}_t$, against $\overline{u}_{in}$ gives the dynamic behavior of the accelerational term. In the same manner, the frictional and restriction terms are expressed by

$$\overline{\Delta p}_{f2} = -\left(\lambda/2d\right)\left(\phi_0^2\right)\left(G_{t0}\right)^2/\varrho_L\cdot\overline{z}_1$$

$$+ \left(\frac{\lambda}{2d}\frac{1}{\varrho_L}\right)\int_0^{\tau_2}\left\{\left(G_{t0}\right)^2\left(\overline{\phi^2}\right) + 2G_{t0}\left(\phi_0^2\right)\overline{G}_t\right\}u_{G0}d\tau_2 \tag{5.131}$$

$$\overline{\Delta p}_{K_{ex}} = K_{ex}\left(G_{t0}\right)^2/\varrho_L\cdot\left(\overline{\phi_{ex}^2}\right) + 2K_{ex}\left(\phi_{ex0}^2\right)G_{t0}\overline{G}_t/\varrho_L \tag{5.132}$$

where $\overline{\phi^2} = (a + 2bx_0)\overline{x}$. Then, finally, we have the dynamic behavior of the total pressure drop in the form

$$\overline{\Delta p}_t = \left(R_{gr1} + R_{a1} + R_{f1} + R_{K_{in}}\right)\overline{u}_{in} + \left(R_{gr2} + R_{a2} + R_{f2} + R_{K_{ex}}\right)\overline{u}_{in}$$

$$= R_1\overline{u}_{in} + R_2\overline{u}_{in} \tag{5.133}$$

where the function R represents the transfer function of each pressure drop term against the inlet velocity perturbation $\overline{u}_{in}$.

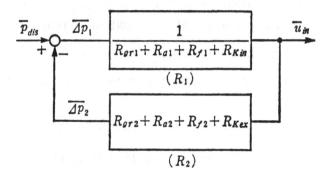

Figure 5.35 Block diagram of boiling-channel system.

On the basis of the above discussion, the block diagram of the feedback system is constructed under the condition of constant pressure drop between the headers, that is, Figure 5.30a, as shown in Figure 5.35. In constructing such a block diagram, the transfer function in the subcooled region is, in general, set as the feed-forward element, and the transfer function in the boiling region is set as the feedback element. Then the transfer function of the whole system, that is, the transfer function of the inlet velocity perturbation against a small perturbation of the pressure disturbance $\bar{p}_{\mathrm{dis}}$, is expressed by

$$W(S) = \frac{1/R_1}{1 + R_2/R_1} = \frac{1}{R_1 + R_2} \qquad (5.134)$$

The characteristic equation then becomes $R_1 + R_2 = 0$. The system stability is examined by drawing the trajectory of the open loop transfer function $R = R_2/R_1$ on the complex plane, which is known as the Nyquist method. That is, the trajectory of the transfer function is drawn against $S = j\omega (j = \sqrt{-1})$ in the range from $\omega = 0$ to ∞, where ω represents angular frequency. When the trajectory passes through the left side of the point $-1 + j0$, then the system is judged to be unstable. Contrariwise, when the trajectory passes through the right side of $-1 + j0$, then the system is stable. Figure 5.36 is an example of the unstable state. By applying the Nyquist method, the system stability is determined and is demonstrated in Figure 5.37. The experimentally determined threshold condition is shown by the shaded region, and the predicted threshold conditions are represented by the dashed lines for the parameters of the drift–flux model. It is noted here that the predicted results in Figure 5.37 have been obtained by taking into account the heat capacity effect of the tube wall, which is not discussed in this chapter. The curve for $C_0 = 1$, $V_{\mathrm{GJ}} = 0$ corresponds to the case of the homogeneous flow model, which under-predicts the threshold heat flux. In other words, the homogeneous

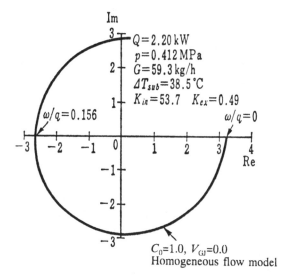

Figure 5.36 Example of Nyquist diagram.

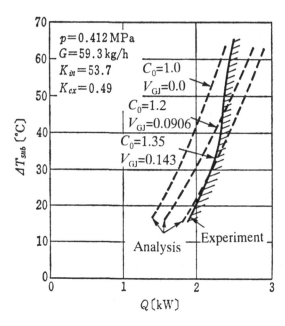

Figure 5.37 Comparison between analytical and experimental results of threshold condition.

flow model can predict the threshold condition with a safety margin. It should be noted that this is not generally true but depends on each system.

The principal feature of the stability analysis was explained on the basis of rather old-fashioned linear stability analysis. We now have very fast and convenient computers and/or work stations that make it possible to calculate directly the conservation equation to obtain the dynamic behavior by using methods such as the finite-difference one. The reader may prefer such computation. The goal of the present chapter is, however, to promote understanding of the functional relationships between each parameter, which is the best approach to the similarity laws shown in Figure 5.32.

5.8.6 Scaling law in flow stability problems

The pressure drop components change in magnitude depending on the system configuration and the operating conditions. When the principal term dominating the dynamics of the system is the frictional component, the stability diagram is drawn in a rather simplified form as seen in Figure 5.32. This is, however, not the general case. In the case when not only frictional but also gravitational terms are comparable in describing the dynamics of the system, we must add an additional scaling parameter to the scale law composed of the phase change number, the subcooling number, and the friction number. This makes it rather difficult to draw such simplified stability diagrams as in Figure 5.32. The effects of various parameters on flow stability are briefly reviewed in Table 5.3 by Bouré and Mihaila (1967).

Table 5.3 shows the influence of the gravitational, inertial, and frictional terms in various parts of the flow channel on the density wave oscillation. The pressure drop in the subcooled region plays the role of damping factor, while that in the boiling region acts as a driving factor. It is suggested that density wave oscillation will occur even in the case when only the gravitational term is a dominant pressure drop in the channel. We recommend that the reader examine such an effect, shown in Table 5.3, by him- or herself.

5.9 Nonlinear analysis of flow instability

In Section 5.8, stability analysis was conducted on the basis of linearized conservation equations. Keeping the principal feature of the treatment, the flow dynamics can be solved numerically by using the nonlinear momentum equations in the time–space domain. One simple way is to apply lumped-parameter modeling. This modeling assumes a linear distribution of velocity, specific volume, and enthalpy in the designated subsections. The conservation equations in the time–space domain can be integrated easily in the corresponding region. The jump condition and/or the continuity relationship between subsequent subsections are obtained by applying the well-known Leibnitz rule. Here this jump condition is briefly explained and a method of analysis is exemplified. Detailed discussion can be found elsewhere (Lahey et al. 1989; Ozawa et al. 1993).

Table 5.3. *Driving and damping terms for density wave oscillations in upward flow (Bouré and Mihaila 1967)*

Terms	External loop upstream	Heated channel		Downstream	Whole channel
		Subcooled region	Boiling region		
Gravity	No influence	Damping factor	Driving factors	Driving factors	Driving or damping
Inertia	Influence only on the frequency		Generally driving factors	Driving factors	Driving
Friction	Always damping factors		Driving or damping factors		Damping or driving
Total	Always damping	Generally damping	Generally driving factors, possibly damping		

Energy equation (5.74) is integrated in the range $z = 0$ to $z = z_1$:

$$\int_0^{z_1} \frac{\partial}{\partial t}\left(\varrho_L i\right) dz + \int_0^{z_1} \frac{\partial}{\partial z}\left(\varrho_L i u_1\right) dz = \int_0^{z_1} q_v \, dz \qquad (5.135)$$

Next, Leibnitz's rule is applied to Eq. (5.135). Leibnitz's rule is expressed as

$$\frac{d}{dt}\int_V \left(\varrho_L i\right) dV = \int_V \left(\frac{\partial \varrho_L i}{\partial t}\right) dV + \int_a \varrho_L i\left(\boldsymbol{u}_a \cdot \boldsymbol{n}\right) da \qquad (5.136)$$

where $\int_V dV$ represents a volume integration and $\int_a da$ a surface integration. The vector $\boldsymbol{u}_a$ represents the moving velocity of the surface and $\boldsymbol{n}$ the unit normal vector to the surface. The scalar product $\boldsymbol{u}_a \cdot \boldsymbol{n}$ is then the moving velocity of boiling boundary dz_1/dt. On the basis of this discussion, Eq. (5.135) is rewritten as

$$\frac{d}{dt}\int_0^{z_1}\left(\varrho_L i\right) dz - \varrho_L i_L \frac{dz_1}{dt} + \varrho_L u_1\left(i_L - i_{in}\right) = q_v z_1 \qquad (5.137)$$

Applying the enthalpy distribution shown in Figure 5.38 to the first term in Eq. (5.137), the movement of the boiling boundary is given by

$$\frac{dz_1}{dt} = 2u_{in} - \frac{2q_v z_1}{\varrho_L\left(i_L - i_{in}\right)} \qquad (5.138)$$

which gives a relationship similar to Eq. (5.109), by linearization and Laplace transformation.

In this sense, the lumped-parameter model is quite similar to the above-mentioned linear stability analysis but has the possibility of providing a more

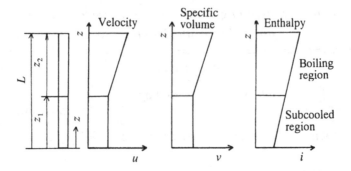

Figure 5.38 Distribution of velocity, specific volume, and enthalpy.

accurate solution including nonlinear effects. The most impressive feature is obtaining the limit cycle oscillation of the flow during the unstable condition. This can be quite interesting for the researcher. The only condition for applying this lumped-parameter modeling is analytical integration along the subsections. This condition, however, makes it difficult to apply the drift–flux model, while the slip–flow model has this possibility.

One of the examples is briefly presented in the following. The boiling channel is subdivided into the subcooled and boiling regions. Then, in each subsection, the parameters are assumed to be distributed linearly as shown in Figure 5.38, which makes the dynamic response, such as the velocity and the enthalpy, against the inlet velocity fluctuation equal to those previously derived in the linearized theory. These equations of dynamic behavior are then substituted into the momentum equation to give a nonlinear equation to obtain the dynamic behavior of the pressure and the flow. This equation system in the time–space domain is then solved by using, for example, the Runge–Kutta method. The result is shown in Figure 5.39, which represents the transient behavior of the inlet velocity and the pressure drop. Based on such numerical simulation, the threshold condition can be determined as shown in Figure 5.40. The stability boundary almost coincides with the experimental one in the case presented. Numerical simulation, including such lumped-parameter modeling, is often available not only in the field of flow instability but in the extensive field of transient behavior. One example has already been demonstrated in predicting the CHF under forced flow oscillation, as shown in Figure 5.6. Recent advances in this field are focused on nonlinear dynamics and chaos (Lahey et al. 1989; Lahey 1992).

Readers, especially students, may have their main interest not in linear stability analysis but in this lumped-parameter nonlinear analysis. We do emphasize

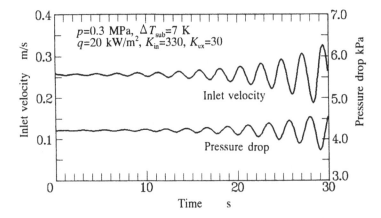

Figure 5.39 Oscillation trace (numerical simulation).

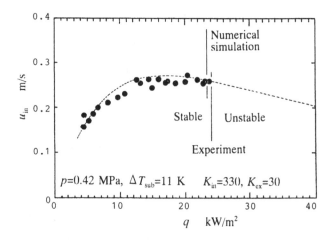

Figure 5.40 Natural-circulation characteristics (liquid nitrogen).

here that analytical understanding is essential, which leads to further understanding of the comprehensive phenomena observed in an actual plant.

5.10 Computer code for flow stability analysis

There exist many analyses published in the last three decades. The most impressive works in this field were found in the Symposium on Two-Phase Flow Dynamics, Eindhoven (1967), which was a significant landmark for research activity. Most of the fundamental aspects can be found in the Proceedings. Other reviews are found in Lahey and Drew (1980) and Bergles and Ishigai (1981). In Table 5.4 some representative computer codes, including even rather old ones,

Table 5.4. *Codes for stability analysis*

Code or authors	Heat flux[a]	Heater dynamics[a]	Subcooled boiling	Two-phase model	Superheated region	Nuclear feedback
Linearized and frequency domain						
LOOP code (Davis and Potter 1967)	Uniform	Not included	Not considered	Homogeneous model	Included	Not considered
NUFREQ code (Lahey and Yadigaroglu 1973)	Arbitrary in S.R., uniform in B.R.	Considered in S.R. but not in B.R.	Not considered	Homogeneous model	Not included	Included
Saha 1974	Uniform	Constant heat input	Considered	Drift–flux model	Not included	Not considered
Nakanishi et al. 1978	Uniform	Considered in S.R. but not in B.R.	Not considered	Drift–flux model, homogeneous if superheated region exists	Included	Not considered
STABLE (Jones 1961)	Arbitrary	Considered	Considered	Slip–flow model	Not included	Not considered
DYNAM code (Efferding 1968)	Arbitrary	Considered	Considered	Slip–flow model	Included	Not considered
Nonlinear and time domain						
HYDNA code (Currin et al. 1961)	Arbitrary	Considered	Considered	Slip–flow model	Included	Not considered
DEW (Takitani and Sakano 1979)	Uniform or heat exchanger mode	Considered	Not considered	Slip–flow model	Included	Not considered

[a] S.R.: single-phase liquid region; B.R.: boiling two-phase region.

are listed for the reader's convenience. Most of the codes are linearized analysis as discussed in this book. The principal feature in these codes is the same as in the present analysis. Thus the reader should have no serious difficulty in understanding them.

5.11 Numerical simulation on thermal hydraulics of two-phase flow

Recent advances in safety research enable us to predict the dynamic behavior of steam-generating plants, especially nuclear reactor plants. This is mainly because of extensive experience in plant construction and operation. Moreover, many incidents, including serious accidents such as those at Three Mile Island (TMI) (1979), Chernobyl (1986), and Mihama (1991), have accelerated this research field. In nuclear power plants, the volumetric heat flux is extremely high compared with that of the conventional oil/gas/coal firing plant. Thus the loss of cooling is the most serious accident estimated at the design stage of the plant (Collier and Hewitt 1987).

The principle in reactor safety is "defense in depth." The first defense is to obtain high quality in design, reliability of equipment and components, and operation. The second defense is to provide protective instrumentation for a variety of faults and to control reactors to bring them to a safe condition. The third defense is to examine a whole range of extreme accidents and to provide additional safeguards. Based on such principles, safety analysis is conducted into two types of incidents, that is, upset conditions and accidents. It is needless to say that experimental verification or examination is of prime importance in such safety research, with or without nuclear reaction. But numerical simulation plays quite an important role in safety analysis. In such analysis, one of the most important scenarios is the loss of coolant accident (LOCA). Before the TMI accident, attention was focused mainly on the large-break LOCA. But the TMI accident emphasized the importance of the small-break LOCA.

The numerical simulation codes in this respect have been developed mainly by the U.S. Nuclear Regulatory Commission, and representative codes are RELAP4 (EG&G 1978a,b), RELAP5 (Ransom et al. 1981, 1985), and TRAC (Taylor 1984; Los Alamos 1986). These codes are applicable to a variety of systems, and include the models of plant control systems, which enable them to apply not only to accidents but also to simulations of dynamic response of power plants. The TRAC code includes one-dimensional neutron dynamics, and is therefore applicable to reactivity accidents as well as other incidents.

At the initial stage of development of such computer codes, the homogeneous flow model with thermal equilibrium is used to describe two-phase flow, and is followed by the slip–flow and drift–flux models. The most recent advanced codes are constructed on the basis of the two-fluid model with thermal nonequilibrium. This model describes two-phase flow by six equations of mass, momentum, and energy conservation. To combine the conservation equations for each phase, it is necessary to have many constitutive equations such as interfacial momentum transfer, interfacial heat transfer, wall friction, wall heat transfer, and flow regime transitions. The interrelationship between constitutive equations and conservation equations is demonstrated in Figure 5.41 (Mishima 1993). By applying such a simulation code to actual events, the time-dependent behavior of various system parameters is obtained, as shown in Figure 5.42, which illustrates the case of the Mihama incident in Japan (Agency of Natural Resources and Energy 1991).

As mentioned above, recent advances in computer systems enable us to simulate such complicated incidents. It should be noted, however, that full understanding is not obtained by such computer simulation but is supported essentially by fundamental understanding on the basis of the two-phase flow theory and experimental and/or operational experience. The lack of such fundamental understanding leads to missing the principles of design as well as operation during tran-

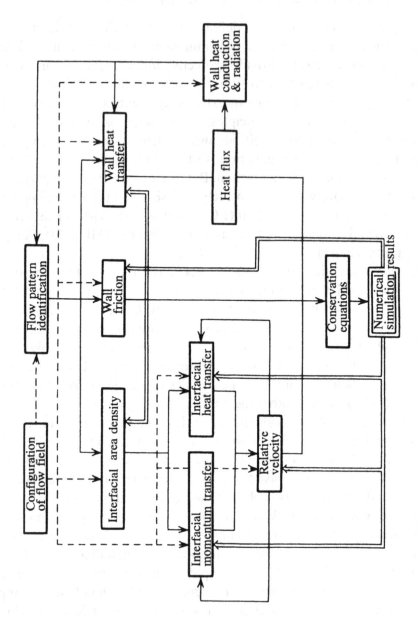

Figure 5.41 Simulation procedure of two-fluid model and constitutive relationship (Mishima 1993).

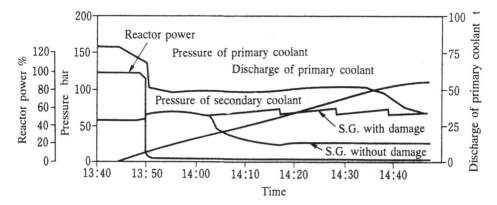

Figure 5.42 Simulation results of Mihama incident (Agency of Natural Resources and Energy 1991).

sient and/or accidental events, which may be followed by damage to power plants or severe accidents.

References

Agency of Natural Resources and Energy. 1991. Report on Mihama-2 Reactor Incident.

Akagawa, K., T. Sakaguchi, M. Kono, et al. 1971: Study on distribution of flow rates and flow stabilities in parallel long evaporators. *Bulletin of the Japan Society of Mechanical Engineers* 14: 837–44.

Bergles, A.E., and S. Ishigai, eds. 1981. *Two-phase flow dynamics*. New York: Hemisphere.

Bouré, J.A., A.E. Bergles, and L.S. Tong. 1971. Review of two-phase flow instability. American Society of Mechanical Engineers paper, 71-HT-42.

Bouré, J.A., and A. Mihaila. 1967. The oscillatory behavior of heated channels. Proceedings of the Symposium on Two-Phase Flow Dynamics, Eindhoven, vol. 1, session 6, pp. 695–720.

Collier, J.G., and G.F. Hewitt. 1987. *Introduction to nuclear power*. New York: Hemisphere. (Japanese edition translated by S. Nakanishi, M. Ozawa, and N. Takenaka includes a detailed description of Mihama accident in Japan.)

Currin, H.B., C.M. Hunin, L. Rivlin, et al. 1961. HYDNA – Digital computer program for hydrodynamic transients in a pressure tube reactor or a closed channel core. CVNA-77, Westinghouse Electric Co.

Davis, A.L., and R. Potter. 1967. Hydraulic stability: An analysis of the causes of unstable flow in parallel channels. Proceedings of the Symposium on Two-Phase Flow Dynamics, Eindhoven, vol. 2, session 9, pp. 1225–66.

Efferding, L.E. 1968. DYNAM, a digital computer program for study of the dynamic stability of once-through boiling flow with steam superheat. GAMD-8656, Gulf General Atomic.

EG&G Idaho, Inc. 1978a. RELAP4/MOD6, a computer code program for transient thermal–hydraulic analysis of nuclear reactors and related system users manual. CDAP-TR-003.

EG&G Idaho, Inc. 1978b. RELAP4/MOD7 Version 2, Users Manual. CDAP-TR-78-036.

Ishii, M., and N. Zuber. 1970. Thermally induced flow instabilities in two-phase mixtures. Proceedings of the 4th International Heat Transfer Conference, Paris, paper no. B5.11.

Jones, A.B. 1961. Hydrodynamic stability of a boiling channel. KAPL-2170, KAPL-2208 (1961), KAPL-2290 (1963), KAPL-3070 (1964).

Lahey, R.T., Jr., and D.A. Drew. 1980. An assessment of the literature related to LWR instability modes. NUREG/CR-1414.

Lahey, RT., Jr., and G. Yadigaroglu. 1973. NUFREQ, a computer program to investigate thermo–hydrodynamic stability. NEDO-13344, G.E.

Lahey, R.T., Jr., A. Clause, and P. DiMarco. 1989. Chaos and non-linear dynamics of density-wave instabilities in a boiling channel. *American Institute of Chemical Engineering Symposium Series* 85, no. 269: 256–61.

Los Alamos National Laboratory, Safety Code Development Group. 1986. TRAC-PF1/MOD1, an advanced best-estimate computer program for pressurized water reactor thermal-hydraulic analysis. NUREG/CR-3858, LA-10157-MS.

Magnus, K. 1976. *Schwingungen.* B.G. Teubner, Stuttgart.

Martinelli, R.C., and D.B. Nelson. 1948. Prediction of pressure drop during forced-circulation boiling of water. *Transactions of the American Society of Mechanical Engineers* 70: 695–702.

Mishima, K. 1993. Constitution equations. In *Numerical simulation of two-phase flow* (ed. Atomic Energy Society of Japan), 76–82. Tokyo: Asakura.

Mishima, K., H. Nishihara, and I. Michiyoshi. 1985. Boiling burnout and flow instabilities for water flowing in a round tube under atmospheric pressure. *International Journal of Heat Mass Transfer* 28: 1115–29.

Nakanishi, S., M. Ozawa, S. Ishigai, et al. 1978. Analytical investigation of density wave oscillation. Technology Report, Osaka University, vol. 28, no. 1421, pp. 243–51.

Nakanishi, S., M. Kaji, and S. Yamauchi. 1986. An approximation method for construction of stability map of density-wave oscillation. *Nuclear Engineering and Design* 95: 55–64.

Ozawa, M., S. Nakanishi, S. Ishigai, et al. 1979a. Flow instabilities in boiling channels, Part 1: Pressure drop oscillation. *Bulletin of the Japan Society of Mechanical Engineers* 22, no. 170: 1113–8.

Ozawa, M., S. Nakanishi, S. Ishigai, et al. 1979b. Flow instabilities in boiling channels, Part 2: Geysering. *Bulletin of the Japan Society of Mechanical Engineers* 22, no. 170: 1119–26.

Ozawa, M., Y. Asao, and N. Takenaka. 1993. Density wave oscillation in a natural circulation loop of liquid nitrogen. In *Instabilities in multiphase flows* (ed. G. Gouesbet and A. Berlemont), 113–24. New York: Plenum Press.

Ransom, V.H., et al. 1981. RELAP5/MOD1 code manual, Vol. 1: System models and numerical methods. NUREG/CR-1826.

Ransom, V.H., et al. 1985. RELAP5/MOD2 code manual, Vol. 1: Code structure, system models, and solution methods. NUREG/CR-4312, EGG-2796.

Saha, P. 1974. Thermally induced two-phase flow instabilities, including the effect of thermal non-equilibrium between the phases. Ph.D. diss., Georgia Institute of Technology, Atlanta.

Takitani, K., and K. Sakano. 1979. Density wave instability in once-through boiling flow system (III) – Distributed parameter model. *Journal of Nuclear Science and Technology* 16: 16–29.

Thom, J.R.S. 1964. Prediction of pressure drop during forced circulation boiling of water. *International Journal of Heat Mass Transfer* 7: 709–24.

Taylor, D. 1984. TRAC-BD1/MOD1, an advanced best estimate computer program for boiling water reactor transient analysis, EG&G Idaho, Inc. NUREG/CR-36233.

Umekawa, H., M. Ozawa, and A. Miyazaki. 1995. CHF in a boiling channel under oscillatory flow condition. In *Advances in multiphase flow*, eds. A. Serizawa, T. Fukano, and J. Bataille. Amsterdam: Elsevier Sci. B.V., 497–506.

van der Pol, B. 1926. On relaxation-oscillations. *Philosophical Magazine* 2: 978–92.

Yadigaroglu, G., and A.E. Bergles. 1972. Fundamental and higher-mode density-wave oscillations in two-phase flow. *Transactions of the American Society of Mechanical Engineers, Journal of Heat Transfer* 94: 189–95.

Zuber, N., and J.A. Findlay. 1965. Average volumetric concentration in two-phase flow system. *Transactions of the American Society of Mechanical Engineers, Journal of Heat Transfer* 87: 453–68.

Appendix

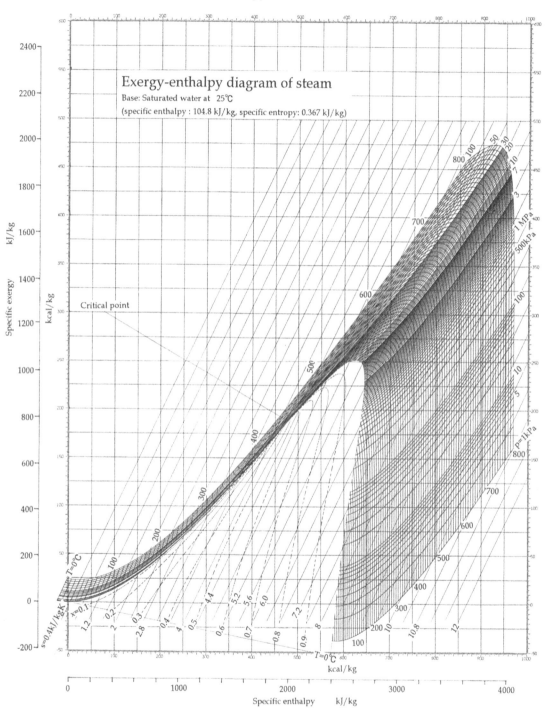

Figure A.1 Exergy–enthalpy diagram of steam.

Index